Section 23

Process Safety

Daniel A. Crowl, Ph.D. *Professor of Chemical Engineering, Michigan Technological University; Fellow, American Institute of Chemical Engineers (Section Editor, Process Safety Introduction, Combustion and Flammability Hazards, Gas Explosions, Vapor Cloud Explosions, Boiling-Liquid Expanding-Vapor Explosions)*

Laurence G. Britton, Ph.D. *Process Safety Consultant; Consulting Scientist, Neolytica, Inc.; Fellow, American Institute of Chemical Engineers; Fellow, Energy Institute; Member, Institute of Physics (U.K.) (Flame Arresters)*

Walter L. Frank, P.E., B.S.Ch.E. *Senior Consultant, ABS Consulting; Fellow, American Institute of Chemical Engineers (Hazards of Vacuum, Hazards of Inerts)*

Stanley Grossel, M.S.Ch.E. *President, Process Safety & Design; Fellow, American Institute of Chemical Engineers (Emergency Relief Device Effluent Collection and Handling, Flame Arresters)*

Dennis Hendershot, M.S.Ch.E. *Principal Process Safety Specialist, Chilworth Technology, Inc.; Fellow, American Institute of Chemical Engineers (Hazard Analysis)*

W. G. High, C.Eng., B.Sc., F.I.Mech.E. *Consultant, Burgoyne Consultants (Estimation of Damage Effects)*

Robert W. Johnson, M.S.Ch.E. *President, Unwin Company; Member, American Institute of Chemical Engineers (Reactivity, Storage and Handling of Hazardous Materials)*

Trevor A. Kletz, D.Sc. *Visiting Professor, Department of Chemical Engineering, Loughborough University (U.K.); Adjunct Professor, Department of Chemical Engineering, Texas A&M University; Fellow, American Institute of Chemical Engineers; Fellow, Royal Academy of Engineering (U.K.); Fellow, Institution of Chemical Engineers (U.K.); Fellow, Royal Society of Chemistry (U.K.) (Inherently Safer and More User-Friendly Design, Incident Investigation and Human Error, Institutional Memory, Key Procedures)*

Joseph C. Leung, Ph.D. *President, Leung Inc.; Member, American Institute of Chemical Engineers (Pressure Relief Systems)*

David A. Moore, MBA, B.Sc. *President, AcuTech Consulting Group; Registered Professional Engineer (FPE, PA); Certified Safety Professional (CSP); ASSE, ASIS, NFPA (Security)*

Robert Ormsby, M.S.Ch.E. *Process Safety Consultant; Fellow, American Institute of Chemical Engineers (Risk Analysis)*

Jack E. Owens, B.E.E. *Electrostatics Consultant, E. I. Dupont de Nemours and Co.; Member, Institute of Electrical and Electronics Engineers; Member, Electrostatics Society of America (Static Electricity)*

Richard W. Prugh, M.S.P.E., C.S.P. *Senior Process Safety Specialist, Chilworth Technology, Inc.; Fellow, American Institute of Chemical Engineers; Member, National Fire Protection Association (Toxicity)*

Carl A. Schiappa, B.S.Ch.E. *Retired, The Dow Chemical Company (Project Review and Audit Processes)*

Richard Siwek, M.S. *Managing Director, President, FireEx Consultant Ltd.; Member, European Committee for Standardization (CENTC305); Member, Association of German Engineers (VDI 2263,3673); Member, International Section for Machine Safety (ISSA) (Dust Explosions, Preventive Explosion Protection, Explosion Protection through Design Measures)*

Thomas O. Spicer III, Ph.D., P.E. *Professor and Head, Ralph E. Martin Department of Chemical Engineering, University of Arkansas; Member, American Institute of Chemical Engineers (Atmospheric Dispersion)*

Angela Summers, Ph.D., P.E. *President, SIS-TECH; Adjunct Professor, Department of Environmental Management, University of Houston—Clear Lake; Senior Member, Instrumentation, Systems and Automation Society; Member, American Institute of Chemical Engineers (Safety Instrumented Systems)*

Ronald Willey, Ph.D., P.E. *Professor, Department of Chemical Engineering, Northeastern University; Fellow, American Institute of Chemical Engineers (Case Histories)*

John L. Woodward, Ph.D. *Senior Principal Consultant, Baker Engineering and Risk Consultants, Inc.; Fellow, American Institute of Chemical Engineers (Discharge Rates from Punctured Lines and Vessels)*

PROCESS SAFETY INTRODUCTION

CASE HISTORIES

HAZARDOUS MATERIALS AND CONDITIONS

INHERENTLY SAFER DESIGN AND OTHER PRINCIPLES

PROCESS SAFETY ANALYSIS

SAFETY EQUIPMENT, PROCESS DESIGN, AND OPERATION

PROCESS SAFETY INTRODUCTION

General References: AICHE/CCPS, *Guidelines for Chemical Process Quantitative Risk Analysis,* 2d ed., American Institute of Chemical Engineers, New York, 2000. AICHE/CCPS, *Guidelines for Hazards Evaluation Procedures,* 2d ed., American Institute of Chemical Engineers, New York, 1992. Crowl and Louver, *Chemical Process Safety: Fundamentals with Applications,* 2d ed., Prentice-Hall, Englewood Cliffs, N.J., 2002. Mannan, *Lees' Loss Prevention in the Process Industries,* 3d ed., Elsevier, Amsterdam.

Process safety differs from the traditional approach to accident prevention in several ways (Mannan, *Lees' Loss Prevention in the Process Industries*, 3d ed., Elsevier, 2005, p. 1/9):

- There is greater concern with accidents that arise out of the technology.
- There is greater emphasis on foreseeing hazards and taking action before accidents occur.
- There is greater emphasis on a systematic rather than a trial-and-error approach, particularly on systematic methods of identifying hazards and of estimating the probability that they will occur and their consequences.
- There is concern with accidents that cause damage to plant and loss of profit but do not injure anyone, as well as those that do cause injury.
- Traditional practices and standards are looked at more critically.

The term *loss prevention* can be applied in any industry but is widely used in the process industries where it usually means the same as *process safety*.

Chemical plants, and other industrial facilities, may contain large quantities of hazardous materials. The materials may be hazardous due to toxicity, reactivity, flammability, or explosivity. A chemical plant may also contain large amounts of energy—the energy either is required to process the materials or is contained in the materials themselves. An *accident* occurs when control of this material or energy is lost. An *accident* is defined as an unplanned event leading to undesired consequences. The consequences might include injury to people, damage to the environment, or loss of inventory and production, or damage to equipment.

A *hazard* is defined as a chemical or physical condition that has the potential for causing damage to people, property, or the environment (AICHE/CCPS, *Guidelines for Chemical Process Quantitative Risk Analysis,* 2d ed., American Institute of Chemical Engineers, New York, 2000, p. 6). Hazards exist in a chemical plant due to the nature of the materials processed or due to the physical conditions under which the materials are processed, i.e., high pressure or temperature. These hazards are present most of the time. An initiating event is required to begin the accident process. Once initiated, the accident follows a sequence of steps, called the event sequence, that results in an incident outcome. The consequences of the accident are the resulting effects of the incident. For instance, a rupture in a pipeline due to corrosion (initiating event) results in leakage of a flammable liquid from the process. The liquid evaporates and mixes with air to form a flammable cloud, which finds an ignition source (event sequence), resulting in a fire (incident outcome). The consequences of the accident are considerable fire damage and loss of production.

Risk is defined as a measure of human injury, environmental damage, or economic loss in terms of both the incident likelihood (probability) and the magnitude of the loss or injury (consequence) (AICHE/CCPS, *Guidelines for Chemical Process Quantitative Risk Analysis,* 2d ed., American Institute of Chemical Engineers, New York, 2000, pp. 5–6). It is important that both likelihood and consequence be included in risk. For instance, seat belt use is based on a reduction in the consequences of an accident. However, many people argue against seat belts based on probabilities, which is an incorrect application of the risk concept.

A good safety program identifies and removes existing hazards. An outstanding safety program prevents the existence of safety hazards in the first place. An outstanding safety program is achieved by company commitment, visibility, and management support. This is usually achieved by a corporatewide safety policy. This safety policy usually includes the following items: (1) the company is very serious about safety, (2) safety cannot be prioritized and is a part of everyone's job function, (3) everyone is responsible for safety, including management.

To ensure that the safety program is working, most companies have a safety policy follow-through. This includes monthly safety meetings, performance reviews, and safety audits. The monthly safety meetings include a discussion of any accidents (and resolution of prevention means), training on specific issues, inspection of facilities, and delegation of work. Performance reviews within the company for all employees must have a visible safety performance component.

Safety audits are a very important means of ensuring that the safety program is operating as intended. Audits are usually done yearly by an audit team. The audit team is comprised of corporate and site safety people and other experts, as needed, including industrial hygiene, toxicology, and/ or process safety experts. The audit team activities include (1) reviewing records (including accident reports, training, monthly meetings), (2) inspecting random facilities to see if they are in compliance, (3) interviewing the employees to determine how they participate in the safety program, (4) making recommendations on

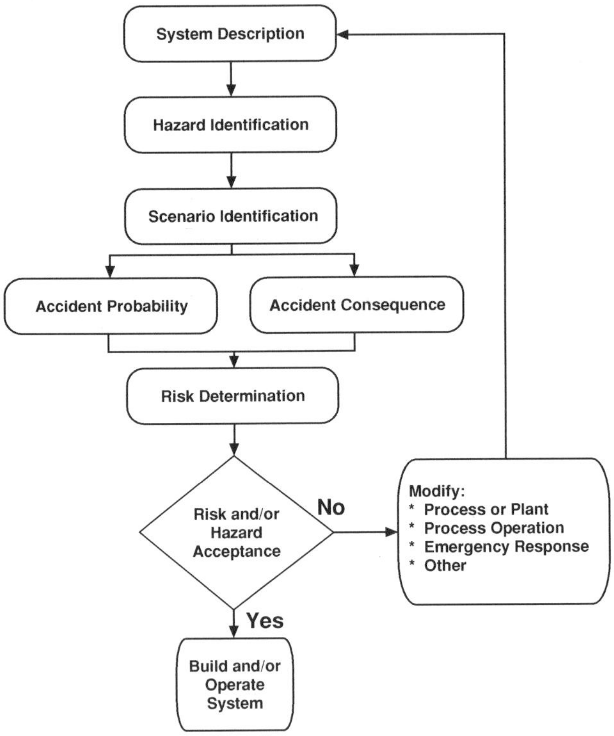

FIG. 23-1 The hazard identification and risk assessment procedure. [Guidelines for Hazards Evaluation Procedures, *Center for Chemical Process Safety (CCPS) of the American Institute of Chemical Engineers (AIChE); copyright 1985 AICHE and reproduced with permission.*]

how the program can be improved, and (5) rating the performance of the unit. The audit results are reported to upper management with the expectation that the designated unit will implement improvements in short order. Many companies perform a combined audit, which may include environmental and quality issues.

Figure 23-1 shows the hazards identification and risk assessment procedure. The procedure begins with a complete description of the process. This includes detailed PFD and P&I diagrams, complete specifications on all equipment, maintenance records, operating procedures, and so forth. A hazard identification procedure is then selected (see Hazard Analysis subsection) to identify the hazards and their nature. This is followed by identification of all potential event sequences and potential incidents (scenarios) that can result in loss of control of energy or material. Next is an evaluation of both the consequences and the probability. The consequences are estimated by using source models (to describe the release of material and energy) coupled with a consequence model to describe the incident outcome. The consequence models include dispersion, fire, and explosion modeling. The results of the consequence models are used to estimate the impacts on people, environment, and property. The accident probability is estimated by using fault trees or generic databases for the initial event sequences. Event trees may be used to account for mitigation and postrelease incidents. Finally, the risk is estimated by combining the potential consequence for each event with the event frequency and summing over all events.

Once the risk is determined, a decision must be made on risk acceptance. This can be done by comparison to a relative or absolute standard. If the risk is acceptable, then the decision is made to build and/or operate the process. If the risk is not acceptable, then something must be changed. This could include the process design, the operation, or maintenance, or additional layers of protection might be added.

CASE HISTORIES

General References: *One Hundred Largest Losses: A Thirty Year Review of Property Damage Losses in the Hydrocarbon Chemical Industry*, 20th ed. (M&M Protection, Consultants, Chicago); Mannan, S., ed., *Lees' Loss Prevention in the Process Industries*, Elsevier, 2005; Kletz, T. A., *Learning from Accidents*, Gulf Professional Publishing, 2001; Kletz, T. A., *What Went Wrong? Case Histories of Process Plant Disasters*, Editions Technip, 1998; and Sanders, R. E., *Chemical Process Safety: Learning from Case Histories*, Editions Technip, 1999.

INTRODUCTION

Engineers must give significant thought to the consequences of their decisions and indecisions. A wise step during conceptual and design phases is to review previous negative experiences of others and within your own organization. Periodically review the status of recent chemical accidents. The U.S. Chemical Safety and Hazards Investigation Board web site, *www.csb.gov*, offers details on many investigations related to chemical industry accidents within the United States. Look for similarities and dissimilarities to your current practice, and carefully make appropriate changes and improvements to avoid repeating similar accidents.

HYDROCARBON FIRES AND EXPLOSIONS

The explosion and fires at the Texaco Refinery, Milford Haven, Wales, 24 July 1994. Reference: Health and Safety Executive (HSE); HSE Books, Her Majesty's Stationary Office, Norwich, England, 1997.

On July 24, 1994, an explosion followed by a number of fires occurred at 13:23 at the Texaco refinery in Milford Haven, Wales, England. Prior to this explosion, around 9 a.m., a severe coastal electrical storm caused plant disturbances that affected the vacuum distillation, alkylation, butamer, and FCC units. The explosion occurred due to a combination of failures in management, equipment, and control systems. Given its calculated TNT equivalent of at least 4 tons, significant portions of the refinery were damaged. That no fatalities occurred is attributed partially to the accident occurring on a Sunday, as well as the fortuitous location of those who were near the explosion.

As the plant attempted adjustments to the upsets caused by the electrical storm, liquid was continuously pumped into a process vessel with a closed outlet valve. The control system indicated that this valve was open. As the unit overfilled, the only means of exit was a relief system designed for vapor. When the liquid reached the relief system, its momentum was high enough to rip apart the ductwork and cause a massive release of hydrocarbons into the environment. Minutes prior to the explosion, operating personnel were responding to 275 alarms of which 80 percent had high priority. An ignition source was found 110 m away. Recommendations from the accident investigation included the necessity of operating personnel having knowledge about simple volumetric and mass balances; that control systems be configured to provide an overview of the condition of the process; that safety critical alarms be distinguishable from other alarms; and that liquid knockout drums exist for relief systems designed for vapor.

DUST EXPLOSIONS

West Pharmaceutical Services Plant in Kinston, North Carolina, 29 January 2003, and CTA Acoustics Manufacturing Plant in Corbin, Kentucky, 20 February 2003. Reference: U.S. Chemical Safety Board (CSB); *www.csb.gov/index.cfm?folder = completed_investigations&page = info&INV_ID=34* and *ID = 35*

On January 29, 2003, the West Pharmaceutical explosion killed six workers and injured dozens more. The CSB determined that fine polyethylene dust particles, released during the production of rubber products, had accumulated above the tiles of a false ceiling, creating an explosion hazard at the plant. A similar incident occurred a few weeks later, at the CTA Acoustics manufacturing plant in Corbin, Kentucky, fatally injuring seven workers and injuring more than 30 others. This facility produced fiberglass insulation for the automotive industry. CSB investigators found that the explosion was fueled by resin dust accumulated in a production area, likely ignited by flames from a malfunctioning oven. The resin involved was a phenolic binder used in producing fiberglass mats.

CSB investigators determined that both disasters resulted from accumulations of combustible dust. Workers and workplaces need to be protected from this insidious hazard. The lesson learned here is the importance of housekeeping. Some companies will allow only $\frac{1}{32}$ in of dust to accumulate before cleaning. Suspended ceilings must be suspected as areas that can accumulate dust. Often the first explosion may be minor, but the dust dislodged can be explosive enough to level the building on the second ignition.

REACTIVE CHEMICALS

Explosion, Morton International, Inc., Paterson, New Jersey, 8 April 1998. Reference: CSB; *www.csb.gov/completed_investigations/docs/MortonInvestigationReport.pdf*

On April 8, 1998, at 20:18, an explosion and fire occurred during the production of Automate Yellow 96 Dye at Morton International, Inc. Yellow 96 dye was produced by mixing and reacting two chemicals, *ortho*-nitrochlorobenzene (*o*-NCB) and 2-ethylhexylamine (2-EHA). The explosion and fire were the consequence of a runaway reaction, which overpressurized a 2000-gal capacity chemical reactor vessel and released flammable material that ignited. Nine employees were injured, including two seriously, and potentially hazardous materials were released into the community. The CSB investigation team determined that the reaction accelerated beyond the heat removal capability of the kettle. The resulting high temperature led to a secondary runaway reaction (decomposition of *o*-NCB). The initial runaway reaction was most likely caused by a combination of the following factors: (1) The reaction was started at a temperature higher than normal, (2) the steam used to initiate the reaction was left on for too long, and (3) the use of cooling water to control the reaction rate was not initiated soon enough. The Paterson facility was not aware of the decomposition reaction. A similar incident occurred with a process using *o*-NCB in Sauget, Illinois, in 1974 (Vincent, G. C., *Loss Prev.* 1971, **5:** 46–52).

MATERIALS OF CONSTRUCTION

Ruptured chlorine hose. Reference: CSB; *www.csb.gov/safety_publications/docs/ChlorineHoseSafetyAdvisory.pdf*

On August 14, 2002, a 1-in chlorine transfer hose (CTH) used in a railcar offloading operation at DPC Enterprises in Festus, Missouri, catastrophically ruptured and initiated a sequence of events that led to the release of 48,000 lb of chlorine into neighboring areas. The material of construction of the ruptured hose was incorrect. The distributor fabricated bulk CTH with Schedule 80 Monel 400 end fittings and a high-density polyethylene spiral guard. Three hoses were shipped directly to the Festus facility from the distributor; two were put into service on June 15, 2002. The hose involved in the incident failed after 59 days in service.

Most plastics react chemically with chlorine because of their hydrocarbon structural makeup. This reactivity is avoided with some plastics in which fluorine atoms have been substituted into the hydrocarbon molecule. The Chlorine Institute recommends that hoses constructed with such an inner lining "have a structural layer braid of polyvinylidene fluoride (PVDF) monofilament material or a structural braid of Hastelloy C-276." An underlying lesson here is material compatibility. Material compatibility tables exist that engineers can consult, including in other sections within this volume.

TOXICOLOGY

Vessel explosion, D. D. Williamson & Co., Inc., Louisville, Kentucky, 11 April 2003. Reference: CSB; *www.csb.gov/completed_investigations/docs/CSB_DDWilliamson Report.pdf*

On April 11, 2003, at approximately 2:10 a.m., a 2200-gal stainless steel spray dryer feed tank at the D. D. Williamson & Co., Inc. (DDW), plant in Louisville, Kentucky, exploded. One operator was killed. The other four men working at the plant at the time of the incident were not injured. The incident was most likely initiated by overheating by a 130-psi steam supply. The feed tank was manually controlled for temperature and pressure. The tank had a maximum working pressure of 40 psi. A concrete block wall to the east separated the feed tank from a 12,000-gal aqua ammonia storage tank (29.4% ammonia). After the explosion, the feed tank's shell split open in a vertical line. It was propelled through the wall and struck the ammonia storage tank, located 15 ft to the west. The ammonia storage tank was knocked off its foundation approximately 10 ft, and piping was ripped loose. This resulted in a 26,000-lb aqua ammonia leak. Metro Louisville Health Department obtained maximum ammonia readings of 50 parts per million (ppm) at the fence line and 35 ppm on a nearby street. No injuries were reported in the area of the ammonia release.

A number of management decisions factor into this case. There was no program to evaluate necessary layers of protection on the spray dryer feed tanks. Likewise, there was no recognition of the need to provide process control and alarm instrumentation on the two feed tanks. Reliance on a single local temperature indicator that must be read by operators is insufficient. On the morning of the incident, the operators were unaware that the system had exceeded normal operating conditions. The feed tanks were installed for use in the spray dryer process without a review of their design versus system requirements. Safety valves on the spray dryer feed tanks had been removed to transport the tanks to Louisville and were never reinstalled. Inadequate hazard analysis systems didn't identify feed tank hazards. The ASME Code, Section VIII (2001 ASME Boiler and Pressure Vessel Code: Design and Fabrication of Pressure Vessels, American Society of Mechanical Engineers, 2001), requires that all vessels having an internal operating pressure exceeding 15 psi be provided with pressure relief devices. Finally, equipment layout should always be considered in the design stage. Methods such as the Dow Fire and Explosion Index (AIChE, 1994) can assist in determining the optimum spacing between critical units.

NITROGEN ASPHYXIATION

Union Carbide Corporation, Hahnville, Louisiana, 27 March 1998. Reference: CSB; *www.csb.gov/completed_investigations/docs/Final Union Carbide Report.pdf* and */SB-Nitrogen-6-11-03.pdf.*

On March 27, 1998, at approximately 12:15 p.m., two workers at Union Carbide Corporation's Taft/Star Manufacturing Plant in Hahnville, Louisiana, were overcome by nitrogen gas while performing a black light inspection at an open end of a 48-in-wide horizontal pipe. One Union Carbide employee was killed, and an independent contractor was seriously injured due to nitrogen asphyxiation. Nitrogen was being injected into a nearby reactor to prevent contamination of a catalyst by oxygen and related materials. The nitrogen also flowed through some of the piping systems connected to the reactors. No warning sign was posted on the pipe opening identifying it as a confined space. Nor was there a warning that the pipe contained potentially hazardous nitrogen.

HAZARDOUS MATERIALS AND CONDITIONS

FLAMMABILITY

Nomenclature

K_G	deflagration index for gases (bar·m/s)
K_{St}	deflagration index for dusts (bar·m/s)
LFL	lower flammability limit (vol % fuel in air)
LOC	limiting oxygen concentration
n	number of combustible species
P	pressure
T	temperature (°C)
t	time (s)
UFL	upper flammability limit (vol. % fuel in air)
V	vessel volume (m^3)
y_i	mole fraction of component i on a combustible basis
z	stoichiometric coefficient for oxygen
ΔH_c	net heat of combustion (kcal/mol)

General References: Crowl and Louvar, *Chemical Process Safety: Fundamentals with Applications,* 2d ed., Prentice-Hall, Upper Saddle River, N.J., 2002, Chaps. 6 and 7. Crowl, *Understanding Explosions,* American Institute of Chemical Engineers, New York, 2003. Eckoff, *Dust Explosions in the Process Industries,* 2d ed., Butterworth-Heinemann, now Elsevier, Amsterdam, 1997. Kinney and Graham, *Explosive Shocks in Air,* 2d ed., Springer-Verlag, New York, 1985. Lewis and von Elbe, *Combustion, Flames and Explosions of Gases,* 3d ed., Academic Press, New York, 1987. Mannan, *Lees' Loss Prevention in the Process Industries,* 3d ed., Elsevier, Amsterdam, 2005, Chap. 16: Fire, Chap. 17: Explosion.

Introduction Fire and explosions in chemical plants and refineries are rare, but when they do occur, they are very dramatic.

Accident statistics have shown that fires and explosions represent 97 percent of the largest accidents in the chemical industry (J. Coco, ed., *Large Property Damage Losses in the Hydrocarbon-Chemical Industry: A Thirty Year Review,* J. H. Marsh and McLennan, New York, 1997).

Prevention of fires and explosions requires

1. An understanding of the fundamentals of fires and explosions
2. Proper experimental characterization of flammable and explosive materials
3. Proper application of these concepts in the plant environment

The technology does exist to handle and process flammable and explosive materials safely, and to mitigate the effects of an explosion. The challenges to this problem are as follows:

1. Combustion behavior varies widely and is dependent on a wide range of parameters.
2. There is an incomplete fundamental understanding of fires and explosions. Predictive methods are still under development.
3. Fire and explosion properties are not fundamentally based and are an artifact of a particular experimental apparatus and procedure.
4. High-quality data from a standardized apparatus that produces consistent results are lacking.
5. The application of these concepts in a plant environment is difficult.

The Fire Triangle The fire triangle is shown in Fig. 23-2. It shows that a fire will result if fuel, oxidant, and an ignition source are present. In reality, the fuel and oxidant must be within certain concentration ranges, and the ignition source must be robust enough to initiate the fire. The fire triangle applies to gases, liquids, and solids. Liquids are volatized and solids decompose prior to combustion in the vapor phase. For dusts arising from solid materials, the particle size, distribution, and suspension in the gas are also important parameters in the combustion—these are sometimes included in the fire triangle.

The usual oxidizer in the fire triangle is oxygen in the air. However, gases such as fluorine and chlorine; liquids such as peroxides and chlorates; and solids such as ammonium nitrate and some metals can serve the role of an oxidizer. Exothermic decomposition, without oxygen, is also possible, e.g., with ethylene oxide or acetylene.

Ignition arises from a wide variety of sources, including static electricity, hot surfaces, sparks, open flames, and electric circuits. Ignition sources are elusive and difficult to eliminate entirely, although efforts should always be made to reduce them.

If any one side of the fire triangle is removed, a fire will not result. In the past, the most common method for fire control was elimination of ignition sources. However, experience has shown that this is not robust enough. Current fire control prevention methods continue with elimination of ignition sources, while focusing efforts more strongly on preventing flammable mixtures.

Definition of Terms The following are terms necessary to characterize fires and explosions (Crowl and Louvar, *Chemical Process Safety: Fundamentals with Applications*, 2d ed. Prentice-Hall, Upper Saddle River, N.J., 2002, pp. 227–229).

Autoignition temperature (AIT) This is a fixed temperature above which adequate energy is available in the environment to provide an ignition source.

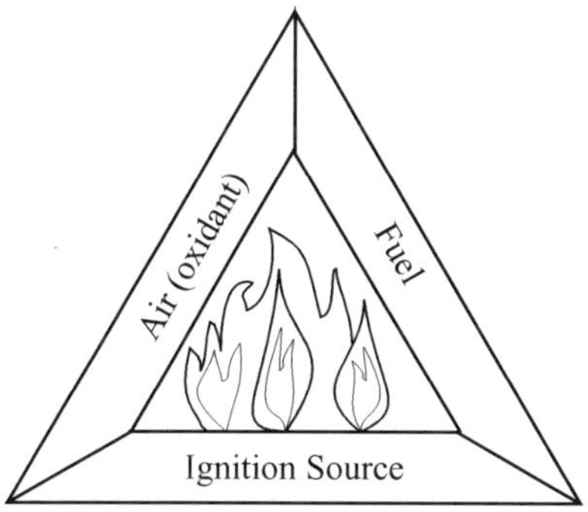

FIG. 23-2 The fire triangle showing the requirement for combustion of gases and vapors. [*D. A. Crowl,* Understanding Explosions, *Center for Chemical Process Safety (CCPS) of the American Institute of Chemical Engineers (AIChE); copyright 2003 AIChE and reproduced with permission.*]

Boiling-liquid expanding-vapor explosion (BLEVE) A BLEVE occurs if a vessel that contains a liquid at a temperature above its atmospheric pressure boiling point ruptures. The subsequent BLEVE is the explosive vaporization of a large fraction of the vessel contents, possibly followed by combustion or explosion of the vaporized cloud if it is combustible. This type of explosion occurs when an external fire heats the contents of a tank of volatile material. As the tank contents heat, the vapor pressure of the liquid within the tank increases, and the tank's structural integrity is reduced because of the heating. If the tank ruptures, the hot liquid volatilizes explosively.

Combustion or fire Combustion or fire is a chemical reaction in which a substance combines with an oxidant and releases energy. Part of the energy released is used to sustain the reaction.

Confined explosion This explosion occurs within a vessel or a building.

Deflagration In this explosion the reaction front moves at a speed less than the speed of sound in the unreacted medium.

Detonation In this explosion the reaction front moves at a speed greater than the speed of sound in the unreacted medium.

Dust explosion This explosion results from the rapid combustion of fine solid particles. Many solid materials (including common metals such as iron and aluminum) become flammable when reduced to a fine powder and suspended in air.

Explosion An explosion is a rapid expansion of gases resulting in a rapidly moving pressure or shock wave. The expansion can be mechanical (by means of a sudden rupture of a pressurized vessel), or it can be the result of a rapid chemical reaction. Explosion damage is caused by the pressure or shock wave.

Fire point The fire point is the lowest temperature at which a vapor above a liquid will continue to burn once ignited; the fire point temperature is higher than the flash point.

Flammability limits Vapor-air mixtures will ignite and burn only over a well-specified range of compositions. The mixture will not burn when the composition is lower than the lower flammable limit (LFL); the mixture is too lean for combustion. The mixture is also not combustible when the composition is too rich, i.e., that is, when it is above the upper flammable limit (UFL). A mixture is flammable only when the composition is between the LFL and the UFL. Commonly used units are volume percent of fuel (percentage of fuel plus air).

Lower explosion limit (LEL) and upper explosion limit (UEL) are used interchangeably with LFL and UFL.

Flash point (FP) The flash point of a liquid is the lowest temperature at which it gives off enough vapor to form an ignitable mixture with air. At the flash point, the vapor will burn but only briefly; inadequate vapor is produced to maintain combustion. The flash point generally increases with increasing pressure.

There are several different experimental methods used to determine flash points. Each method produces a somewhat different value. The two most commonly used methods are open cup and closed cup, depending on the physical configuration of the experimental equipment. The open-cup flash point is a few degrees higher than the closed-cup flash point.

Ignition Ignition of a flammable mixture may be caused by a flammable mixture coming in contact with a source of ignition with sufficient energy or by the gas reaching a temperature high enough to cause the gas to autoignite.

Mechanical explosion A mechanical explosion results from the sudden failure of a vessel containing high-pressure, nonreactive gas.

Minimum ignition energy This is the minimum energy input required to initiate combustion.

Overpressure The pressure over ambient that results from an explosion.

Shock wave This is an abrupt pressure wave moving through a gas. A shock wave in open air is followed by a strong wind; the combined shock wave and wind is called a blast wave. The pressure increase in the shock wave is so rapid that the process is mostly adiabatic.

Unconfined explosion Unconfined explosions occur in the open. This type of explosion is usually the result of a flammable gas spill. The gas is dispersed and mixed with air until it comes in contact with an ignition source. Unconfined explosions are rarer than confined explosions because the explosive material is frequently diluted below the LFL by

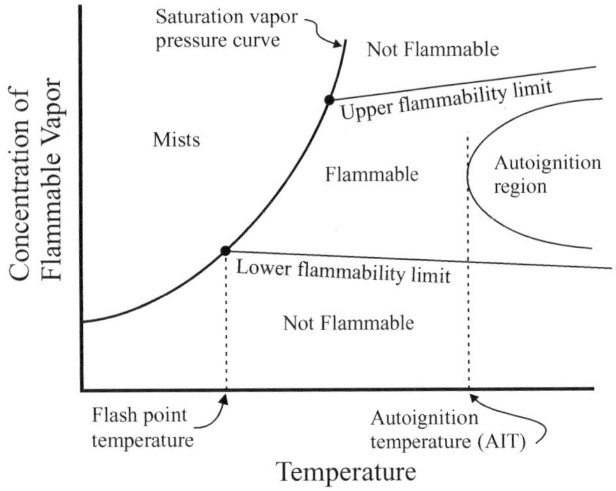

FIG. 23-3 The relationship between the various flammability properties. (*D. A. Crowl and J. F. Louvar,* Chemical Process Safety: Fundamentals with Applications, *2d ed., © 2002. Adapted by permission of Pearson Education, Inc., Upper Saddle River, N.J.*)

wind dispersion. These explosions are destructive because large quantities of gas and large areas are frequently involved.

Figure 23-3 is a plot of concentration versus temperature and shows how several of these definitions are related. The exponential curve in Fig. 23-3 represents the saturation vapor pressure curve for the liquid material. Typically, the UFL increases and the LFL decreases with temperature. The LFL theoretically intersects the saturation vapor pressure curve at the flash point, although experimental data are not always consistent. The autoignition temperature is actually the lowest temperature of an autoignition region. The behavior of the autoignition region and the flammability limits at higher temperatures are not well understood.

The flash point and flammability limits are not fundamental properties but are defined only by the specific experimental apparatus and procedure used.

Section 2 provides flammability data for a number of compounds.

Combustion and Flammability Hazards

Vapor Mixtures Frequently, flammability data are required for vapor mixtures. The flammability limits for the mixture are estimated by using LeChatelier's rule [LeChatelier, "Estimation of Firedamp by Flammability Limits," *Ann. Mines* (1891), ser. 8, 19: 388–395, with translation in *Process Safety Progress*, 23(3): 172].

$$\text{LFL}_{\text{mix}} = \frac{1}{\sum_{i=1}^{n} y_i/\text{LFL}_i} \tag{23-1}$$

where LFL_i = lower flammability limit for component i (in volume %)
y_i = mole fraction of component i on a combustible basis
n = number of combustible species

An identical equation can be written for the UFL.

Note that Eq. (23-1) is only applied to the combustible species, and the mole fraction is computed using only the combustible species.

LeChatelier's rule is empirically derived and is not universally applicable. Mashuga and Crowl [Mashuga and Crowl, "Derivation of LeChatelier's Mixing Rule for Flammable Limits," *Process Safety Progress*, 19(2): 112–118 (2000)] determined that the following assumptions are present in LeChatelier's rule:

1. The product heat capacities are constant.
2. The number of moles of gas is constant.
3. The combustion kinetics of the pure species is independent of and unchanged by the presence of other combustible species.
4. The adiabatic temperature rise at the flammability limit is the same for all species.

These assumptions were found to be reasonably valid at the LFL and less so at the UFL.

Liquid Mixtures Flash point temperatures for mixtures of liquids can be estimated if only one component is flammable and the flash point temperature of the flammable component is known. In this case the flash point temperature is estimated by determining the temperature at which the vapor pressure of the flammable component in the mixture is equal to the pure component vapor pressure at its flash point. Estimation of flash point temperatures for mixtures of several flammable components can be done by a similar procedure, but it is recommended that the flash point temperature be measured experimentally.

Flammability Limit Dependence on Temperature In general, as the temperature increases, the flammability range widens, i.e., the LFL decreases and the UFL increases. Zabetakis et al. (Zabetakis, Lambiris, and Scott, "Flame Temperatures of Limit Mixtures," *7th Symposium on Combustion*, Butterworths, London, 1959) derived the following empirical equations, which are approximate for many hydrocarbons:

$$\text{LFT}_T = \text{LFL}_{25} - \frac{0.75}{\Delta H_c}(T - 25)$$
$$\text{UFL}_T = \text{UFL}_{25} + \frac{0.75}{\Delta H_c}(T - 25) \tag{23-2}$$

where ΔH_c is the net heat of combustion (kcal/mol) and T is the temperature (°C).

Flammability Limit Dependence on Pressure Pressure has little effect on the LFL except at very low pressures (<50 mmHg absolute) where flames do not propagate.

The UFL increases as the pressure is increased. A very approximate equation for the change in UFL with pressure is available for some hydrocarbon gases (Zabetakis, "Fire and Explosion Hazards at Temperature and Pressure Extremes," *AICHE Inst. Chem. Engr. Symp.*, ser. 2, pp. 99-104, 1965):

$$\text{UFL}_P = \text{UFL} + 20.6\,(\log P + 1) \tag{23-3}$$

where P is the pressure (megapascals absolute) and UFL is the upper flammability limit (vol % fuel in air at 1 atm).

Estimating Flammability Limits There are a number of very approximate methods available to estimate flammability limits. However, for critical safety values, experimental determination as close as possible to actual process conditions is always recommended.

Jones [Jones, "Inflammation Limits and Their Practical Application in Hazardous Industrial Operations," *Chem. Rev.*, 22(1): 1–26 (1938)] found that for many hydrocarbon vapors the LFL and UFL can be estimated from the stoichiometric concentration of fuel:

$$\text{LFL} = 0.55C_{\text{st}}$$
$$\text{UFL} = 3.50C_{\text{st}} \tag{23-4}$$

where C_{st} is the stoichiometric fuel concentration (vol % fuel in air).

For a stoichiometric combustion equation of the form

$$(1)\ C_mH_xO_y + z\ O_2 \rightarrow m\ CO_2 + (x/2)\ H_2O \tag{23-5}$$

it follows that

$$z = m + \frac{x}{4} - \frac{y}{2} \tag{23-6}$$

and furthermore that

$$C_{\text{st}} = \frac{100}{1 + z/0.21} \tag{23-7}$$

TABLE 23-1 Limiting Oxygen Concentrations (Volume Percent Oxygen Concentrations above Which Combustion Can Occur)

Gas or vapor	N_2 / Air	CO_2 / Air
Methane	12	14.5
Ethane	11	13.5
Propane	11.5	14.5
n-Butane	12	14.5
Isobutane	12	15
n-Pentane	12	14.5
Isopentane	12	14.5
n-Hexane	12	14.5
n-Heptane	11.5	14.5
Ethylene	10	11.5
Propylene	11.5	14
1-Butene	11.5	14
Isobutylene	12	15
Butadiene	10.5	13
3-Methyl-1-butene	11.5	14
Benzene	11.4	14
Toluene	9.5	—
Styrene	9.0	—
Cyclopropane	11.5	14
Gasoline		
(73/100)	12	15
(100/130)	12	15
(115/145)	12	14.5
Kerosene	10 (150°C)	13 (150°C)
JP-1 fuel	10.5 (150°C)	14 (150°C)
Natural gas	12	14.5
Acetone	11.5	14
t-Butanol	NA	16.5 (150°C)
Carbon disulfide	5	7.5
Carbon monoxide	5.5	5.5
Ethanol	10.5	13
Ethyl ether	10.5	13
Hydrogen	5	5.2
Hydrogen sulfide	7.5	11.5
Isobutyl formate	12.5	15
Methanol	10	12
Methyl acetate	11	13.5

Data from NFPA 68, *Venting of Deflagrations* (Quincy, Mass.: National Fire Protection Association, 1994).

Equation (23-7) can be used with (23-4) to estimate the LFL and UFL.

Suzuki [Suzuki, "Empirical Relationship between Lower Flammability Limits and Standard Enthalpies of Combustion of Organic Compounds," *Fire and Materials*, 18: 333–336 (1994); Suzuki and Koide, "Correlation between Upper Flammability Limits and Thermochemical Properties of Organic Compounds," *Fire and Materials*, 18: pp. 393–397 (1994)] provides more detailed correlations for the UFL and LFL in terms of the heat of combustion.

Flammability limits can also be estimated by using calculated adiabatic flame temperatures and a chemical equilibrium program [Mashuga and Crowl, "Flammability Zone Prediction Using Calculated Adiabatic Flame Temperatures," *Process Safety Progress*, 18 (3) (1999)].

Limiting Oxygen Concentration (LOC) Below the limiting oxygen concentration it is not possible to support combustion, independent of the fuel concentration. The LOC is expressed in units of volume percent of oxygen. The LOC is dependent on the pressure and temperature, and on the inert gas. Table 23-1 lists a number of LOCs, and it shows that the LOC changes if carbon dioxide is the inert gas instead of nitrogen.

The LOC can be estimated for many hydrocarbons from

$$\text{LOC} = z\ \text{LFL} \tag{23-8}$$

where z is the stoichiometric coefficient for oxygen [see Eq. (23-5)] and LFL is the lower flammability limit.

Flammability Diagram Figure 23-4 shows a typical flammability diagram. Point A shows how the scales are oriented—at any point on the diagram the concentrations must add up to 100 percent. At point A we have 60% fuel, 20% oxygen, and 20% nitrogen. The air line represents all possible combinations of fuel and air—it intersects the nitrogen axis at 79% nitrogen which is the composition of air. The stoichiometric line represents all stoichiometric combinations of fuel and oxygen. If the combustion reaction is written according to Eq. (23-5), then the intersection of the stoichiometric line with the oxygen axis is given by

$$100\left(\frac{z}{1+z}\right) \tag{23-9}$$

The LFL and UFL are drawn on the air line from the fuel axis values.

The flammability zone for most hydrocarbon vapors is shown as drawn in Fig. 23-4. Any concentration within the flammability zone is defined as flammable.

The LOC is the oxygen concentration at the very nose of the flammability zone. It is found from a line drawn from the nose of the flammability zone to the oxygen axis.

Crowl (*Understanding Explosions*, American Institute of Chemical Engineers, New York, 2003, App. A) derived a number of rules for using flammability diagrams:

1. If two gas mixtures *R* and *S* are combined, the resulting mixture composition lies on a line connecting points *R* and *S* on the flammability diagram. The location of the final mixture on the straight line depends on the relative moles in the mixtures combined: If mixture *S* has more moles, the final mixture point will lie closer to point *S*. This is identical to the lever rule used for phase diagrams.
2. If a mixture *R* is continuously diluted with mixture *S*, the mixture composition follows along the straight line between points *R* and *S* on the flammability diagram. As the dilution continues, the mixture composition moves closer and closer to point *S*. Eventually, at infinite dilution the mixture composition is at point *S*.
3. For systems having composition points that fall on a straight line passing through an apex corresponding to one pure component, the other two components are present in a fixed ratio along the entire line length.

Figure 23-5 shows how nitrogen can be used to avoid the flammable zone during the vessel preparation for maintenance. In this case nitrogen is pumped into the vessel until a concentration is reached at point S. Then air can be pumped in, arriving at point R. Figure 23-6 shows the reverse procedure. Now nitrogen is added until point S is reached, then fuel is pumped in until point R is reached. In both cases the flammable zone is avoided.

A complete flammability diagram requires hundreds of tests in a combustion sphere [Mashuga and Crowl, "Application of the Flammability Diagram for the Evaluation of Fire and Explosion Hazards of Flammable Vapors," *Proc. Safety Prog.*, 17 (3): 176–183 (1998)]. However, an approximate diagram can be drawn by using the LFL, UFL, LOC, and flammability limits in pure oxygen. The following procedure is used:

1. Draw the flammability limits in air as points on the air line, using the fuel axis values.
2. Draw the flammability limits in pure oxygen as points on the oxygen scale, using the fuel axis values. Table 23-2 provides a number of values for the flammability limits in pure oxygen. These are drawn on the oxygen axis using the fuel axis concentrations.
3. Use Eq. (23-9) to draw a point on the oxygen axis, and then draw the stoichiometric line from this point to the 100 percent nitrogen apex.
4. Locate the LOC on the oxygen axis. Draw a line parallel to the fuel axis until it intersects the stoichiometric line. Draw a point at the intersection.
5. Connect the points to estimate the flammability zone.

In reality, not all the data are available, so a reduced form of the above procedure is used to draw a partial diagram (Crowl, *Understanding Explosions*, American Institute of Chemical Engineers, New York, 2003, p. 27).

Ignition Sources and Energy Table 23-3 provides a list of the ignition sources for major fires. As seen in Table 23-3, ignition sources are very common and cannot be used as the only method of fire prevention.

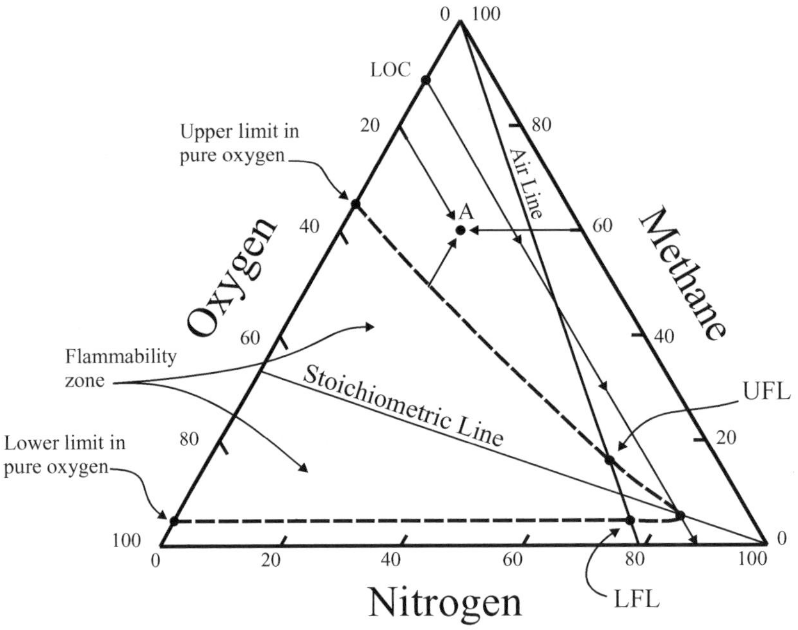

FIG. 23-4 Flammability diagram for methane at an initial temperature and pressure of 25°C and 1 atm. [*C. V. Mashuga and D. A. Crowl, "Application of the Flammability Diagram for Evaluation of Fire and Explosion Hazards of Flammable Vapors,"* Process Safety Progress, *vol. 17, no. 3; copyright 1998 American Institute of Chemical Engineers (AIChE) and reproduced with permission.*]

The minimum ignition energy (MIE) is the minimum energy input required to initiate combustion. All flammable materials (including dusts) have an MIE. The MIE depends on the species, concentration, pressure, and temperature. A few MIEs are provided in Table 23-4. In general, experimental data indicate that

1. The MIE increases with increasing pressure.
2. The MIE for dusts is, in general, at energy levels somewhat higher than that of combustible gases.
3. An increase in nitrogen concentration increases the MIE.

Most hydrocarbon vapors have an MIE of about 0.25 mJ. This is very low—a static spark that you can feel is greater than about 20 mJ. Dusts typically have MIEs of about 10 mJ. In both the vapor and dust cases, wide variability in the values is expected.

Aerosols and Mists The flammability behavior of vapors is affected by the presence of liquid droplets in the form of aerosols or mists. Aerosols are liquid droplets or solid particles of size small

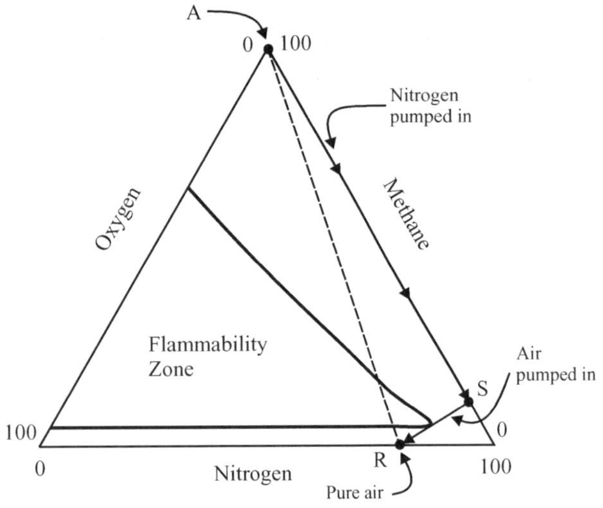

FIG. 23-5 A procedure for avoiding the flammability zone for taking a vessel out of service. [*D. A. Crowl,* Understanding Explosions, *Center for Chemical Process Safety (CCPS) of the American Institute of Chemical Engineers (AIChE); copyright 2003 AIChE and reproduced with permission.*]

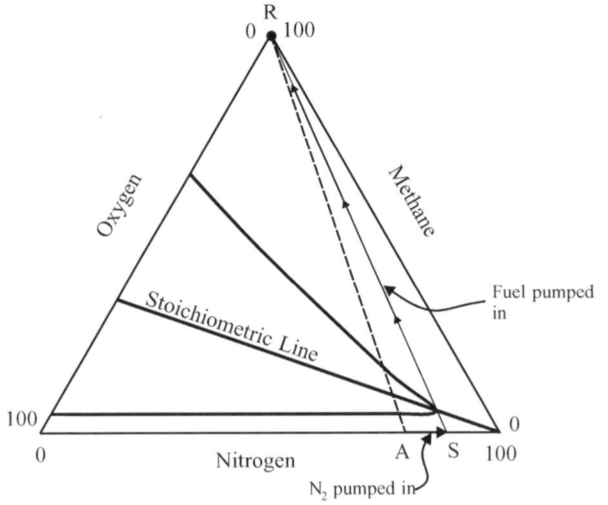

FIG. 23-6 A procedure for avoiding the flammability zone for placing a vessel into service. [*D. A. Crowl,* Understanding Explosions, *Center for Chemical Process Safety (CCPS) of the American Institute of Chemical Engineers (AIChE); copyright 2003 AIChE and reproduced with permission.*]

TABLE 23-2 Flammability Limits in Pure Oxygen

Compound	Formula	Limits of flammability in pure oxygen	
		Lower	Upper
Hydrogen	H_2	4.0	94
Carbon monoxide*	CO	15.5	94
Ammonia	NH_3	15.0	79
Methane	CH_4	5.1	61
Ethane	C_2H_6	3.0	66
Ethylene	C_2H_4	3.0	80
Propylene	C_3H_6	2.1	53
Cyclopropane	C_3H_6	2.5	60
Diethyl ether	$C_4H_{10}O$	2.0	82
Divinyl ether	C_4H_6O	1.8	85

*The limits are insensitive to p_{H_2O} above a few millimeters of mercury.
Data from B. Lewis and G. von Elbe, *Combustion, Flames and Explosions of Gases* (New York: Harcourt Brace Jovanovich, 1987).

enough to remain suspended in air for prolonged periods. Mists are suspended liquid droplets produced by condensation of vapor into liquid or by the breaking up of liquid into a dispersed state by splashing, spraying, or atomizing.

For liquid droplets with diameters less than 0.01 mm, the LFL is virtually the same as the substance in vapor form. For mechanically formed mists with drop diameters between 0.001 and 0.2 mm, the LFL decreases as the drop diameter increases. In experiments with larger drop diameters the LFL was less than one-tenth of the vapor LFL. Thus, suspended droplets have a profound effect on flammability.

Explosions

Introduction Gas explosions depend on a large number of parameters, including temperature, pressure, gas composition, ignition source, geometry of surroundings, turbulence in the gas, mixing, time before ignition, and so forth. Thus, gas explosions are difficult to characterize and predict.

An explosion occurs when energy is released into the gas phase in a very short time, typically milliseconds or less. If the energy is released into the gas phase, the energy causes the gases to expand very rapidly, forcing back the surrounding gas and initiating a pressure wave that moves rapidly outward from the blast origin. The pressure wave contains energy which causes damage to the surroundings. A prediction of the damage effects from an explosion requires a clear understanding of how this pressure wave behaves.

TABLE 23-3 Ignition Sources of Major Fires*

Electrical (wiring of motors)	23%
Smoking	18
Friction (bearings or broken parts)	10
Overheated materials (abnormally high temperatures)	8
Hot surfaces (heat from boilers, lamps, etc.)	7
Burner flames (improper use of torches, etc.)	7
Combustion sparks (sparks and embers)	5
Spontaneous ignition (rubbish, etc.)	4
Cutting and welding (sparks, arcs, heat, etc.)	4
Exposure (fires jumping into new areas)	3
Incendiarism (fires maliciously set)	3
Mechanical sparks (grinders, crushers, etc.)	2
Molten substances (hot spills)	2
Chemical action (processes not in control)	1
Static sparks (release of accumulated energy)	1
Lightning (where lightning rods are not used)	1
Miscellaneous	1

**Accident Prevention Manual for Industrial Operations* (Chicago: National Safety Council, 1974).

TABLE 23-4 Minimum Ignition Energy (MIE) for Selected Gases

Chemical	Minimum ignition energy, mJ
Acetylene	0.020
Benzene	0.225
1,3-Butadiene	0.125
n-Butane	0.260
Cyclohexane	0.223
Cyclopropane	0.180
Ethane	0.240
Ethene	0.124
Ethylacetate	0.480
Ethylene oxide	0.062
n-Heptane	0.240
Hexane	0.248
Hydrogen	0.018
Methane	0.280
Methanol	0.140
Methyl acetylene	0.120
Methyl ethyl ketone	0.280
n-Pentane	0.220
2-Pentane	0.180
Propane	0.250

Data from I. Glassman, *Combustion*, 3d ed. (New York: Academic Press, 1996).

Detonation and Deflagration The difference between a detonation and deflagration depends on how fast the pressure wave moves out from the blast origin. If the pressure wave moves at a speed less than the speed of sound in the ambient gas, then a deflagration results. If the pressure wave moves at a speed greater than the speed of sound in the ambient gas, then a detonation results.

For ideal gases, the speed of sound is a function of temperature and molecular weight only. For air at 20°C the speed of sound is 344 m/s (1129 ft/s).

For a detonation, the reaction front moves faster than the speed of sound, pushing the pressure wave or shock front immediately ahead of it. For a deflagration, the reaction front moves at a speed less than the speed of sound, resulting in a pressure wave that moves at the speed of sound, moving away from the reaction front. A noticeable difference is found in the resulting pressure-time or pressure-distance plots.

The difference in behavior between a detonation and deflagration results in a significant difference in the damage. For a detonation, the damage is usually localized. However, for a deflagration, the damage is more widespread.

For high explosives, such as TNT, detonations are the normal result. However, for flammable vapors, deflagrations are more common.

Confined Explosions A confined explosion occurs in a building or process. Empirical studies on deflagrations (Tang and Baker, "A New Set of Blast Curves from Vapor Cloud Explosions," 33d Loss Prevention Sympsoium, AICHE, 1999; Mercx, van Wees, and Opschoor, "Current Research at TNO on Vapour Cloud Explosion Modeling," *Plant/Operations Progress,* October 1993) have shown that the behavior of the explosion is highly dependent on the degree of confinement. Confinement may be due to process equipment, buildings, storage vessels, and anything else that impedes the expansion of the reaction front.

These studies have found that increased confinement leads to flame acceleration and increased damage. The flame acceleration is caused by increased turbulence which stretches and tears the flame front, resulting in a larger flame front surface and an increased combustion rate. The turbulence is caused by two phenomena. First, the unburned gases are pushed and accelerated by the combustion products behind the reaction front. Second, turbulence is caused by the interaction of the gases with obstacles. The increased combustion rate results in additional turbulence and additional acceleration, providing a feedback mechanism for even more turbulence.

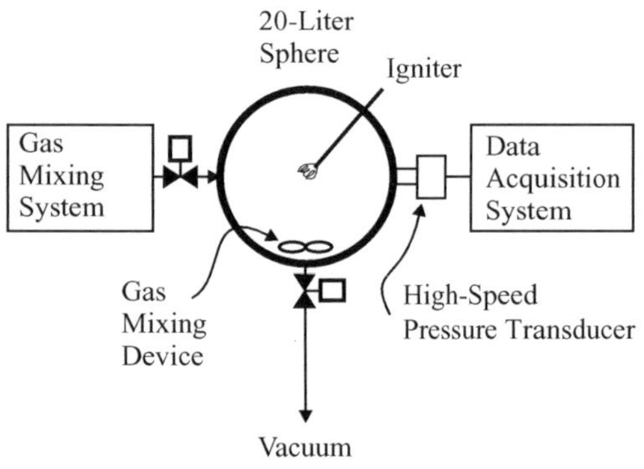

FIG. 23-7 An apparatus for collecting explosion data for gases and vapors. [*D. A. Crowl, Understanding Explosions, Center for Chemical Process Safety (CCPS) of the American Institute of Chemical Engineers (AIChE); copyright 2003 AIChE and reproduced with permission.*]

Characterizing Explosive Behavior for Vapors and Dusts Figure 23-7 is a schematic of a device used to characterize explosive vapors. This vessel is typically 3 to 20 L. It includes a gas handling and mixing system (not shown), an igniter to initiate the combustion, and a high-speed pressure transducer capable of measuring the pressure changes at the millisecond level.

The igniter can be of several types, including a fuse wire, spark, or chemical ignition system. A typical energy for ignition is 10 J, although gases can be ignited at much lower levels.

The gas is metered into the chamber to provide a mixture of a known composition. At a specified time the igniter is activated, and data are collected from the pressure transducer.

A typical pressure time plot is shown in Fig. 23-8. After ignition, the pressure increases rapidly, reaches a peak, and then diminishes as the reaction products are consumed and the gases are quenched and cooled by the vessel wall.

The experiment is repeated over a range of concentrations. A plot of the maximum pressure versus fuel concentration is used to determine the flammability limits, as shown in Fig. 23-9. A pressure increase of 7 percent over initial ambient pressure is used to define the flammability limits (ASTM E918, *Standard Procedure for Determining Flammability Limits at Elevated Temperature and Pressure*, American Society for Testing and Materials, Philadelphia, 1992).

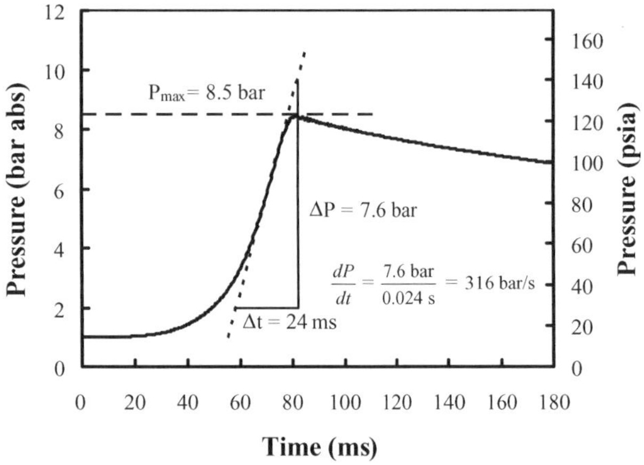

FIG. 23-8 Typical pressure versus time data obtained from gas explosion apparatus shown in Fig. 23-9. (*Daniel A. Crowl and Joseph F. Louvar,* Chemical Process Safety: Fundamentals with Applications, *2d ed., © 2002. Adapted by permission of Pearson Education, Inc., Upper Saddle River, N.J.*)

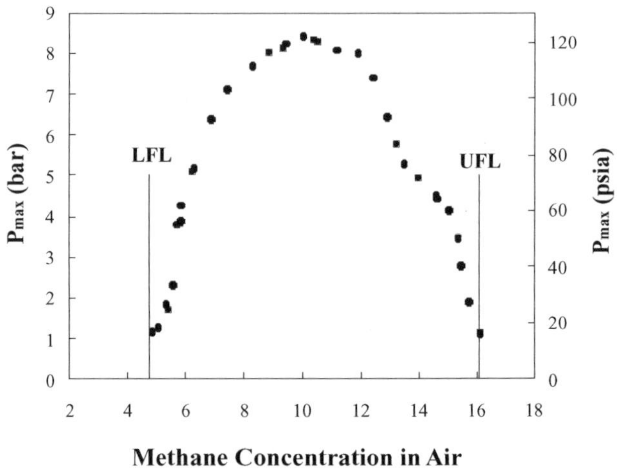

FIG. 26-9 Maximum pressure as a function of volume percent concentration for methane in air in a 20-L test sphere. The initial temperature and pressure are 25°C and 1 atm. The stoichiometric concentration is 9.51% methane. [*C. V. Mashuga and D. A. Crowl, "Application of the Flammability Diagram for Evaluation of Fire and Explosion Hazards of Flammable Vapors,"* Process Safety Progress, *vol. 17, no. 3; copyright 1998 American Institute of Chemical Engineers (AIChE) and reproduced with permission.*]

Figure 23-10 shows a device used to characterize the combustion of dusts. In this case, the dusts are initially contained in a small carrier external to the vessel, and the dust is blown in just prior to ignition. A typical pressure-time curve for the dust apparatus is shown in Fig. 23-11. The vessel is initially at a pressure less than atmospheric, but the pressure increases to atmospheric after the dust sample is blown in. After the dust is blown in, a delay time occurs in order for the dust to become quiescent but still suspended in the gas. The results are highly dependent on the delay time.

Two parameters are used to characterize the combustion for both the vapor and dust cases. The first is the maximum pressure during the combustion process, and the second is the maximum rate of

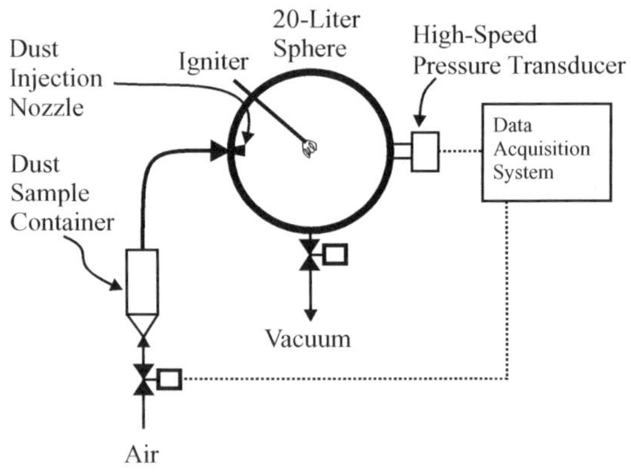

FIG. 23-10 An apparatus for collecting explosion data for dusts. [*D. A. Crowl, Understanding Explosions, Center for Chemical Process Safety (CCPS) of the American Institute of Chemical Engineers (AIChE); copyright 2003 AIChE and reproduced with permission.*]

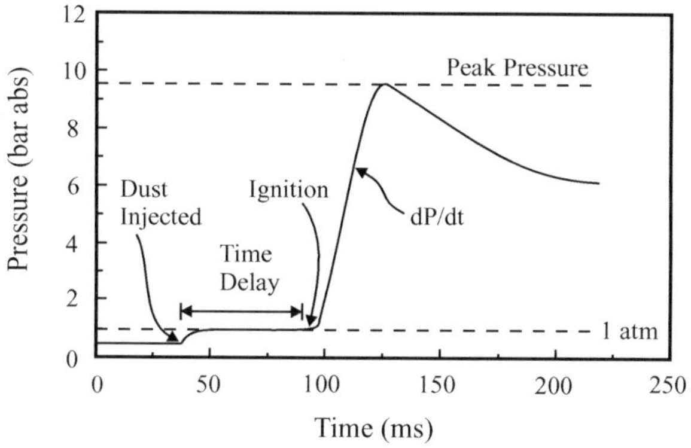

FIG. 23-11 Pressure data from dust explosion device. (*Daniel A. Crowl and Joseph F. Louvar,* Chemical Process Safety: Fundamentals with Applications, *2d ed., © 2002. Adapted by permission of Pearson Education, Inc., Upper Saddle River, N.J.*)

pressure increase. Empirical studies have shown that a deflagration index can be computed from the maximum rate of pressure increase:

$$K_G \text{ or } K_{St} = \left(\frac{dP}{dt}\right)_{max} V^{1/3} \qquad (23\text{-}10)$$

where K_G = deflagration index for gases (bar·m/s)
K_{St} = deflagration index for dusts (bar·m/s)
P = pressure (bar)
t = time (s)
V = is the vessel volume (m^3)

The higher the value of the deflagration index, the more robust the combustion. Table 23-5 contains combustion data for gases while Table 23-6 contains combustion data for dusts.

Vapor Cloud Explosions A vapor cloud explosion (VCE) occurs when a large quantity of flammable material is released, is mixed with enough air to form a flammable mixture, and is ignited. Damage from a VCE is due mostly to the overpressure, but significant damage to equipment and personnel may occur due to thermal radiation from the resulting fireball.

A VCE requires several conditions to occur (*Estimating the Flammable Mass of a Vapor Cloud,* American Institute of Chemical Engineers, New York, 1999):

1. The released material must be flammable.
2. A cloud of sufficient size must form prior to ignition.
3. The released material must mix with an adequate quantity of air to produce a sufficient mass in the flammable range.
4. The speed of the flame propagation must accelerate as the vapor cloud burns. This acceleration can be due to turbulence, as discussed in the section on confined explosions. Without this acceleration, only a flash fire will result.

Most VCEs involving flammable liquids or gases result only in a deflagration—detonations are unlikely. As the confinement of the vapor cloud increases, due to congestion from process equipment, the flame accelerates and higher overpressures are achieved. The higher overpressures may approach the severity of a detonation.

Four methods are available to estimate the damage from a VCE: TNT equivalency, TNO Multi-Energy method, Baker-Strehlow-Tang method, and computational fluid dynamics. The TNT equivalency method is discussed in the Estimation of Damage Effects section. The other methods are discussed elsewhere (*Guidelines for Evaluating the Characteristics of Vapor Cloud Explosions, Flash Fires and BLEVES,* American Institute of Chemical Engineers, New York, 1994; *Guidelines for Chemical Process Quantitative Risk Analysis,* 2d ed., American Institute of Chemical Engineers, 2000).

Boiling-Liquid Expanding-Vapor Explosions A boiling-liquid expanding-vapor explosion, commonly called a BLEVE (pronounced ble-vee), occurs when a vessel containing a liquid stored at a temperature above its normal boiling point fails catastrophically. After failure, a fraction of the liquid flashes almost instantaneously into vapor. Damage may be caused, in part, by the rapid expansion of the vapor and fragments from the failing vessel. The liquid may be water.

TABLE 23-5 Maximum Pressures and Deflagration Indices for a Number of Gases and Vapors

	Maximum pressure P_{max}, barg			Deflagration index K_G, bar·m/s		
Chemical	NFPA 68 1997	Bartknecht 1993	Senecal 1998	NFPA 68 1997	Bartknecht 1993	Senecal 1998
Acetylene	10.6			109		
Ammonia	5.4			10		
Butane	8.0	8.0		92	92	
Carbon disulfide	6.4			105		
Diethyl ether	8.1			115		
Ethane	7.8	7.8	7.4	106	106	78
Ethyl alcohol	7.0			78		
Ethyl benzene	6.6	7.4		94	96	
Ethylene			8.0			171
Hydrogen	6.9	6.8	6.5	659	550	638
Hydrogen sulfide	7.4			45		
Isobutane			7.4			67
Methane	7.05	7.1	6.7	64	55	46
Methyl alcohol		7.5	7.2		75	94
Methylene chloride	5.0			5		
Pentane	7.65	7.8		104	104	
Propane	7.9	7.9	7.2	96	100	76
Toluene		7.8			94	

Data selected from:
NFPA 68: Venting of Deflagrations (Quincy, Mass.: National Fire Protection Association, 1997).
W. Bartknecht, *Explosionsschutz: Grundlagen und Anwendung* (New York: Springer-Verlag, 1993).
J. A. Senecal and P. A. Beaulieu, "K_G: Data and Analysis," *31st Loss Prevention Symposium* (New York: American Institute of Chemical Engineers, 1997).

TABLE 23-6 Combustion Data for Dust Clouds*

Dust	Median particle size, μm	Minimum explosive dust conc., g/m³	P_{max}, barg	K_{St}, bar·m/s	Minimum ignition energy, mJ
Cotton, Wood, Peat					
Cotton	44	100	7.2	24	—
Cellulose	51	60	9.3	66	250
Wood dust	33	—	—	—	100
Wood dust	80	—	—	—	7
Paper dust	<10	—	5.7	18	—
Feed, Food					
Dextrose	80	60	4.3	18	—
Fructose	200	125	6.4	27	180
Fructose	400	—	—	—	>4000
Wheat grain dust	80	60	9.3	112	—
Milk powder	165	60	8.1	90	75
Rice flour	—	60	7.4	57	>100
Wheat flour	50	—	—	—	540
Milk sugar	10	60	8.3	75	14
Coal, Coal Products					
Activated carbon	18	60	8.8	44	—
Bituminous coal	<10	—	9.0	55	—
Plastics, Resins, Rubber					
Polyacrylamide	10	250	5.9	12	—
Polyester	<10	—	10.1	194	—
Polyethylene	72	—	7.5	67	—
Polyethylene	280	—	6.2	20	—
Polypropylene	25	30	8.4	101	—
Polypropylene	162	200	7.7	38	—
Polystyrene (copolymer)	155	30	8.4	110	—
Polystyrene (hard foam)	760	—	8.4	23	—
Polyurethane	3	<30	7.8	156	—
Intermediate Products, Auxiliary Materials					
Adipinic acid	<10	60	8.0	97	—
Naphthalene	95	15	8.5	178	<1
Salicylic acid	—	30	—	—	—
Other Technical, Chemical Products					
Organic dyestuff (blue)	<10	—	9.0	73	—
Organic dyestuff (red)	<10	50	11.2	249	—
Organic dyestuff (red)	52	60	9.8	237	—
Metals, Alloys					
Aluminum powder	<10	60	11.2	515	—
Aluminum powder	22	30	11.5	110	—
Bronze powder	18	750	4.1	31	—
Iron (from dry filter)	12	500	5.2	50	—
Magnesium	28	30	17.5	508	—
Magnesium	240	500	7.0	12	—
Silicon	<10	125	10.2	126	54
Zinc (dust from collector)	<10	250	6.7	125	—
Other Inorganic Products					
Graphite (99.5% C)	7	<30	5.9	71	—
Sulfur	20	30	6.8	151	—
Toner	<10	60	8.9	196	4

St Classes for Dusts

Deflagration index K_{St}, bar·m/s	St class
0	St-0
1–200	St-1
200–300	St-2
>300	St-3

*Data selected from R. K. Eckoff, *Dust Explosions in the Process Industries* (Oxford, England: Butterworth-Heinemann, 1997).

The most damaging BLEVE occurs when a vessel contains a flammable liquid stored at a temperature above its normal boiling point. The vessel walls below the liquid level are maintained at a low temperature due to the rapid heat transfer to the liquid. However, the vessel walls exposed to the fire above the liquid level will heat rapidly due to the much lower heat transfer to the vapor. The vessel wall temperature will increase to a point where the strength of the vessel wall is significantly reduced. The vessel wall will fail catastrophically, resulting in the flashing of a large quantity of flammable liquid into vapor. Since a fire is already present, the resulting vapor cloud will ignite almost immediately. Overpressures from the vessel failure may result, but most of the damage is caused by radiation from the resulting large fireball.

Dust Explosions

Nomenclature

A	Area of a vent opening, m^2
A_W	Effective vent area, m^2
A_K	Geometric vent area, m^2
$(dP/dt)_{max}$	Maximum rate of pressure rise, bar·s^{-1}
K_{max}	Maximum explosion constant, m·bar·s^{-1}
K_{min}	Minimum explosion constant, m·bar·s^{-1}
ℓ	Length of a pipe, m
L_D	Length of vent duct, m
L_{DS}	Maximum length of vent duct, m
L_F	Flame length, m
P_a	Activation overpressure, bar
P_{ext}	Maximum external peak of overpressure, bar
P_{max}	Maximum explosion overpressure, bar
P_{red}	Reduced explosion overpressure, bar
$P_{red,max}$	Maximum reduced explosion overpressure, bar
P_{stat}	Static activation overpressure, bar
R	Distance to vent area, m
T_{max}	Maximum permissible surface temperature, °C
V	Vessel volume, m^3
$\mathbf{v}_c$	Relative circumferential speed, m·s^{-1}
W_F	Width of flame, m
α	Angle between the horizontal and vertical axes of vent area, deg

GENERAL REFERENCES: Bartknecht, *Dust Explosions*, Springer, New York, 1989. Bartknecht, *Explosionsschutz* (Explosion Protection), Springer, Berlin, 1993. Crowl and Louvar, *Chemical Process Safety*, Prentice-Hall, New Jersey, 1990. "Dust Explosions," 28th Annual Loss Prevention Symposium, Atlanta, 1994. Eckhoff, *Dust Explosions in the Process Industries*, 2d ed., Butterworth-Heinemann, London, 1997. *Health, Safety and Loss Prevention in the Oil, Chemical and Process Industries*, Butterworth/Heinemann, Singapore, 1993. NFPA 69, *Standard on Explosion Prevention Systems*, Quincy, Mass., 1997. Hattwig and Steen, eds., *Handbook of Explosion Prevention and Protection*, Wiley-Vch Verlag GmbH & Co. KGaA, Weinheim, 2004. VDI-Report 1601, *Safe Handling of Combustible Dust*, VDI-Verlag GmbH, Düsseldorf, 2001. VDI-Guideline 2263, *Dust Fires and Dust Explosions*, Beuth Verlag, Berlin, 1992. European Standard EN 1127-1, *Explosives atmospheres: Explosion prevention and protection*, Pt. 1: *Basic Concepts and Methodology*, 1997.

Definition of Dust Explosion A dust explosion is the rapid combustion of a dust cloud. In a confined or nearly confined space, the explosion is characterized by relatively rapid development of pressure with flame propagation and the evolution of large quantities of heat and reaction products. The required oxygen for this combustion is mostly supplied by the combustion air. The condition necessary for a dust explosion is a simultaneous presence of a dust cloud of proper concentration in air that will support combustion and a suitable ignition source.

Explosions are either deflagrations or detonations. The difference depends on the speed of the shock wave emanating from the explosion. If the pressure wave moves at a speed less than or equal to the speed of sound in the unreacted medium, it is a deflagration; if it moves faster than the speed of sound, the explosion is a detonation.

The term *dust* is used if the maximum particle size of the solids mixture is below 500 μm.

In the following, dusts are called combustible in the airborne state only if they require oxygen from the air for exothermic reaction.

Glossary

Activation overpressure P_a That pressure threshold, above the pressure at ignition of the reactants, at which a firing signal is applied to the suppressor(s).

Cubic low The correlation of the vessel volume with the maximum rate of pressure rise. $V^{1/3} \cdot (dP/dt)_{max} = \text{constant} = K_{max}$.

Dust explosions class, St Dusts are classified in accordance with the K_{max} values.

Explosion Propagation of a flame in a premixture of combustible gases, suspended dust(s), combustible vapor(s), mist(s), or mixtures thereof, in a gaseous oxidant such as air, in a closed or substantially closed vessel.

Explosion pressure resistant (EPR) Design of a construction following the calculation and construction directions for pressure vessels.

Explosion pressure-shock resistant (EPSR) Design of a construction allowing greater utilization of the material strength than the EPR design.

Maximum reduced explosion overpressure $P_{red,max}$ The maximum pressure generated by an explosion of a dust-air mixture in a vented or suppressed vessel under systematically varied dust concentrations.

Minimum ignition energy (MIE) Lowest electrical energy stored in a capacitor which, upon discharge, is just sufficient to effect ignition of the most ignitable atmosphere under specified test conditions.

Minimum ignition temperature of a dust cloud (MIT_C) The lowest temperature of a hot surface on which the most ignitable mixture of the dust with air is ignited under specified test conditions.

Minimum ignition temperature of a dust layer (MIT_L) The lowest temperature of a hot surface on which a dust layer is ignited under specified test conditions.

Static activation overpressure P_{stat} Pressure at which the retaining element breaks or releases such that the venting element is able to open when the rate of pressure rise is ≤ 0.1 bar/min.

Vent area A Area of an opening for explosion venting.

Venting efficiency EF Ratio of the effective vent area A_w to the geometric vent area A_k.

Venting element That part of vent area device that covers the vent area and opens under explosion conditions.

Vessel length-to-diameter ratio (L/D) The ratio of the longest linear dimension L (length, height) of a round vessel to its geometric or equivalent diameter D.

Prevention and Protection Concept against Dust Explosions Explosion protection encompasses the measures implemented against explosion hazards in the handling of combustible substances and the assessment of the effectiveness of protective measures for the avoidance or dependable reduction of these hazards. The explosions protection concept is valid for all mixtures of combustible substances and distinguishes between

1. Measures that prevent or restrict formation of a hazardous explosive atmosphere
2. Measures that prevent the ignition of a hazardous, explosive atmosphere
3. Constructional measures that limit the effects of an explosion to a harmless level

From a safety standpoint, priority must be given to the measures in item 1. Item 2 cannot be used as a sole protective measure for flammable gas or solvent vapors in industrial practice with sufficient reliability, but can be applied as the sole protective measure when only combustible dusts are present if the minimum ignition energy of the dusts is high (≥10 mJ) and the operating area concerned can easily be monitored.

If the measures under items 1 and 2, which are also known as preventive measures, cannot be used with sufficient reliability, the constructional measures must be applied.

In practice, in most cases it is sufficient to determine and judge systematically the explosion risk with a sequence of specific questions, shown in Fig. 23-12.

During the evaluation it is assumed that ignition of an existing combustible atmosphere is always possible. The assessment is thus independent of the question of whether ignition sources are present.

Preventive Explosion Protection The principle of preventive explosion protection comprises the reliable exclusion of one of the requirements necessary for the development of an explosion. An explosion can thus be excluded with certainty by

- Avoiding the development of explosive mixtures
- Replacing the atmospheric oxygen by *inert gas*, working in a vacuum, or using *inert dust*
- Preventing the occurrence of effective ignition sources

Avoidance or Reduction of Explosive Combustible Fuel-Air Mixtures

Flammable gas or vapor-air mixtures This can be achieved if the flammable substance can be replaced by a nonflammable substance or the concentration of flammable substance can be kept so low that the gas or vapor-air mixture is too lean for an explosion.

The development of a hazardous atmosphere can also be prevented by ventilation measures. A distinction is made here between *natural* ventilation, which is usually sufficient only in the open air, and *artificial*

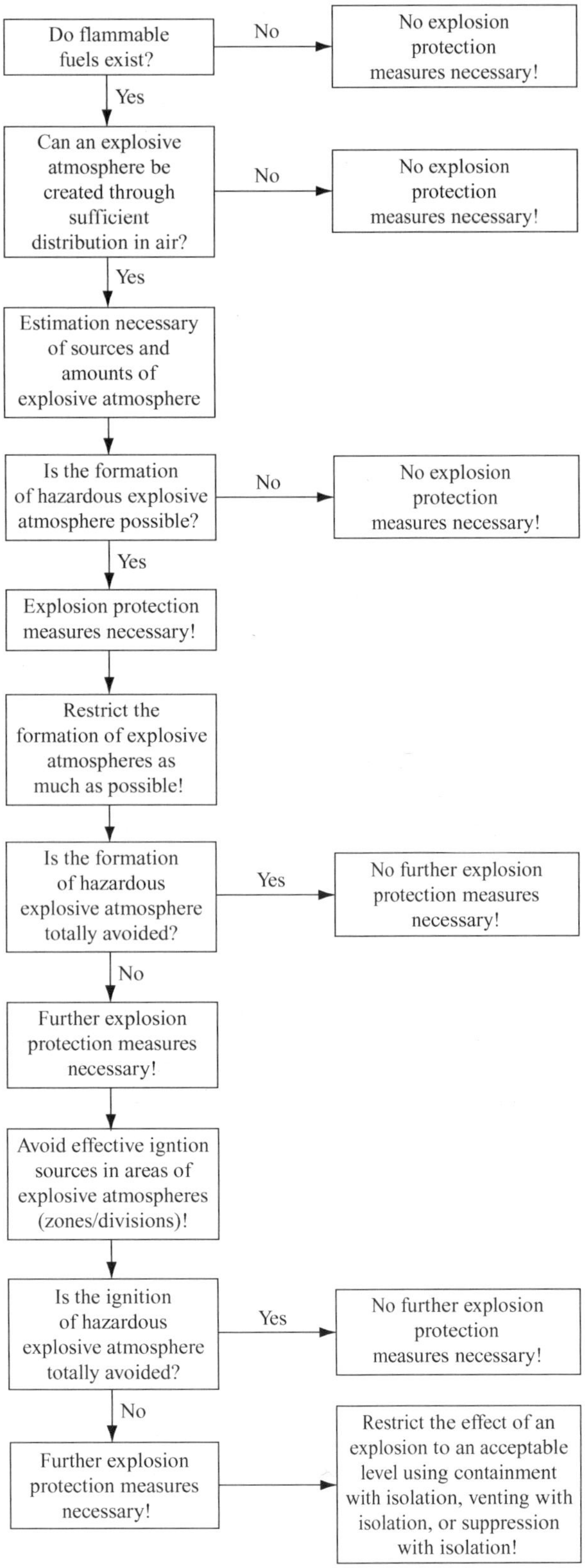

FIG. 23-12 Evaluation flowchart for preventing and/or limiting explosions.

ventilation (IEC 60079-10, "Electrical Apparatus for Explosive Atmospheres. Part 10: Classification of Hazardous Areas"). *Artificial* ventilation permits the use of greater amounts of air and the selective circulation of air in areas surrounding the equipment. Its use and the calculation of the minimum volume flow rate for the supply and exhaust air are subject to certain requirements.

The use of *gas alarm devices* in connection with, e.g., ventilation measures is also possible. The factors influencing the decision regarding setting of the required gas detectors include the relative density of the flammable gases and vapors. With gases and vapors that are heavier than air, the sensors and the waste air openings should be installed near the floor, with those that are lighter than air near the ceiling.

Combustible dust-air mixtures These mixtures can be avoided or restricted if the combustible dust can be replaced by a noncombustible dust or the dust concentration can be kept so low that an explosive dust-air mixture is never actually formed.

The explosibility limits do not have the same meaning as with flammable gases and flammable vapors, owing to the interaction between dust layers and suspended dust. This protective measure can, e.g., be used when dust deposits are avoided in operating areas or in the airstream of clean air lines after filter installation, where in normal operation the lower explosion limit (LEL) is not reached. However, dust deposits must be anticipated with time. When these dust deposits are whirled up in the air, an explosion hazard can arise. Such a hazard can be avoided by regular cleaning. The dust can be extracted directly at its point of origin by suitable ventilation measures.

Avoidance of Explosions through Inerting The introduction of inert gas in the area to be protected against explosions lowers the oxygen volume content below the limiting oxygen concentration (LOC) so that ignition of the mixture can no longer take place. This process is called *inerting* (CEN/TR 15281, "Guidance on Inerting for the Prevention of Explosions").

One has to be aware of the danger of asphyxiation from gases in inerted equipment. This is also important for surrounding areas in case of major leaks.

Inerting is not a protective measure to avoid exothermic decomposition. For the avoidance of (smoldering) fires, oxygen concentrations lower than the LOC must usually be adhered to and must be determined from case to case. In addition to the nitrogen normally used, all nonflammable gases which do not support combustion or react with the combustible dust can be considered for use as the inert gas. The inerting effect generally decreases in the following order: carbon dioxide → water vapor → flue gas → nitrogen → noble gases. In special gases, liquid nitrogen or dry ice is used.

The LOC depends upon the combustible material and the type of inert gas used. It decreases with increased temperature and pressure. A distinction has to be made between the determined LOC value and the concentration which results by subtracting a safety margin.

The maximum allowable oxygen concentration (MAOC), which is, in general, 2 vol % below the LOC, has to include the following considerations: Fluctuation in oxygen concentrations due to process and breakdown conditions per time and location, as well as the requirement for protective measures or emergency measures to become effective. In addition, a concentration level for an alarm has to be set below the MAOC.

Explosive dusts can also be changed into mixtures which are no longer explosive by the addition of inert dusts (e.g., rock salt, sodium sulfate). In general, inert dust additions of more than 50 wt % are necessary here. It is also possible to replace flammable solvents and cleaning agents by nonflammable halogenated hydrocarbons or water, or flammable pressure transmission fluids by halocarbon oils.

Avoidance of Effective Ignition Sources Explosions can be prevented if ignition sources capable of igniting combustible material-air mixtures can successfully be avoided. A distinction is made between trivial ignition sources (e.g., welding, smoking, cutting) and mechanically generated spark, mechanically generated hot surfaces, lumps of smoldering material, and static electricity. Trivial ignition sources can also reliably be excluded by organizational measures such as the systematic employment of permits. This measure should always be employed, even when constructional measures are applied, unless the explosive atmosphere is avoided with certainty.

TABLE 23-7 Zone Classification versus Minimum Extent of Preventive Measures against Danger of Ignition

Explosion hazard zone	Avoid ignition sources which can occur:
20/0	During normal operation, during frequent malfunction, and during rare malfunction
21/1	During normal operation and during frequent malfunction
22/2	During normal operation

Hazardous places are classified in terms of zones (divisions) on the basis of the frequency and duration of the occurrence of an explosive atmosphere (IEC 60079-10:200X, "Classification of Hazardous Areas"; European Standard EN 50281-3:200X, "Classification of Areas Where Combustible Dusts Are or May Be Present").

Fundamentally all 13 kinds of ignition sources mentioned and also described in detail in EN 1127-1 must be considered. In EN 1127-1, the descriptions refer to both the ignition mechanisms of the different kinds of ignition source and the necessary scope of protection (material- and zone-dependent). The ignition sources should be classified according to the likelihood of their occurrence in the following manner:

1. Sources of ignition which can occur *continuously* or *frequently*
2. Sources of ignition which can occur in *rare* situations
3. Sources of ignition which can occur in *very rare* situations

In terms of the equipment, protective systems, and components used, this classification must be considered equivalent to

1. Sources of ignition which can occur during *normal operation*
2. Sources of ignition which can occur solely as a result of *malfunctions*
3. Sources of ignition which can occur solely as a result of *rare malfunctions*

Ignition sources at devices, protective systems, and components must be avoided, depending on zones (Table 23-7); i.e., the zone determines the minimum extent of preventive measures against danger of ignition.

Flammable gas or vapor-air mixtures Due to their low minimum ignition energies (<1 mJ), avoidance of effective ignition sources in flammable gas or vapor-air mixtures is in principle possible only in exceptional cases. For *hot surfaces* a maximum permissible surface temperature T_{max} must be specified, with the help of the minimum ignition temperature (MIT) of flammable gases, vapors, or liquids, so that the temperature of all surfaces is not exceeded (Table 23-8).

Combustible dust-air mixtures For every installation a check has to be made to determine which ignition source may become effective and whether it can be prevented with a sufficient degree of safety. With more sensitive products and complex installations, it becomes more and more difficult to exclude ignition sources with ample safety.

With dusts the avoidance of effective ignition sources depends on the ignition sensitivity, i.e., on the temperature-related minimum ignition energy (MIE). Because a low MIE means that both the number of effective ignition sources rises and the probability of the release of an ignition of a dust-air mixture rises. Safe handling of particularly or extremely ignition-sensitive dusts requires in general increased use of preventive measures in comparison to normally ignitable sensitive dusts.

It is therefore usual to combine the evaluation of the ignition sensitivity (MIE) of the dusts with the type and extent of protective measures in accordance with Table 23-9.

TABLE 23-8 Measures against Ignition of Flammable Gas, Vapors, and Liquids by Hot Surfaces

Zone	Protection by limiting the hot surface temperature (EN 1127-1)
0	$T_{max} \leq 0.8$ MIT
1	$T_{max} \leq 0.8$ MIT
2	$T_{max} \leq$ MIT

TABLE 23-9 Ignition Sensitivity of Dusts as a Function of Protective Measures

Ignition sensitivity of dusts	Protective measures
Normally sensitive (MIE ≥ 10 mJ)	Avoidance of effective ignition sources.
Easily sensitive (3 mJ ≤ MIE < 10 mJ)	Expert decision **or** Avoidance of effective ignition sources **and** additional protective measures.
Extremely sensitive (MIE < 3 mJ)	Avoidance of effective ignition sources **and** additional protective measures.

Protective measures: Inerting or constructional explosion protection (e.g., containment).

For hot surfaces a maximum permissible surface temperature T_{max} must be specified, with the help of the minimum ignition temperature of a dust cloud MIT_C and the minimum ignition temperature of a dust layer MIT_L (usually 5-mm dust layer = glow temperature), so that the temperature of all surfaces is not exceeded. Independent of the zone the T_{max} for dust clouds is ⅔ MIT_C and for dust layers $MIT_L - 75$ K.

Mechanically generated sparks and resultant *hot surfaces* together are regarded as one of the more important causes of ignition in industrial practice. The hot surfaces show considerably better incendivity in comparison with the short-lived, mechanically generated sparks. Neither ignition source appears in industrial practice from the normal metallic materials of construction rubbing against each other or against stone if the relative circumferential speeds v_c are less than or equal to 1 m/s and the power requirement is no more than 4 kW (Table 23-10). This is not valid for cerium-iron, titanium, and zirconium.

Regarding *electrostatic ignition sources,* see the subsection "Static Electricity," later. In addition, see CENELEC CLC/TR 50404, "Electrostatics—Code of Practice for the Avoidance of Hazards due to Static Electricity," June 2003.

Explosion Protection through Design Measures Design measures which restrict the effects of an explosion to a safe level are always necessary when the goal of avoiding explosions cannot be achieved—or at least not with sufficient reliability—through the use of preventive explosion protection. This ensures that people are not injured and further that the protected equipment is usually ready for operation a short time after an explosion. In applying design measures, the possibility of an explosion is not prevented. Therefore, all exposed equipment has to be built to be explosion-pressure-resistant in order to withstand the anticipated explosion pressure. The anticipated explosion pressure may be the maximum explosion overpressure or the maximum reduced explosion overpressure. In addition, *any propagation of an explosion to other parts or process areas has to be prevented.* Depending on the anticipated explosion pressure, a distinction is made between the following explosion-pressure-resistant designs:

- Those capable of withstanding the maximum explosion overpressure
- Those capable of withstanding an explosion overpressure reduced by explosion suppression or explosion venting

The strength of the protected vessels or apparatus may be either explosion-pressure-resistant or explosion pressure shock-resistant.

Containment Containment is understood to mean the possibility of designing vessels and equipment for the full *maximum explosion overpressure*, which is generally from $P_{max} = 7$ to 10 bar. The explosion-resistant vessel can then be designed as explosion-pressure-resistant or

TABLE 23-10 Influence of Relative Circumferential Speed v_c on Danger of Ignition for Combustible Dusts

*$v_c \leq 1$ m/s	There is no danger of ignition.
$v_c > 1$–10 m/s	Every case has to be judged separately, considering the product and material-specific characteristics.
$v_c > 10$ m/s	In every case there is danger of ignition.

* In addition, low power requirements W ≤ 4 kW.

TABLE 23-11 Explosion Venting System Design Parameters

Explosion hazard definition	Venting system definition
Volume of vessel (free volume)	Type of venting device
Shape of vessel (cubic or elongated vessel)	Detection method for triggering a shutdown
Length-to-diameter ratio of vessel	Static activation overpressure P_{stat} of venting device
Strength of vessel	Venting capability of venting device
Type of dust cloud distribution (ISO method/ pneumatic-loading method)	Location of venting device on the vessel
Dust explosibility characteristics:	Position of equipment to be protected in the building
maximum explosion overpressure P_{max}	Length and shape of relief pipe if existent
maximum explosion constant K_{max}	Recoil force during venting
toxicity of the product	Duration of recoil force
Maximum flame length	Total transferred impulse
Pressure outside the vent areas	

explosion pressure shock-resistant. This protective measure is generally employed when small vessel volumes need to be protected, such as small filter units, fluidized-bed dryers, cyclones, rotary vales, or mill housings.

One has to consider that all connected devices must also withstand the maximum explosion overpressure. The NFPA 69 Standard, *Explosion Prevention System,* 1997; European Standard prEN 14460, *Explosion Resistant Equipment,* 2005; and Kirby and Siwek, "Preventing Failures of Equipment Subject to Explosions," *Chemical Engineering,* June 23, 1986, provide excellent guidance on the practice of containment.

Explosion venting The concept of explosion venting encompasses all measures used to open the originally closed vessels and equipment either briefly or permanently in a nonhazardous direction following an explosion. *Explosion venting is inadmissible when the escape of toxic or corrosive, irritating, carcinogenic, harmful-to-fruit, or genetically damaging substances* is anticipated. In contrast to containment, explosions in a vented vessel are characterized by the maximum reduced explosion overpressure $P_{red,max}$ instead of the maximum explosion overpressure P_{max} and by the maximum reduced rate of pressure rise $(dP/dt)_{red,max}$ instead of the maximum rate of pressure rise $(dP/dt)_{max}$.

By this method in general, the expected inherent maximum explosion overpressure on the order of P_{max} = 7 to 10 bar will be reduced to a value of $P_{red,max}$ < 2 bar. In this case, the static activation overpressure of the venting device is $P_{stat} \leq 0.1$ bar. The resulting $P_{red,max}$ may not exceed the design pressure of the equipment. The explosion as such is not prevented; only the dangerous consequences are limited. However, subsequent fires must be expected.

Among other things, one prerequisite to calculate the pressure relief openings needed on the apparatus is knowledge about the explosion threat definition and venting system hardware definition. The various factors are summarized in Table 23-11.

The inertia, the opening behavior of a bursting disk or of the movable cover of an explosion device, and its arrangement (horizontal, vertical) can affect the venting efficiency and may result in a higher maximum reduced explosion overpressure inside the protected vessel (Fig. 23-13). This venting efficiency is mainly dependent upon the specific mass of the venting device.

If the specific mass of a venting device is < 0.5 kg/m², then it has a venting efficiency of EF = 1 and is called inertia-free and does not impede the venting process (European Standard prEN 14797-2006, "Explosion Venting Devices"). For such explosion venting devices EF testing is therefore not required. Explosion venting devices with venting elements with a specific mass > 0.5 kg/m² can influence the venting process by their opening and release behavior. Experiments have shown that explosion venting devices with a specific mass > 0.5 kg/m² and ≤ 10 kg/m² can be considered as inertia-free, which means having a venting efficiency EF = 1 provided that

$$\frac{A}{V} < 0.07 \qquad (23\text{-}11)$$

The equations are only valid for

- Vessel volumes of 0.1 m³ ≤ V ≤ 10,000 m³
- Static activation overpressure of venting device $P_{stat} \leq 0.1$ bar

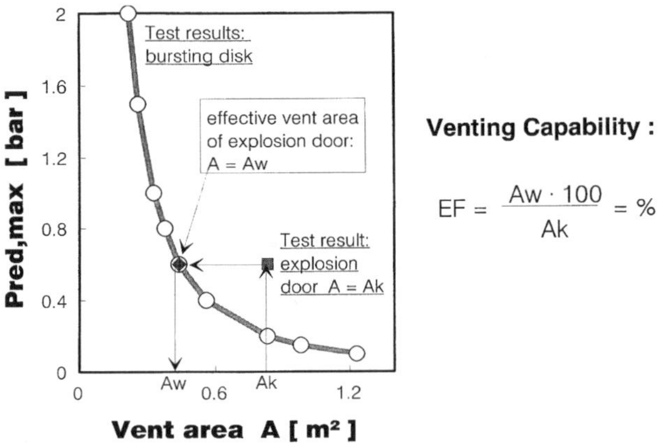

FIG. 23-13 Definition of the venting capability EF of an explosion door in comparison with a plastic foil rupture disk.

- Vessel strength (= $P_{red,max}$) of 0.1 bar < P ≤ 2 bar
- $P_{red,max} > P_{stat}$

For all other conditions EF has to be determined by tests (Fig. 23-13). EF and therefore the effective vent area A_W of a non-inertia-free explosion device are smaller than the venting efficiency of an inertia-free vent device (specific mass < 0.5 kg/m²) with the same vent area. Therefore, such devices need testing to determine the mechanical strength before actual use, and the EF or the pressure rise, respectively, has to be chosen relative to the $P_{red,max}$ of the rupture disk of the same area.

When explosion doors that close the vent area after the explosion are in use, the cooling of the hot gases of combustion may create a vacuum in the vessel, resulting in its deformation. To prevent this from happening, vacuum breakers have to be provided.

Sizing of vent areas The empirical equation (23-12) can be used to calculate the required vent area for flammable gas or solvent vapor explosions. The equation is valid for flammable gas-air mixtures which have been ignited in a quiescent state (nonturbulent) with an ignition source of E = 10 J.

$$A = [(0.1265 \log K_{max} - 0.0567) P_{red,max}^{-0.5817} + 0.1754 P_{red,max}^{-0.5722} (P_{stat} - 0.1)] \times V^{2/3} + \frac{[AK_{max}(L/D - 2)^2]}{750} \qquad (23\text{-}12)$$

The equation is valid for

- Vessel volumes 0.1 m³ ≤ V ≤ 1000 m³
- Vessel length-to-diameter ratio 1 ≤ L/D ≤ 5
- Static activation overpressure of rupture disk 0.1 bar ≤ P_{stat} ≤ 0.5 bar
- Maximum reduced explosion overpressure 0.1 bar ≤ $P_{red,max}$ ≤ 2 bar
- $P_{red,max} > P_{stat}$ + 0.05 bar

- Maximum explosion constant 50 bar·m/s $\le K_{max} \le$ 550 bar·m/s
- Gas-air mixtures ignited at zero turbulence
- Venting efficiency EF = 1

If it is necessary to locate equipment with explosion vents inside buildings, vent ducts should be used to direct vented material from the equipment to the outdoors. Vent ducts will significantly increase the pressure development in the equipment during venting. They require at least the same cross section as the vent area and the same design pressure as the protected equipment.

The use of vent ducts results in an increase in $P_{red,max}$. The maximum reduced explosion overpressure $P'_{red,max}$ caused by the downstream vent duct can be calculated with Eqs. (23-13) and (23-14).

Length of vent duct = 0 m $< L_D \le$3 m:

$$P'_{red,max} = 1.24P_{red,max}^{0.8614} \tag{23-13}$$

Length of vent duct = 3 m $< L_D \le$ 6 m:

$$P'_{red,max} = 2.48P_{red,max}^{0.5165} \tag{23-14}$$

where $P'_{red,max}$ = maximum explosion overpressure with vent duct, bar, and $P_{red,max}$ = maximum explosion overpressure without vent duct, bar.

The following empirical equations, Eqs. (23-15) to (23-17), allow the calculation of the size of a vent area A for combustible dust-air-mixture explosions.

For $P_{red,max} < 1.5$ bar

$$A = B\left(1 + C\log\frac{L}{D}\right) \quad \text{m}^2 \tag{23-15}$$

For $P_{red,max} \ge 1.5$ bar

$$A = B \qquad \text{m}^2$$

with

$$B = [3.264\cdot10^{-5}\cdot P_{max}\,K_{max}\,P_{red,max}^{-0.569} + 0.27(P_{stat} - 0.1)P_{red,max}^{-0.5}]\,V^{0.753} \tag{23-16}$$

$$C = -4.305\log P_{red,max} + 0.758 \tag{23-17}$$

The equations are valid for

- Vessel volumes 0.1 m$^3 \le V \le$ 10,000 m^3
- Static activation overpressure of venting device 0.1 bar $\le P_{stat} \le$ 1 bar
- Maximum reduced explosion overpressure 0.1 bar $< P_{red,max} \le$ 2 bar
- $P_{red,max} > P_{stat}$,
- Maximum explosion overpressure 5 bar $\le P_{max} \le$ 10 bar for a maximum explosion constant 10 bar·m/s $\le K_{max} \le$ 300 bar·m/s
- Maximum explosion overpressure 5 bar $\le P_{max} \le$ 12 bar for a maximum explosion constant 300 bar·m/s $\le K_{max} \le$ 800 bar·m/s
- Length-to-diameter ratio of the vessel $L/D \le 20$
- L/D limited in that the maximum vent area shall not be greater than the cross-sectional area of the equipment
- Venting efficiency EF = 1

The required area for pressure venting increases with increased length (height) to diameter ratio of the vessel, in comparison with the area requirement for L/D = 1 vessel. For low $P_{red,max}$, the required effective vent area will be markedly influenced by the ratio L/D. Such influence diminishes with increasing reduced explosion overpressure and ceases at $P_{red,max}$ = 1.5 bar as per experimental results. However, with $P_{red,max} \ge 1.5$ bar, no influence of the L/D ratio of the vessel can be noticed.

The influence of the vent duct upon the pressure increase is most pronounced when the flame propagation from the secondary explosion in the vent duct reaches the velocity of sound. This is valid for vent ducts of

$$L_D = L_{DS} = 4.564P_{red,max}^{-0.37} \qquad \text{m} \tag{23-18}$$

Vent ducts with a length of $L_D > L_{DS}$ have no additional effect upon the pressure increase. Therefore, L_{DS} will be the maximum vent duct length that has to be considered. *The above-mentioned equation (23-18) is not valid for metal dusts.*

Independent of the location of the vent duct, the maximum reduced explosion overpressure $P'_{red,max}$ caused by the downstream vent duct can be calculated for vessels having an L/D ratio of 1 with Eq. (23-19).

Vessel ratio L/D = 1 (longitudinal und transversal):

$$P'_{red,max} = P_{red,max}[1 + 17.3L_D(AV^{-0.753})^{1.6}] \tag{23-19}$$

where $P'_{red,max}$ = maximum explosion overpressure with vent duct, bar
$P_{red,max}$ = maximum explosion overpressure without vent duct, bar
A = vent area, m^2, without vent duct
V = volume of protected vessel, m^3
L_D = length of vent duct, m

The equation is valid for

- Vessel volumes 0.1 m$^3 \le V \le$ 10,000 m^3
- Static activation overpressure of the venting device 0.1 bar $\le P_{stat} \le$ 1 bar
- Maximum reduced explosion overpressure 0.1 bar $< P_{red,max} \le$ 2 bar, with $P_{red,max} > P_{stat}$
- Maximum explosion overpressure 5 bar $\le P_{max} \le$ 12 bar and a maximum explosion constant 10 m·bar/s $\le K_{max} \le$ 800 m·bar/s
- Vessel L/D = 1
- Vent duct length $L_D \le L_{DS}$

Experimental studies have proved that the influence of vent duct with longitudinal arrangement—*located on the roof*—decreases markedly with increased vessel length-to-diameter ratio. The increase of the maximum explosion overpressure is at its maximum if vessel ratio L/D = 1.

Hazard due to flame and pressure in the surroundings The maximum outside range of a flame L_F originating from a vessel *increases* with increased volume of the vented vessel.

Pressure and blast effects external to a vent arise from pressures generated by the vented explosion inside the plant and the explosion area outside the vent.

Explosion suppression During a suppression of an explosion, not products, residues from combustion, residues from gases, or flames can escape from the protected vessel, because an explosion suppression system reduces the effects of these explosions to a harmless level, by restricting the action of the flames during the initial phase of the explosion. This prevents the installation in question from being destroyed and people standing in the area of the installation from being injured. A further benefit of explosion suppression systems is that they can be deployed for combustible products with toxic properties and can be used irrespective of the equipment location.

An explosion can generally be considered suppressed if the expected maximum explosion pressure P_{max} at the optimum concentration of the combustible product (7 to 10 bar)—assuming the explosion suppression system has an activation overpressure P_a^* of 0.1 bar—is reduced to a maximum reduced explosion overpressure $P_{red,max}$ $\le$ 1 bar. This means that a vessel safeguarded in this way needs to be designed so that it is secured against explosions of up to 1 bar (equivalent to $P_{red,max}$). The activation overpressure P_a is that pressure at which an explosion suppression system will be activated.

To initiate an explosion suppression system, a detector is used to sense either an overpressure generated by, or a flame of, an incipient explosion. It is important to locate the detector in a position that ensures sufficient time for the suppression system to sense and activate the devices to extinguish the explosion.

Optical detectors shall be used in more open configurations where pressure buildup due to the incipient explosion is limited. Optical detectors shall not be used where high dust concentrations limit the reliability of the suppression system. Both uv and ir detectors are available for optical detection. The use of daylight-sensitive sensors shall be avoided to avoid spurious activation. The sensor shall be mounted such that the angle of vision allows it to cover all the protected hazard area. The performance of an optical detector will also be affected by any obstacles within its vision, and this shall be overcome by the introduction of more detectors. Optical detectors shall be fitted with air shields to keep the optical lens clean.

TABLE 23-12 Suppression System Design Parameters

Explosion hazard definition	Suppression system hardware definition
Volume of vessel (free volume V) Shape of vessel (area and aspect ratio) Type of dust cloud distribution (ISO method/pneumatic-loading method) Dust explosibility characteristics: Maximum explosion overpressure P_{max} Maximum explosion constant K_{max} Minimum ignition temperature MIT	Type of explosion suppressant and its suppression efficiency Type of HRD suppressors: number and free volume of HRD suppressors and the outlet diameter and valve opening time Suppressant charge and propelling agent pressure Fittings: elbow and/or stub pipe and type of nozzle Type of explosion detector(s): dynamic or threshold pressure, UV or IR radiation, effective system activation overpressure P_a Hardware deployment: location of HRD suppressor(s) on vessel

Pressure detection shall be used for closed enclosure applications. Threshold detectors provide an electric signal when a preset overpressure is exceeded. Dynamic detectors provide an electric signal to the control and indicating equipment (CIE). Typically they have both rate-of-rise and pressure threshold triggering points that can be configured specifically to the application conditions. Although this type of detector minimizes spurious activation of the isolation system (due to pressure fluctuations other than explosion pressure rise), care shall be taken to set up such detectors to meet appropriate detection response criteria for the particular application and protected enclosure geometry.

The suppressants deployed in suppression systems are water and dry and liquid chemicals. Apart from the effectiveness of the suppressant used, the compatibility of the suppressant with the process shall be considered. A suppressant is regarded as being very effective when an increase of the activation pressure P_a of the explosion suppression system leads to a small increase in the maximum reduced explosion overpressure $P_{red,max}$. The application of a suppressant is dependent upon how effective it is at suppressing an explosion. Testing shall be used to determine the effectiveness and performance of the suppressant, thus quantifying the applicability of the suppressant. The following parameters shall be considered when selecting a suppressant:

- Any adverse reaction with the process products
- The toxicity levels of the suppressant relating to occupational exposure limits
- The temperature stability of the suppressant

In addition, the following properties shall be taken into account where necessary:

- Will the suppressant have to be food-compatible?
- Will the suppressant cause the onset of corrosion?
- Is the suppressant environmentally friendly?
- Can the suppressant be easily removed from the process?

Suppression system design parameters fall into the two categories of explosion hazard definition and suppression system hardware definition. The various influences are summarized in Table 23-12.

Comparison of explosion protection design measures In Table 23-13, comparison is made of the explosion protection design measures of containment, explosion venting and explosion suppression. In addition all three design measures are in combination with explosion isolation. Regarding the effectiveness of the different explosion design measures, all three techniques are equal if the design of these measures is performed properly and the design measures are inspected by a component person at least once a year or more often depending on the process and/or environmental conditions.

Explosion Isolation For all equipment systems protected by design safety measures it is also necessary to prevent the propagation of an explosion from these protected vessels into operating areas or equipment connected via interconnecting pipeline. Such an approach is referred to as *explosion isolation.*

To prevent an explosion occurring in, e.g., a constructional protected installation from spreading through a pipeline ($\ell > 6$ m) to part of the installation fitted with preventive explosion protection, explosion isolation measures (see Fig. 23-14) must be implemented. As explosions are generally propagated by *flames* and not by pressure waves, it is especially important to detect, extinguish, or block this flame front at an early stage, i.e., to isolate or disengage the explosion. If there is no explosion isolation, then the flame issuing from the equipment, e.g., from the equipment protected through design (equipment part 1), through the connecting pipeline comes into contact with a highly turbulent recompressed mixture in the equipment with preventive protection (equipment part 2). The mixture will ignite in an instant and explode; a large increase in the rate of combustion reaction and, naturally, in the reduced explosion overpressure is the result. The equipment in question may be destroyed.

The isolation can be done with very different systems, which have in common that they become effective only by an explosion.

Since the action of the isolation systems requires the physical effects of an explosion, in the selection of a suitable system consideration must be given to process engineering and machine boundary

TABLE 23-13 Comparison of Explosion Protection Design Measures

	Containment with isolation	Explosion venting with isolation	Explosion suppression with isolation
Pressure resistance P	7–10 bar	Without relief pipe, up to 2 bar With relief pipe, up to 4 bar	St 1 up to 0.5 bar St 2+3 up to 1.0 bar
Location	Independent	Dependent	Independent
Limits of application	Products which decompose spontaneously	Toxic products and products which decompose spontaneously	Products which decompose spontaneously, metal dust hazard
Environmentally friendly	Yes	No (flame, pressure, and product)	Yes
Loss of material*	+++++	+++++	++
Maintenance requirements†	++++	++++	+++++

*The loss of material by using containment and explosion venting is always much greater than that by using explosion suppression.

†To ensure the reliability of explosion protection devices, regular servicing and maintenance are required. The nature and time intervals of these activities depend on technical specifications and on the plant situation. Normally, after commissioning of the plant, inspections are carried out in comparatively short intervals, e.g., every month. Positive experience may subsequently provide for longer service intervals (every three months). It is recommended to contract service and maintenance to reliable, specialized companies.

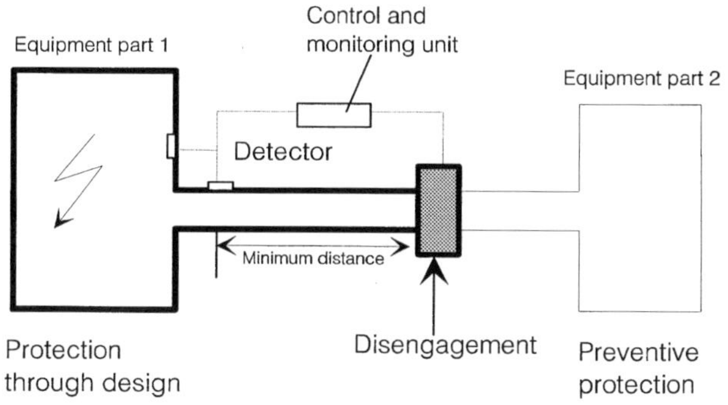

FIG. 23-14 Principle of the constructional measure explosion isolation.

conditions, particularly since the function and operability of these systems are not generally unrestricted.

Dust-carrying pipelines are different from gas-carrying pipelines—isolation devices can only be used which do not lose their function by the presence of the dust.

Today it is usual to divide the assigned isolation system in accordance with its mode of operation in *passive* isolation systems and *active* isolation systems. The passive isolation systems work without additional control units; i.e., the function (release) is determined by the physical effects of the explosion. Active isolation systems, however, are dependent on additional control and/or release mechanisms, without which they are nonfunctioning. Table 23-14 summarizes the different isolation systems.

Proper installation is dependent on the existing explosion protection measures and on the corresponding isolation system to be used. This guarantees operability. In addition the application and installation remarks of the manufacturer as well as the limits of application in accordance with the Type-Examination Certificate are to be obeyed in every detail. In Table 23-15 the most frequently used isolation systems for explosion protection measures are summarized.

An optical flame sensor installed at the beginning of the pipeline is the most suitable device for such an isolation system, since the propagating flame from the explosion has to be detected and extinguished. Pressure detectors alone are, in principle, not suited to the case on hand because there is no distinct separation between the pressure and flame fronts for explosion in pipelines. Optical ir sensors that have a relatively low sensitivity to daylight are normally chosen and have proved themselves amply in industrial practice. Therefore, daylight into the pipe in the vicinity of the sensor must be avoided. It is necessary to flush the optical lens with gas (e.g., nitrogen, air) to keep it dust-free.

TABLE 23-14 Different Isolation Devices

Type of isolation	Suitable for:
Rotary air lock (active)	Dusts
Extinguishing barrier (active)	Gas, dusts, and hybrid mixtures
Explosion protection valve (passive or active)	Gas, dusts, and hybrid mixtures
Explosion diverter (passive)	Gas, dusts, and hybrid mixtures
Double slide valve (active)	Gas, dusts, and hybrid mixtures
Product layer as a barrier (active)	Dusts and solvent humid products
Screw conveyer (passive)	Dusts and solvent humid products
Extinguishing barrier or explosion protection valve in combination with explosion diverter (active)	Gas, dusts, and hybrid mixtures
Flame arresters	Gas (see subsection "Flame Arresters," later)

TABLE 23-15 Most Frequently Used Isolation Systems as Related to the Installed Construction Explosion Protection Measure

Isolation system	Containment	Explosion venting	Explosion suppression
Double slide valve system	✔	✔	✔
Explosion diverter	✔	✔	✔
Explosion protection valve system	✔	✔	✔
Extinguishing barrier	—	✔	✔
Extinguishing barrier in combination with explosion diverter	—	✔	✔
Explosion protection valve in combination with explosion diverter	✔	✔	✔
Product layer as a barrier	—	✔	✔
Screw conveyer	✔	✔	✔
Rotary valve	✔	✔	✔

In difficult situations, it is recommended to install both types of sensors—pressure detector in the vessel and flame detector in the pipe—and they must be switched in an OR-logic to activate the isolation device (see Fig. 23-14).

The pressure detector in the vessel provides the earliest detection of an explosion in the interconnected vessel, whereas the flame detector in the pipeline provides assured detection, even for lazy flames, as they propagate down the pipeline toward the extinguishing barrier.

For the explosion protection valves always two installation distances are to be indicated, namely, the *minimum* and *maximum* installation distances. The *minimum installation distance* ensures that the given isolation system can react in time to prevent an explosion propagation beyond the installation place. The *maximum installation distance* ensures that a detonation in the piping cannot be formed up to the isolation system and/or that excessive pressure loads are avoided at the isolation system.

With extinguishing barriers the minimum and the maximum installation distances and the *extinguishing distance* must be considered after the extinguishing barrier. This *extinguishing distance* is specified as the minimum length after the extinguishing barrier to guarantee proper function of that extinguishing barrier. Only after this extinguishing distance may the piping connect to other equipment.

The minimum installation distance between the detector which activates the isolation system and the isolation system itself depends essentially on

- Type of detectors (pressure/flame)
- Volume of the vessel to be protected
- Maximum explosion overpressure P_{max} or maximum reduced explosion overpressure $P_{red,max}$ in the protected vessel
- Diameter of the pipeline
- Reaction time of the isolation system
- Maximum explosion overpressure P_{max} of the fuel
- Maximum explosion constant K_{max} of the fuel
- Minimum explosion constant K_{min} of the fuel (typically $\frac{1}{3}$ K_{max}, but minimum 50 m·bar/s)
- Forward air velocity in the pipeline
- Minimum ignition temperature of a dust cloud (only for extinguishing barriers)
- Type of suppressant (only for extinguishing barriers)

Finally, it must be pointed out that all explosion protection devices or systems used in practice may be used only when their pressure rating, flameproof, and functional testing have been proved by competent bodies and their test results including the limits of application are documented in a type test certificate.

Static Electricity

Nomenclature

C	Capacitance
C/kg	Unit of charge density
C/m^2	Unit surface charge density
F	Farad
J	Unit of energy (joules)
K_e	Relative dielectric constant, dimensionless
kV/m	Unit of electric field intensity
m	Meter
MIE	Minimum ignition energy, mJ
mJ	Millijoule
Ω^2	Resistivity value, ohms per square, usually used for fabrics and films
pS	Unit of conductivity (picosiemens)
pS/m	Unit of electrical conductivity of liquid
RH	Relative humidity, %
S	Siemen, formerly mho
T	Time, s
V	Electric potential, V
V/m	Unit of electrical field intensity

GENERAL REFERENCES: Gibson and Lloyd, "Incendivity of Discharges from Electrostatically Charged Plastics," *Brit. J. of Appl. Phys.* **16**, pp. 1619–1631, 1965. *Plant/Operations Progress* **7,** No. 1, Jan. 1988. Entire issue devoted to papers on static electricity, presented at AIChE meeting, Minneapolis, Minn., August 1987. "Protection Against Ignitions Arising Out of Static, Lightning and Stray Currents," American Petroleum Institute Recommended Practice 2003, 1998. M. Glor, "Electrostatic Ignition Hazards Associated with Flammable Substances in the Form of Gases, Vapors, Mists, and Dusts," *Inst. Phys. Conf. Ser. No. 163,* London, pp. 199–206, 1999. National Fire Protection Association, "Recommended Practice on Static Electricity," NFPA 77, Quincy, Mass., 2007. National Fire Protection Association, "Standard for the Manufacture of Organic Coatings," NFPA 35, 2005. "Safety of Machinery; Guidance and Recommendations for the Avoidance of Hazards due to Static Electricity," Cenelec Report RO 44-001, European Committee for Electrotechnical Standardization, Brussels, 1999.

Introduction Spark ignition hazards must be considered whenever static charges may accumulate in an environment that contains a flammable gas, liquid, or dust. The need for electrical bonding and grounding of conductive process equipment in hazardous (classified) locations is widely recognized. Less well understood are the ignition hazards associated with static charges on poorly conductive, flammable liquids, solids, and powders. Static charges, generated on these materials by normal handling and processing, cannot be conducted to ground quickly, and may cause hazardous charge accumulations. The electric fields associated with these charges may stress the surrounding air sufficiently to cause breakdown by some type of electrical discharge.

Electrical discharges from poorly conductive materials take several forms, each differing in its ability to ignite flammable mixtures. It is not possible to calculate the incendivity of most of these discharges, because of their varying time and spatial distributions. Several engineering rules of thumb for estimating the relative hazard of these discharges are discussed below.

An analysis of static ignition hazard should start with data on the ignition sensitivity of the particular flammable material at its most flammable concentration in air, i.e., its MIE. This is especially important for dusts. It is prudent to determine this value on fines of the specific dust of interest, rather than to rely on published data. *Hybrid mixtures*, i.e., mixtures of dust and vapor for which vapor concentrations may be below their lower flammable limit, can be ignited by smaller discharge energies than might be expected.

The key to safe operation is to provide an adequate means of charge dissipation from charged materials to ground. This requires mobility of charges in or on the charged material *plus* electrical continuity from the material to ground.

Definitions

Bonding A method of providing electrical continuity between two or more conductive objects to prevent electrical sparking between them.

Charge relaxation time The time required for a charge in a liquid or on a solid material to dissipate to 36.8 percent of its initial value when the material is grounded.

Electrical discharge A current flow that occurs when the electrical field strength exceeds the dielectric breakdown value of a medium such as air.

Flammable mixture A mixture of a gas, vapor, mist, or dust in air which is within its flammable range.

Grounding A special form of bonding in which a conductive object is connected to (earth) ground.

Incendive discharge Any electrical discharge that has sufficient energy to ignite a specified flammable mixture.

Minimum ignition energy The smallest amount of spark energy that has been found capable of igniting a specified flammable mixture in a standard test.

Static-dissipative (antistatic) material One with an electrical resistivity that is low enough to make it incapable of accumulating hazardous concentrations of static charges when grounded.

Electrostatic Charging The primary cause of electrostatic charging is contact electrification, which takes place when two different materials are brought into contact and separated. Other causes include induction charging, the formation of sprays, and impingement of charged mist or particles on an ungrounded conductor.

Contact electrification involves the contact and separation of solid-solid, solid-liquid, or liquid-liquid surfaces. Pure gases do not cause charging unless they carry droplets or dust particles.

Efforts to quantify the magnitude and polarity of contact charging have had limited success, because minute variations in the types and concentration of contaminants exert a large influence on charge separation. Even like solid-solid surfaces can produce significant charge separation. The charge density on separated surfaces is usually very nonuniform. Each surface may contain both + and − charges, with more of one polarity than of the other. After separation, the charges dissipate slowly or rapidly, depending upon the electrical resistivity of the material and the presence of a path to ground.

Contact electrification at liquid-liquid and liquid-solid surfaces is attributed to the adsorption of ions of one polarity by one surface. Ions of opposite polarity form a diffuse layer near the interface. If the diffuse layer is carried along by moving liquid, or in a pipeline, the flowing charges (called a *streaming current*) may create a sparking hazard downstream. One protective measure is to keep the charged liquid in a closed, grounded system (a *relaxation chamber*) long enough to allow for safe dissipation of the charges.

The magnitude of the streaming current in any given situation is not readily calculated. Equations, derived experimentally, for some liquids (Bustin and Dukek, *Electrostatic Hazards in the Petroleum Industry*, Research Studies Press, Letchworth, England, 1983) show that flow velocity and filters have the greatest influence on pipeline charging. Streaming currents can usually be limited to safe levels by limiting velocities to less than 1 m/s.

Charge induction takes place when a conducting object is exposed to electric fields from other objects. Examples include the induction charging of a human body by charged clothing, the charging of a conductive liquid in a charged plastic container, and the charging of the conductive coating on one-side-metallized film by static charges on the uncoated surface.

Although charge induction can take place whether or not the conductive object is grounded, a sparking hazard is present only if the conductor is not grounded. This phenomenon can convert a relatively innocuous charge buildup on a nonconductor to a serious sparking hazard by raising the potential of the conductor aboveground (Owens, "Spark Ignition Hazards Caused by Charge Induction," *Plant/Operations Progress* **7,** no. 1, pp. 37–39, 1988).

Droplets, formed by **spray nozzles,** tend to be highly charged, even if the conductivity of the liquid is high. Because there is no path to ground from the droplets, their charges can accumulate on an ungrounded conductor to cause sparking. If flammable vapor is present, as in some tank cleaning operations, it is essential that the spray nozzle and the tank be bonded or separately grounded. Other precautions include the use of a nonflammable cleaning solvent or the use of an inert gas.

Although charged **mists** are unable to cause ignition of flammable vapor by self-generated sparking, it is important that the mist not impinge on an ungrounded conductor.

Charge Dissipation It is an experimental fact that charged objects exert a force on other charged objects. This behavior is explained by the presence of an *electric field*, i.e., electric lines of force, each of which emanates from a + charge and terminates on a − charge. The *magnitude* of the field is defined as the force on a unit test charge, placed at the point of interest. The *direction* of the field is the direction of the force on a + test charge placed at that point.

Static charge generation causes an ignition hazard only if the accumulated charges create an electric field that is sufficient to produce an electrical discharge in a flammable atmosphere. In most processes, this means that the electric field intensity at some location must reach the dielectric breakdown strength of air (nominally 3×10^6 V/m). The objective of static control measures is to ensure that electric field intensities cannot reach this value.

Bonding and grounding are the primary means of dissipating charges from **conductive objects.** Bonding clamps should be of the single-point type, which bites through oxide or enamel coatings to make contact with the bare metal. Owing to the sturdy construction of bonding clamps and cables, their initial resistance is less than 1 Ω. It is good practice to visually inspect the condition of bonding cables and clamps during each use, and to measure the resistance of temporary bonding cables at least annually, to confirm that it is less than, say, 25 Ω.

Charge-dissipative materials allow static charges to dissipate without causing hazardous accumulations. Charge dissipation normally takes place by conduction along the material to ground. The charge-dissipating behavior of such materials is measured at a controlled temperature and relative humidity, in terms of Ω^2 (ohms per square) of *electrical surface resistivity*. The maximum safe resistivity depends, in part, upon the rate of charge generation, but is typically in the range of 10^8 to 10^{11} Ω^2 for fabrics and films [ASTM Standard Test Method D257-99 (2005), "DC Resistance or Conductance of Insulating Materials"].

An alternate test for charge-dissipating performance is the *charge decay* test, in which the time of charge decay is measured after a potential of 5 kV has been applied to the specimen (Federal Test Method Standard 101C, Method 4046.1). For many purposes, a charge decay time of 0.5 s to 500 V, measured at the RH in end use, indicates acceptable performance.

The electrical surface resistivity and the charge decay time of most materials vary substantially with RH. It is important that test specimens be conditioned and tested at the lowest RH expected during use. Items that are acceptable at 50 percent RH may not be safe at 20 percent RH.

Some fabrics contain a small percentage of conductive or antistatic fibers or staple, which limit charge accumulation by *air ionization*. These static-dissipative fabrics do not depend upon electrical conduction of static charges, and may not pass the resistivity or the charge decay test. Their performance is not humidity-dependent. Antistatic performance is determined by measuring the charge transferred in electrical discharges from the charged fabric, and by the ability of these discharges to ignite flammable mixtures having a known MIE.

The rate of dissipation of charges in **liquid,** assuming that its conductivity and dielectric properties are constant, can be expressed as

$$T = 8.85K_e/C \qquad (23\text{-}20)$$

where T = time required for charge density to dissipate to 36.8% of its initial value, s
K_e = relative dielectric constant of liquid, dimensionless
C = electrical conductivity of liquid, pS/m

Flammable liquids are considered particularly static-prone if their electrical conductivity is within the range of 0.1 to 10 pS/m. If no particulates or immiscible liquids are present, these liquids are considered safe when their conductivity has been raised to 50 pS/m or higher. Blending operations or other two-phase mixing may cause such a high rate of charging that a conductivity of at least 1000 pS/m is needed for safe charge dissipation (British Standard 5958, part 1, "Control of Undesirable Static Electricity," para. 8, 1991).

Electrostatic Discharges An electrostatic discharge takes place when a gas- or a vapor-air mixture is stressed electrically to its breakdown value. Depending upon the specific circumstances, the breakdown usually appears as one of four types of discharges, which vary greatly in origin, appearance, duration, and incendivity.

Spark discharges are most common between solid conductors, although one electrode may be a conductive liquid or the human body. They appear as a narrow, luminous channel and carry a large peak current for a few microseconds or less. Sparks are the only form of discharge for which a maximum energy can be calculated, by using the expression

$$J = 0.5CV^2 \qquad (23\text{-}21)$$

where J = total energy dissipated, J
C = capacitance of charged system, F
V = initial potential difference between electrodes, V

Incident investigations often require that estimates be made of the possible spark energy from an ungrounded conductor. If the discharge path contains significant resistance, some of the stored energy is dissipated in the resistance, thereby lowering the energy in the spark gap.

A **corona** is generated when a highly nonuniform electric field of sufficient strength terminates on a conductor that has a small radius of curvature, i.e., a point, wire, or knife edge. The luminous (breakdown) region is confined to a small volume near the corona electrode. Because of their small peak currents and long duration, corona discharges do not have sufficient energy to ignite most flammable materials found in industry, i.e., materials having an MIE greater than 0.2 mJ. For this reason, nonpowered devices that employ corona discharges for static neutralization can be used safely in most hazardous (classified) locations. Corona discharges *can* ignite hydrogen-air and oxygen-enriched gas mixtures.

Brush discharges take place between conductors and charged nonconductors, where the radius of curvature of the conductor is too large for corona generation. The name refers to the brushlike appearance of the discharge, which spreads from the conductor to discrete areas on the nonconductor. The brush discharge may have a hot "stem" near the conductor, which may cause ignition by raising the temperature of the flammable mixture to its *autoignition* value.

Brush discharges from − charged nonconductors have been found more incendive that those from + charged nonconductors. Brush discharges may ignite flammable mixtures that have an MIE of less than 4 mJ. This limitation arises because charges from a small area on the nonconductor are able to participate in the discharge. Most dust-air mixtures cannot be ignited by brush discharges, because their MIE exceeds 4 mJ.

Surface charge densities cannot exceed the theoretical value of 2.7×10^{-5} C/m², set by air breakdown, and are normally less than 1.5×10^{-5} C/m².

Propagating brush discharges are much less common than brush discharges. They may occur when a nonconductive film or plastic layer acquires a double layer of charges, i.e., + charges on one surface and − charges on the opposite surface. With the electric fields within the film, surface charge densities can be large, because they are not limited by air breakdown.

The double layer can be formed by contact (triboelectric) charging on one surface of the nonconductor while the opposite surface is in contact with a conductor, e.g., a nonconductive coating on a metal chute or a plastic-lined, metal pipe for powders. A less frequent cause is contact charging of one surface of a nonconductor while air ions collect on the opposite surface.

Investigations by Glor ("Discharges and Hazards Associated with the Handling of Powders," *Electrostatics 1987*, Inst. Phys. Conf. Ser. no. 85, pp. 207–216, 1987) and others conclude that propagating brush discharges require surface charge densities above 2.7×10^{-4} C/m². In addition, the dielectric breakdown voltage of the insulating layer must exceed 4 kV for a thickness of 10 μm, or 8 kV for a thickness of 200 μm.

If a conductor approaches the charged surface, the electric field will produce air ionization at the surface, which creates a semiconductive layer, thereby allowing charges from a large area to participate in a single discharge. Because these discharges can have energies of 1 J or more, they are very hazardous in a flammable environment. They may also cause severe shocks to operators who reach into a nonconductive container that is receiving charged powder, pellets, or fibers.

Causes of Hazardous Discharges with Liquids **Self-generated discharges** in vapor-air mixtures can be ignited by static discharges from highly charged liquids. Such liquids are said to "carry their own match." Typical causes of such charging for poorly conductive (< 50 pS/m) liquids include

1. High-velocity flow
2. Free-fall/splashing
3. Filtering
4. Spraying
5. Agitation with air
6. Blending with powder
7. Settling of an immiscible component, e.g., water in gasoline
8. Liquid sampling from pressurized lines, using ungrounded or nonconductive containers

Conductive liquids in nonconductive containers may cause sparking if the outside of the container is charged by rubbing.

External causes of incendive static discharges include

1. Sparks from ungrounded conductors or persons
2. Brush discharges from flexible intermediate bulk containers (FIBCs), plastic bags, stretch wrap, or other plastic film
3. Propagating brush discharges from metal-backed, plastic film or linings

Powders Contact charging of powders occurs whenever particles move relative to one another or to a third surface. Significant charging is most often generated by pneumatic transfer. Maximum charge densities (C/kg) on airborne powders increase as particle size decreases, because of larger surface/mass ratios. Dry fines can be expected to charge more highly than those containing moisture. While suspended in air, charged powder poses an ignition risk only if nonconductive piping is used in the conveying lines, or if conductive piping is not bonded. The collected powder may accumulate so much charge per unit volume that the associated electric fields cause breakdown of the surrounding air in the form of a corona or a brush discharge. For receiving containers larger than about 1 m^3, *bulking* or *cone* discharges may be present. These discharges typically have energies of less than 10 mJ, but the value is sometimes higher. The ignition hazard from bulking discharges can be minimized by, e.g., using a rotary valve or bag filter to prebulk small volumes of charged powder prior to its collection in a large receiver.

Personnel and Clothing Sparks from ungrounded persons pose a serious ignition hazard in flammable gas-air, vapor-air, and some dust-air mixtures, because the body is a conductor and can store energies as high as 40 mJ. Induction of static charges on a person's ungrounded body by charged clothing is a common cause of personnel electrification. Even at the *threshold of shock sensation*, the stored energy is about 1 mJ.

It is essential that persons be grounded in hazardous (classified) locations. For most chemical operations, the resistance from skin to ground should not exceed 100 MΩ. A lower allowable resistance may be specified for locations where the presence of primary explosives, hydrogen-air mixtures, oxygen-enriched mixtures, or certain solid-state devices requires faster charge dissipation.

The combination of conductive flooring and conductive (ESD) footwear is the preferred method of grounding. Untreated concrete flooring with conductive footwear is usually adequate, but the resistance to ground should be measured. Where this method is impractical, personnel-grounding devices are available.

In most chemical plants, grounded persons can wear any type of clothing safely. For the unusually sensitive environments noted above, charge-dissipative or conductive clothing should be worn, and personnel should be grounded. Removal of outer garments in a flammable location can cause hazardous discharges and should be avoided. Although most gloves used in the chemical industry show a resistance of less than 100 MΩ from the wearer's palm to a handheld electrode, this value should be verified.

CHEMICAL REACTIVITY

GENERAL REFERENCES: ASTM E 1231, "Practice for Calculation of Hazard Potential Figures-of-Merit for Thermally Unstable Materials," ASTM International, West Conshohocken, Penn. ASTM E 1445, "Standard Terminology Relating to Hazardous Potential of Chemicals," ASTM International. ASTM E 2012, "Standard Guide for the Preparation of a Binary Chemical Compatibility Chart," ASTM International. Bartknecht, *Explosions: Course, Prevention, Protection,* Springer-Verlag, Berlin, 1981. Barton and Rogers, eds., *Chemical Reaction Hazards: A Guide to Safety,* 2d ed., Gulf Publishing Co., Houston, Tex., 1997. Beever, "Hazards in Drying Thermally Unstable Powders," *Inst. Chem. Engr., Hazards X,* Symp. Ser. 115, 1989. Beever, "Scaling Rules for Prediction of Thermal Runaway," *Intl. Symp. on Runaway Reactions,* CCPS-AIChE, Cambridge, Mass., 1989. Bollinger et al., *Inherently Safer Chemical Processes: A Life Cycle Approach,* CCPS-AIChE, New York, 1996. Bowes, "A General Approach to the Prediction and Control of Potential Runaway Reaction," *Inst. Chem. Eng., Symp. Ser. 68,* 1981. Bowes, *Self-heating: Evaluating and Controlling the Hazards,* Elsevier, 1984. *Bretherick's Reactive Chemical Hazards Database,* Version 3.0, Butterworth-Heinemann, Oxford, 1999. Brogli, et al., "Assessment of Reaction Hazards by Means of a Bench Scale Calorimeter," *Inst. Chem. Eng., Symp. Ser. 68,* 1981. Brogli, Gygax, and Myer, "Differential Scanning Calorimetry—A Powerful Screening Method for the Estimation of the Hazards Inherent in Industrial Chemical Reactions," *Therm. Analy.,* 6th Symp., vol. 1, 549, 1980. Center for Chemical Process Safety (CCPS), *Guidelines for Chemical Reactivity Evaluation and Application to Process Design,* CCPS-AIChE, New York, 1995. CCPS, *Guidelines for Safe Storage and Handling of Reactive Materials,* CCPS-AIChE, New York, 1995. CCPS, *Guidelines for Process Safety in Batch Reaction Systems,* CCPS-AIChE, New York, 1999. CCPS, Safety Alert, "Reactive Material Hazards: What You Need To Know," CCPS-AIChE, New York, 2001. Coates, "The ARC™ in Chemical Hazard Evaluation," *Thermochim. Acta 85:* 369–372, 1984. Davies, *Organic Peroxides,* Butterworths, London, 1961. ESCIS, *Thermal Process Safety: Data, Assessment Criteria, Measures,* Safety Series, Booklet 8, Expert Commission for Safety in the Swiss Chemical Industry, Basel, Switzerland, April 1993. Fenlon, "Calorimetric Methods Used in the Assessment of Thermally Unstable or Reactive Chemicals," *Intl. Symp. for the Prevention of Major Chemical Accidents,* AIChE, 1987. Fisher et al., "Emergency Relief System Design Using DIERS Technology: The Design Institute for Emergency Relief Systems (DIERS) Project Manual," AIChE, New York, 1992. Frank and Kamenetskii, *Diffusion and Heat Transfer in Chemical Kinetics,* Plenum Press, 1969. Frurip et al., "A Review of Chemical Compatibility Issues," *AIChE Loss Prevention Symp.,* Houston, Tex., 1997. Grewer, "Direct Assessment and Scaling of Runaway Reaction and Decomposition Behavior Using Oven and DEWAR Tests," *Inst. Chem. Eng., Symp. Ser., 68,* 1981. Grewer, "Use of DEWAR Calorimetry," *Proc. Runaway Chem. Reac. Haz. Symp.,* IBC, Amsterdam, November 1986. Grewer, *Thermal Hazards of Chemical Reactions,* Elsevier Science, London, 1994. Halk, "SEDEX Versatile Instrument for Investigating Thermal Stability," *Inst. Chem. Eng., Symp. Ser. 68,* 1981. Hazard Investigation, "Improving Reactive Hazard Management," Report no. 2001-01-H, U.S. Chemical Safety and Hazard Investigation Board, Washington, October 2002. Hendershot, "A Checklist for Inherently Safer Chemical Reaction Process Design and Operation," *Intl. Symp. on Risk, Reliability and Security,* CCPS-AIChE, New York, October 2002. HSE, *Designing and Operating Safe Chemical Reaction Processes,* U.K. Health and Safety Executive, 2000. "How to Prevent Runaway Reactions, Case Study: Phenol-Formaldehyde Reaction Hazards," EPA 550-F99-004, U.S. Environmental Protection Agency, Washington, August 1999. Hub, "Adiabatic Calorimetry and SIKAREX Technique," *Inst. Chem. Eng., Symp. Ser. 68,* 1981. Hub, "Experiences with the TSC 500 and RADEX Calorimeter," *Proc. Runaway Chem. Reac. Haz Symp.,* IBC, Amsterdam, November 1986. Hub, "Heat Balance Calorimetry and Automation of Testing Procedures," *Proc. Runaway Chem. Reac. Haz. Symp.,* IBC, Amsterdam, November 1986. *International Conference/Workshop on Modeling and Mitigating the Consequences of Accidental Releases of Hazardous Materials,* CCPS-AIChE, New York, 1991. *International Conference on Hazard Identification and Risk Analysis, Human Factors and Human Reliability in Process Safety,* CCPS-AIChE, New York, 1992. *International Conference on Managing Chemical Reactivity Hazards and High Energy Release Events,* CCPS-AIChE, New York, 2003. *International Symp. on Runaway Reactions and Pressure Relief Design,* DIERS-AIChE, New York, 1995. Johnson et al., *Essential Practices for Managing Chemical Reactivity Hazards,* CCPS-AIChE, New York, 2003. Johnson and Lodal, "Screen Your Facilities for Chemical Reactivity Hazards," *Chem. Eng. Prog.* 99(8): 50–58, 2003. Johnson and Unwin, "Addressing Chemical Reactivity Hazards in Process Hazard Analysis," *Proc. Intl. Conf. and Workshop on Managing Chemical Reactivity Hazards and High Energy Release Events,* CCPS-AIChE, New York, 2003. Kohlbrand, "Reactive Chemical Screening for Pilot Plant Safety," *Chem. Eng. Prog. 81*(4): 52, 1985. Kohlbrand, "The Relationship Between Theory and Testing in the Evaluation of Reactive Chemical Hazards," *Intl. Symp. for the Prevention of Major Chemical Accidents,* AIChE, New York, 1987. Leggett, "Chemical Reaction Hazard Identification and Evaluation: Taking the First Steps," *AIChE Loss Prevention Symp.,* New Orleans, March 2002.

Lewis, *Sax's Dangerous Properties of Industrial Materials,* 11th ed., Wiley, New York, 2004. Mannan, *Lees' Loss Prevention in the Process Industries,* 3d ed., Elsevier, Amsterdam, 2005. Manufacturing Chemists Association (now American Chemistry Council), *Selected Case Histories,* Washington, 1951–1973. Mosley et al., "Screen Reactive Chemical Hazards Early in Process Development," *Chem. Eng. Prog.* 96(11): 51–65, 2000. NFPA 430, "Code for the Storage of Liquid and Solid Oxidizers," National Fire Protection Association, Quincy, Mass. NFPA 432, "Code for the Storage of Organic Peroxide Formulations," National Fire Protection Association, Quincy, Mass. NFPA 49, "Hazardous Chemicals Data," and NFPA 491, "Hazardous Chemical Reactions," in *Fire Protection Guide to Hazardous Materials,* 13th ed., National Fire Protection Association, Quincy, Mass., 1997. NFPA 704, "Standard System for the Identification of the Hazards of Materials for Emergency Response," National Fire Protection Association, Quincy, Mass., 2001. NOAA, Chemical Reactivity Worksheet, Version 1.5, U.S. National Oceanic and Atmospheric Administration, *response. restoration. noaa.gov/chemaids/react.html,* 2002. Pohanish and Green, *Wiley Guide to Chemical Incompatibilities,* 2d ed., Wiley, New York, 2003. *Second Intl. Symp. on Runaway Reactions, Pressure Relief Design and Effluent Handling,* DIERS-AIChE, New York, 1998. Stull, *Fundamentals of Fire and Explosion,* AIChE, New York, 1976. Stull, "Linking Thermodynamics and Kinetics to Predict Real Chemical Hazards," *Loss Prevention 7,* p. 67, AIChE, 1976. Thomas, "Self Heating and Thermal Ignition—A Guide to Its Theory and Application," ASTM STP 502, pp. 56–82, ASTM International, 1972. Townsend, "Accelerating Rate Calorimetry," *Inst. Chem. Eng., Symp. Ser.* 68, 1981. Townsend and Kohlbrand, "Use of ARC™ to Study Thermally Unstable Chemicals and Runaway Reactions," *Proc. Runaway Chem. Reac. Haz. Symp.,* IBC, Amsterdam, November 1986. Townsend and Tou, "Thermal Hazard Evaluation by an Accelerating Rate Calorimeter," *Thermochem. Acta 73:* 1–30 and references therein, 1980. Urben, ed., *Bretherick's Handbook of Reactive Chemical Hazards,* 6th ed., Butterworth-Heinemann, Oxford, 1999. Yoshida et al., *Safety of Reactive Chemicals and Pyrotechnics,* Elsevier Science, London, 1995.

Introduction *Chemical reactivity* is the tendency of substances to undergo chemical change. A *chemical reactivity hazard* is a situation with the potential for an uncontrolled chemical reaction that can result directly or indirectly in serious harm to people, property, or the environment. A chemical reaction can get out of control whenever the reaction environment is not able to safely absorb the energy and products released by the reaction. The possibility of such situations should be anticipated not only in the reaction step of chemical processes but also in storage, mixing, physical processing, purification, waste treatment, environmental control systems, and any other areas where reactive materials are handled or reactive interactions are possible.

The main business of most chemical companies is to manufacture products by means of controlled chemical reactions. The reactivity that makes chemicals useful can also make them hazardous. Chemical reactions are usually carried out without mishap, but sometimes they get out of control because of problems such as the wrong or contaminated raw material being used, changed operating conditions, unanticipated time delays, failed equipment, incompatible materials of construction, or loss of temperature control. Such mishaps can be worse if the chemistry under both normal and abnormal conditions is not fully understood. Therefore, it is essential that chemical process designers and operators understand the nature of the reactive materials and chemistry involved and what it takes to control intended reactions and avoid unintended reactions throughout the entire life cycle of a process facility.

Life-Cycle Considerations

Considering Chemical Reactivity during Process Development Decisions made at the early development stages of a process facility, including conceptual and research phases, will in large part determine the nature and magnitude of the chemical reactivity hazards that will need to be contained and controlled throughout the entire life cycle of the facility. For this reason, chemical reactivity hazards should be considered from the outset of process development, including creative thinking regarding feasible alternatives to the use of reactive materials or the employment of highly energetic reactive systems. What may seem reasonable to the research chemist—handling materials in very small quantities—will have vast implications to the design and ongoing operation of a full-scale facility that must safely control the intended chemical reactions and avoid unintended reactions throughout the entire facility lifetime.

Mosley et al. describe a chemistry hazard and operability (CHAZOP) analysis approach, similar to a HAZOP study but applied at the early development stages of a new process.

Many companies designate a particular person or position as the "owner" of the process chemistry; this responsibility is likely to change as the life cycle progresses from development to design, construction, and operation. Data on the hazardous properties of the chemical reactions to be employed and the materials to be handled should begin to be assembled into a formal documentation package. Screening tests (described later in this section) may also need to be performed early in the development process to identify consequences of abnormal reactions and of deviations such as exceeding the normal reaction temperature. This documentation package will then form part of the information base upon which safeguards can be developed to control chemical reactivity hazards.

Considering Inherently Safer Approaches Specific to Reactivity Hazards The basic concepts of inherently safer plants, and the general strategies for making a facility inherently safer, are detailed in the later subsection on Inherently Safer and More User-Friendly Design. Strategies that focus on chemical reactivity hazards, and steps to conduct a review of these strategies, are highlighted in that section.

Instead of choosing to receive and store a highly reactive raw material, it may be possible to use a less hazardous material that is one step farther along in the formulation or synthesis chain. Alternatively, a decision may be made to generate the material on demand and eliminate all or most storage and handling of the material. Many reactive materials can be handled in dilute solutions, dissolved in less hazardous solvents, or otherwise handled under inherently safer conditions. (For some reactive materials such as benzoyl peroxide, handling as a dilute paste or solution is essential to the safe handling of the material.)

Inherently safer facilities with respect to chemical reactivity hazards must focus on the magnitude of stored chemical energy, the kinetics of how fast the energy could be released, and the possible reaction products that may themselves have hazardous properties such as toxicity or flammability.

With respect to kinetics, a slower reaction might be considered at first glance to be the inherently safer option as compared to a rapid reaction. This may indeed be the case, if the energy and products of the slower reaction can always be dissipated safely without causing harm or loss. However, this is often not the case, for two important reasons. First, regardless of the speed of the reaction, the same potential chemical energy is still thermodynamically present if only the kinetics is changed, and may be available under abnormal conditions such as an external fire or the introduction of a catalytic contaminant. Second, a slower reaction may allow unreacted material to accumulate. Hence faster reactions are generally more desirable, as discussed in the general reaction considerations below.

Finally, with respect to reaction products, a chemical reaction that does not generate hazardous reaction products or by-products is inherently safer than one that does. Thought must be given not only to hazardous reaction products, however. The generation of any kind of noncondensible gases can cause a vessel rupture due to internal overpressurization, if not adequately vented or relieved.

The following is a typical agenda for an inherent safety review at the concept or development stage of a new facility involving reactivity hazards (Johnson et al. 2003):

1. Review what is known of the chemical reactivity hazards (as well as other hazards) that will need to be contained and controlled in the proposed process. This existing level of knowledge might come from past experience, suppliers, literature reviews, incident reports, etc.
2. Based on the level of knowledge of chemical reactivity hazards, determine if additional screening of reactivity hazards is necessary. Having reactive functional groups might indicate the need to perform literature searches, access databases, or run differential scanning calorimetry.
3. Discuss possible process alternatives and their relative hazards, including discussions on such topics as alternative solvents and possible incompatibilities to avoid.
4. Brainstorm and discuss possible ways to reduce the hazards.
5. Obtain consensus on significant unknowns that will need to be addressed.

6. Document the review, including attendees, scope, approach, and decisions.

7. Assign follow-up items, with responsibilities, goal completion dates, and a closure mechanism such as reconvening after a designated number of weeks.

Scale-up Considerations A key consideration when scaling up a reactive process, such as from a pilot plant to a full-scale facility, is to ensure adequate heat removal for normal or abnormal exothermic reactions. Heat generation is proportional to *volume* (mass) in a reactive system, whereas heat removal is only proportional to *area* (surface area) at best. Even though the reaction temperature can be easily controlled in the laboratory, this does not mean that it can be adequately controlled in a plant-scale reactor. Increasing the size of a reactor, or of another process or storage vessel where, e.g., polymerization or slow degradation can occur, without adequately considering heat transfer can have disastrous effects. The careful design of the agitation or recirculation system is likewise important when scaling up, and the combined effects on the design of the emergency relief system must be taken into account.

Scale-up can also have a significant effect on the basic process control system and safety systems in a reactive process. In particular, a larger process will likely require more temperature sensors at different locations in the process to be able to rapidly detect the onset of out-of-control situations. Consideration should be given to the impact of higher-temperature gradients in plant-scale equipment compared to a laboratory or pilot plant reactor (Hendershot 2002).

Designing Processes for Control of Intended Chemical Reactions

General Considerations The following should be taken into account whenever designing or operating a chemical process that involves intended chemical reactions (Hendershot 2002). CCPS (1999) also details many key issues and process safety practices to consider that are oriented toward the design and operation of batch reaction systems.

- *Know the heat of reaction for the intended and other potential chemical reactions.*
- *Calculate the maximum adiabatic temperature for the reaction mixture.*
- *Determine the stability of all individual components of the reaction mixture at the maximum adiabatic reaction temperature.* This might be done through literature searching, supplier contacts, or experimentation.
- *Understand the stability of the reaction mixture at the maximum adiabatic reaction temperature.* Are there any chemical reactions, other than the intended reaction, that can occur at the maximum adiabatic reaction temperature? Consider possible decomposition reactions, particularly those which generate gaseous products.
- *Determine the heat addition and heat removal capabilities of the reactor.* Don't forget to consider the reactor agitator as a source of energy—about 2550 Btu/(h·hp).
- *Identify potential reaction contaminants.* In particular, consider possible contaminants, which are ubiquitous in a plant environment, such as air, water, rust, oil, and grease. Think about possible catalytic effects of trace metal ions such as sodium, calcium, and others commonly present in process water.
- *Consider the impact of possible deviations from intended reactant charges and operating conditions.* For example, is a double charge of one of the reactants a possible deviation, and, if so, what is the impact?
- *Identify all heat sources connected to the reaction vessel and determine their maximum temperature.*
- *Determine the minimum temperature to which the reactor cooling sources could cool the reaction mixture.*
- *Understand the rate of all chemical reactions.* Thermal hazard calorimetry testing can provide useful kinetic data.
- *Consider possible vapor-phase reactions.* These might include combustion reactions, other vapor-phase reactions such as the reaction of organic vapors with a chlorine atmosphere, and vapor-phase decomposition of materials such as ethylene oxide or organic peroxide.
- *Understand the hazards of the products of both intended and unintended reactions.*
- *Rapid reactions are desirable.* In general, you want chemical reactions to occur immediately when the reactants come into contact.
- *Avoid batch processes in which all the potential chemical energy is present in the system at the start of the reaction step.*
- *Avoid using control of reaction mixture temperature as a means for limiting the reaction rate.*
- *Avoid feeding a material to a reactor at a higher temperature than the boiling point of the reactor contents.* This can cause rapid boiling of the reactor contents and vapor generation.

Exothermic Reactions and "Runaway Reactions" The term *runaway reaction* is often improperly used to refer to any uncontrolled chemical reaction. As properly used, it refers to loss of control of a kinetically limited, exothermic reaction that proceeds at a stable, controlled rate under normal conditions and that includes adequate removal of the heat of reaction (Fig. 23-15). When the situation changes such that the heat of reaction is not adequately removed, the excess heat increases the temperature of the reaction mass, which in turn increases the reaction rate and thus the rate of heat release as an exponential function of reaction temperature. If not limited by some means such as (1) the limiting reactant being exhausted, (2) a solvent removing the heat of reaction by boiling off, or (3) quenching or inhibiting the reaction, this "bootstrap" situation can result in an exponential temperature rise that can reach as high as hundreds of degrees Celsius per minute. The resulting temperature increase, generation of gaseous reaction products, and/or boiloff of evaporated liquid can easily exceed a pressure and/or thermal limit of the containment system, if not adequately relieved. The elevated temperatures may also initiate a secondary or side reaction that is even more rapid or energetic.

This runaway situation can be understood by comparing Fig. 23-15 with Fig. 23-16, which has two new lines added, for two possible upset conditions in a process with a cooling coil or other heat exchanger being used to absorb the heat of an exothermic reaction. The temperature of the cooling medium might increase (shift from line 1 to line 2), or the heat-transfer coefficient might decrease, such as by heat exchanger fouling (shift from line 1 to line 3). When one of these shifts gets past point T_{NR} (temperature of no return), the heat removal line no longer crosses the heat generation line, and stable operation is no longer possible. The heat of reaction causes the system temperature to increase, which further increases the rate of heat generation, which further increases the system temperature, etc.

Many possible abnormal situations can initiate a runaway reaction. These include

- Loss of flow of cooling medium to/from the reactor
- An increase in the temperature of the cooling medium

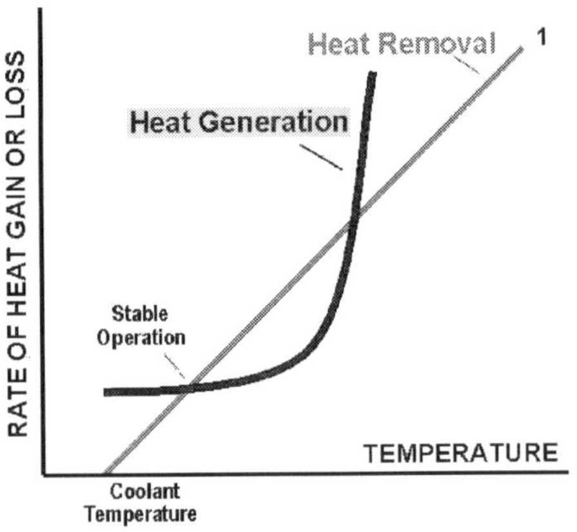

FIG. 23-15 For stable operation, all heat generated by an exothermic reaction is transferred to the surroundings, by whatever means (conduction, evaporation, etc.).

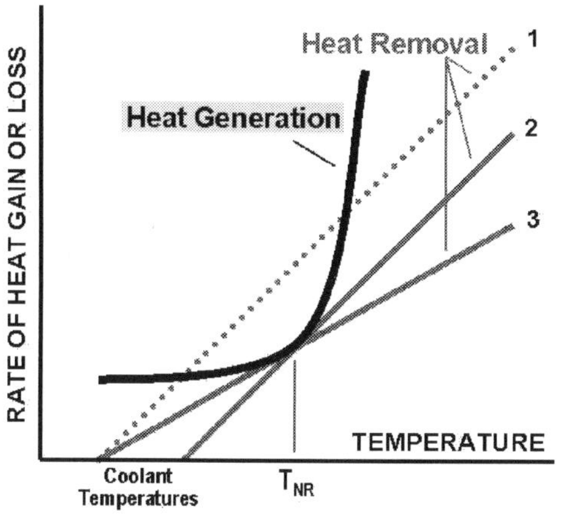

FIG. 23-16 For an exothermic reaction system with heat removal, e.g., to a vessel jacket and cooling coil, the limit of stable operation is reached as the reaction temperature increases to T_{NR} (temperature of no return), beyond which the rate of heat generation, which increases exponentially with increasing temperature, exceeds the capability of the system to remove the heat of reaction (see text).

- A general increase in the temperature of the storage or process configuration, such as due to an extreme ambient condition or loss of refrigeration
- Abnormal heat addition to the reactive material or mixture, such as by an external fire or the injection of steam to a vessel jacket or directly into the material or mixture
- Intentional heating of a vessel containing thermally sensitive material, due to lack of recognition of the runaway hazard or other reason
- Gradual fouling of the heat exchange surfaces to the point that maximum coolant flow is no longer sufficient to remove the heat of reaction
- Loss of agitation or circulation of the reactant mass or other reduction in the heat-transfer coefficient or contact with the heat exchange surface
- Insulation of the system, resulting in less heat dissipation
- Addition of a contaminant or excess catalyst which would increase the reaction rate
- Excess or too rapid addition of a limiting reactant
- Blockage of a vapor line or other means of increasing the system pressure
- Loss of a moderating diluent or solvent
- Inadequate inhibitor concentration in a storage container, or inadequate mixing of the inhibitor (including due to freezing of the material)
- Transfer of the reactive material or mixture to a location not capable of removing the heat of reaction

As can be seen from the above list, runaway reactions do not occur by a single mechanism. They can take place not only in reactors but also in raw material and product storage containers and vessels, purification systems, and anywhere else exothermic reactive systems and self-reacting materials (as described below) are involved.

Historical perspective An analysis of thermal runaways in the United Kingdom (Barton and Nolan, "Incidents in the Chemical Industry due to Thermal Runaway Chemical Reactions," *Hazards X: Process Safety in Fine and Specialty Chemical Plants, IChem 115:* 3–18) indicated that such incidents occur because of the following general causes:

- Inadequate understanding of the process chemistry and thermochemistry
- Inadequate design for heat removal
- Inadequate control systems and safety systems
- Inadequate operational procedures, including training

Semibatch reactions The inherently safer way to operate exothermic reaction processes is to determine a temperature at which the reaction occurs very rapidly (Hendershot 2002). The reactor can be operated at this temperature, while feeding at least one of the reactants gradually to limit the potential energy contained in the reactor. This type of gradual addition process is often called *semibatch*. A physical limit to the possible rate of addition of the limiting reactant is desirable—e.g., a metering pump, flow limited by using a small feed line, or a restriction orifice. Ideally, the limiting reactant should react immediately, or very quickly, when it is charged. The reactant feed can be stopped if necessary, if there is any kind of a failure (e.g., loss of cooling, power failure, loss of agitation), and the reactor will contain little or no potential chemical energy from unreacted material. Some means to confirm actual reaction of the limiting reagent is also desirable. A direct measurement is best, but indirect methods such as monitoring of the demand for cooling from an exothermic batch reactor can also be effective.

Design of Emergency Relief and Effluent Treatment Systems Containment systems are only rarely designed with sufficient pressure and temperature rating to fully contain a runaway reaction. For this reason, overpressure protection is of obvious critical importance as a last line of defense against loss events that can result from runaway reactions. The latter sections in this chapter on Pressure Relief Systems and on Emergency Relief Device Effluent Collection and Handling address design basis selection, relief calculations, and effluent treatment system configurations for reactive system overpressure protection.

Endothermic Reactions An endothermic reaction process is generally easier to bring to a safe state if an out-of-control situation is detected. Discontinuing the heat input is usually the primary line of defense to stop the operation. In this regard, the endothermic reaction is inherently safer than an exothermic reaction.

The following should especially be taken into account:

- The final product of an endothermic chemical reaction has a greater energy content than the starting materials. For this reason, materials with net positive heats of formation are often termed *endothermic compounds*. (Most explosives, e.g., are endothermic compounds.) This energy content can potentially be released in an uncontrolled manner if sufficient energy is again added to the material, such as by heating it to a decomposition temperature.
- Likewise, if control is lost of an endothermic reaction process, such as by a heating control valve opening too far or by a steam leak directly into the reaction mass, a degradation reaction or other secondary or side reaction may be initiated that can be exothermic and can lead to a thermal runaway.
- Some endothermic compounds can gradually degrade, decompose, become more concentrated, or become sensitized over time.

Designing Facilities for Avoidance of Unintended Reactions

General Considerations The following general design and operational considerations for avoiding unintended chemical reactions are summarized from a CCPS Safety Alert (2001):

- Train all personnel to be aware of reactivity hazards and incompatibilities and to know maximum storage temperatures and quantities.
- Design storage and handling equipment with all compatible materials of construction.
- Avoid heating coils, space heaters, and all other heat sources for thermally sensitive materials.
- Avoid confinement when possible; otherwise, provide adequate emergency relief protection.
- Avoid the possibility of pumping a liquid reactive material against a closed or plugged line.
- Locate storage areas away from operating areas in secured and monitored locations.
- Monitor material and building temperatures where feasible with high-temperature alarms.
- Clearly label and identify all reactive materials and what must be avoided (e.g., heat, water).
- Positively segregate and separate incompatible materials, using dedicated equipment if possible.
- Use dedicated fittings and connections to avoid unloading a material to the wrong tank.
- Rotate inventories for materials that can degrade or react over time.

- Pay close attention to housekeeping and fire prevention around storage/handling areas.

Identifying Potential Reactions The U.S. Chemical Safety and Hazard Investigation Board's Hazard Investigation "Improving Reactive Hazard Management" (2002) highlighted the importance of identifying chemical reactivity hazards as a result of an examination of 167 previous reactive incidents. CCPS has published a preliminary screening methodology for identifying where reactive hazards are likely to exist (Johnson et al., 2003). The flowchart for the preliminary screening methodology is shown later in the Hazard Analysis subsection.

The following paragraphs break down the types of reactive materials and reactive interactions that an engineer may need to address in the design of a chemical process or other facility such as a warehouse where reactive materials are handled (Johnson and Lodal, 2003; Johnson et al., 2003). These can be considered to be in three larger categories:

- Self-reactive substances (polymerizing, decomposing, rearranging)
- Substances that are reactive with ubiquitous substances such as air (spontaneously combustible/pyrophoric, peroxide-forming), water (water-reactive), or ordinary combustibles (oxidizers)
- Incompatible materials

Polymerizing, Decomposing, and Rearranging Substances Most of these substances are stable under normal conditions or with an added inhibitor, but can energetically self-react with the input of thermal, mechanical, or other form of energy sufficient to overcome its activation energy barrier (see Sec. 4, Reaction Kinetics, Reactor Design, and Thermodynamics). The rate of self-reaction can vary from imperceptibly slow to violently explosive, and is likely to accelerate if the reaction is exothermic or self-catalytic.

The tendency of a material such as acrylic acid or styrene to *polymerize* is usually recognized, and the material safety data sheet should be checked and the supplier can be contacted as to whether hazardous polymerization might be expected. A less energetic means of self-reaction is by molecular *rearrangement* such as by isomerizing, tautomering, disproportionating, or condensing.

The *decomposition* of some materials into smaller, more stable molecules can be initiated by mechanical shock alone, and they are known as *shock-sensitive.* Many commercially important chemicals are *thermally sensitive* and decompose with the addition of heat. For storage situations, the critical temperature at which the thermal energy is sufficient to start an uncontrolled reaction in a particular storage configuration for a specified time is known as the *self-accelerating decomposition temperature* (SADT), as described in NFPA 49.

Decomposing materials are sometimes referred to as *unstable,* and generally they have a positive heat of formation such that energy will be released when the decomposition reaction occurs. Self-reactive materials can often be recognized by the presence of certain chemical structures that tend to confer reactivity. These include

- Carbon-carbon double bonds not in benzene rings (e.g., ethylene, styrene)
- Carbon-carbon triple bonds (e.g., acetylene)
- Nitrogen-containing compounds (NO_2 groups, adjacent N atoms, etc.)
- Oxygen-oxygen bonds (peroxides, hydroperoxides, ozonides)
- Ring compounds with only three or four atoms (e.g., ethylene oxide)
- Metal- and halogen-containing complexes (metal fulminates, halites, halates, etc.).

A more complete list is given by CCPS (1995), and specific compounds can be investigated in resources such as Urben (1999).

General considerations for avoiding unintended reactions with self-reacting substances include knowing the mechanisms and boundaries of what will initiate a self-reaction, maintaining diluents or inhibitors to extend the boundaries where feasible and avoiding the mechanisms (such as shock and overtemperature) that would initiate the self-reaction, and having reliable controls and last-resort safety systems in place to detect and deal with an incipient out-of-control condition.

Specific design considerations for a few substances including acrylic acid, styrene, organic peroxides, ethylene oxide, and 1,3-butadiene are given in CCPS (1995) on the basis of an industry-practice survey. Detailed information for other substances is distributed by industry user groups. These include methacrylic acid and methacrylate esters (*www.mpausa.org*) and ethylene oxide (*www.ethyleneoxide.com*).

Spontaneously Combustible and Pyrophoric Substances *Spontaneously combustible* substances will readily react with the oxygen in the atmosphere, igniting and burning even without an ignition source. Ignition may be immediate, or may result from a self-heating process that may take minutes or hours (hence, some spontaneously combustible substances are known as *self-heating* materials).

Pyrophoric materials ignite spontaneously on short exposure to air under ordinary ambient conditions. Some materials that are considered pyrophoric require a minimum relative humidity in the atmosphere for spontaneous ignition to occur. The potential of pyrophoric materials to exhibit this behavior is usually well known due to the extreme care required for their safe handling.

Pyrophoric and other spontaneously combustible substances will generally be identified as such on their product literature, material safety data sheets (MSDSs), or International Chemical Safety Cards (ICSCs). If transported, these substances should be identified as DOT/UN Hazard Class 4.2 materials for shipping purposes and labeled as spontaneously combustible. For pyrophoric substances, the NFPA 704 diamond for container or vessel labeling has a red (top) quadrant with a rating of 4, indicating the highest severity of flammability hazard (NFPA 704, 2001). Note that pyrophoric materials often exhibit one or more other reactivity hazards as well, such as water reactivity.

A scenario that has resulted in many fires and explosions in petroleum refineries involves iron sulfide. An impure, pyrophoric sulfide is formed when streams containing hydrogen sulfide or other volatile sulfur compounds are processed in ferrous equipment. Oxidation of moist iron sulfide is highly exothermic. Opening sulfide-containing equipment without adequate purging can result in rapid self-heating and ignition of the iron sulfide, which can then ignite other residual flammable gases or liquids in the equipment.

Many scenarios involving spontaneous combustion involve a combination of materials exposed to sufficient air, often in an insulating situation that prevents heat from a slow oxidation reaction from dissipating, which results in a self-heating situation.

Lists of pyrophoric materials that include less common chemicals, including metals, can be found in volume 2 of *Bretherick's Handbook of Reactive Chemical Hazards* (Urben, 1999). Other spontaneously combustible substances are tabulated by their proper shipping names and UN/NA numbers in the U.S. Dept. of Transportation regulation 49 CFR 172.101.

Possible causes of uncontrolled reactions associated with pyrophoric and other spontaneously combustible materials are listed in Johnson et al. (2003).

Peroxide Formers *Peroxide formers* will react with the oxygen in the atmosphere to form unstable peroxides, which in turn might explosively decompose if concentrated. Peroxide formation, or peroxidation, usually happens slowly over time, when a peroxide-forming liquid is stored with limited access to air.

Substances that are peroxide formers will often have an inhibitor or stabilizer added to prevent peroxidation. They are often not easily identifiable as peroxide formers by using MSDSs or ICSCs. Rather, they are frequently identified by another characteristic, such as flammability, for storage and shipping purposes. Examples of peroxide formers include 1,3-butadiene, 1,1-dichloroethylene, isopropyl ether, and alkali metals. Johnson et al. (2003) tabulate other chemical structures susceptible to peroxide formation.

The total exclusion of air from vessels and equipment containing peroxide formers, and the establishment and observing of strict shelf life limitations, are basic strategies for managing peroxide-forming hazards.

Water-Reactive Substances *Water-reactive substances* will chemically react with water, particularly at normal ambient conditions. For fire protection purposes, a material is considered water-reactive if a gas or at least 30 cal/g (126 kJ/kg) of heat is generated when it is mixed with water (NFPA 704, 2001), using a two-drop mixing calorimeter.

Water reactivity can be hazardous by one or more of several mechanisms. The heat of reaction can cause thermal burns, ignite combustible materials, or initiate other chemical reactions. Flammable, corrosive or toxic gases are often formed as reaction products. The violence of some reactions may disperse hazardous materials. Even slow reactions can generate sufficient heat and off-gases to overpressurize and rupture a closed container.

Substances that are water-reactive will nearly always be identified as such on their MSDSs or ICSCs. They may be identified as DOT/UN Hazard Class 4.3 materials for shipping purposes and labeled as dangerous when wet. However, some water-reactive materials are classified otherwise. Acetic anhydride is designated Class 8; it may also be identified as a combustible liquid.

The total exclusion of water from vessels and equipment containing water-reactive substances, and the maintenance of the primary containment integrity over time, are the obvious design and operational considerations when handling water-reactive substances. Drying of equipment prior to start-up and careful design of provisions for cleaning and purging of equipment are also essential.

Oxidizers and Organic Peroxides An *oxidizer* is any material that readily yields oxygen or other oxidizing gas, or that readily reacts to promote or initiate combustion of combustible materials (NFPA 430, 2000). Thus, most oxidizers can be thought of as being reactive with ordinary combustible liquids or solids, which are commonly used as process, packaging, general use, or structural materials. They can also react with many other reducing substances.

Oxidizers will nearly always be identified as such on their MSDSs or ICSCs. They may be identified as DOT/UN Hazard Class 5.1 materials for shipping purposes and labeled as oxidizers. However, some oxidizers are classified otherwise.

Volume 2 of *Bretherick's Handbook of Reactive Chemical Hazards* (Urben, 1999) lists many structures and individual chemical compounds having oxidizing properties. NFPA 432 can be consulted for typical organic peroxide formulations. Note, however, that some organic peroxide formulations burn with even less intensity than ordinary combustibles and present no chemical reactivity hazard.

NFPA 430 contains safety provisions for the storage of liquid and solid oxidizers. NFPA 432 contains safety provisions for the storage of organic peroxide formulations.

Incompatible Materials In this context, *incompatible* refers to two materials not able to contact each other without undesired consequences. ASTM E 2012 gives a method for preparing a binary compatibility chart for identifying incompatibilities. The NOAA Chemical Reactivity Worksheet uses a group compatibility method to predict the results of mixing any binary combination of the 6080 chemicals in the CAMEO database, including many common mixtures and solutions. Materials to be considered include not only raw materials and products but also by-products, waste products, cleaning solutions, normal and possible abnormal materials of construction, possible contaminants and degradation products, material that could be left in the process from a previous batch or cleanout, and materials in interconnected piping, heat-transfer systems, waste collection systems, or colocated storage.

The essence of the ASTM E 2012 approach is to determine incompatibility *scenarios* that could foreseeably occur by examining all possible binary combinations. It may be necessary to review a process by using a systematic method such as a process hazard analysis (PHA) to identify all incompatibility scenarios that have a significant likelihood of occurrence and severity of consequences. The same review can then be used to evaluate whether adequate safeguards exist or whether further risk reduction is warranted.

Where the consequences of combining two or more materials under given conditions of temperature, confinement, etc., are unknown and cannot be predicted with certainty, testing may need to be performed to screen for potential incompatibilities. Two common test methods used for this purpose are differential scanning calorimetry and mixing cell calorimetry (described later in this section).

Design considerations to avoid contact of incompatible materials include total exclusion of an incompatible substance from the facility; quality control and sampling of incoming materials; approval procedures for bringing new chemicals and materials of construction onsite; dedicated fittings and unloading spots; vessel, piping, and container labeling; dedicated or segregated storage; segregated diking, drainage, and vent systems; quality control of raw materials and of materials of construction (both initial construction and ongoing maintenance and modifications); sealless pumps, double tube sheets, and other means of excluding seal fluid, heat-transfer fluid, and other utility substances; positive isolation of interconnections by physical disconnects, blinding, or double block and bleed valves; avoidance of manifolds with flexible connections; and use of compatible purge gases, cleaning solutions, heat-transfer fluids, insulation, fire-extinguishing and suppression agents whenever possible; and removal of unused materials from the site. These design considerations will always need to be accompanied by procedure training, hazard awareness, and operating discipline for them to be effective on an ongoing basis.

Designing Mitigation Systems to Handle Uncontrolled Reactions (From CCPS, *Guidelines for Safe Storage and Handling of Reactive Materials*, 1995, Chap. 5.) Last-resort safety systems are intended to be used in many reactive chemical storage and handling operations as last-ditch efforts to avert a loss event such as an explosion or a hazardous material release, if the operation exceeds safe operating limits and it is not possible to regain control by using the operation's normal control mechanisms.

Inhibitor Injection Inhibitor injection systems are primarily used with polymerizing materials such as vinyl acetate. If the material begins to self-react in an uncontrolled manner, then injection of a polymerization inhibitor can interfere with the reaction before sufficient pressure and temperature have built up to cause a release from the storage/handling containment. The type of inhibitor needed will depend on the nature of the polymerization reaction; e.g., a free-radical scavenger may be used as an inhibitor for a material that reacts by free-radical polymerization. The inhibitor is often the same inhibitor used for normal storage stability requirements, but injected in a much larger quantity. If a different inhibitor is used that is designed to quickly kill the reaction, it is generally called a *short-stop* system.

Inhibitor injection systems need to be carefully designed and maintained to provide a highly reliable last-resort safety system. Since the inhibitor injection system is on standby and may not be used for months, attention must be paid to how the system components can be functionally and effectively tested on a periodic basis, such as once a month, without excessive disruption of normal operations. CCPS' *Guidelines for Engineering Design for Process Safety* (CCPS-AIChE, New York, 1993, pp. 273–275) discusses testing of continuous-process safety systems. This functional testing is important not only for the checking of adequate inhibitor supply and properly functioning delivery system, but also as the means of detecting an out-of-control situation and actuating the inhibitor injection system. Such systems, as well as other last-resort safety systems, are likely to be considered *safety instrumented systems* (SISs); and they should be selected, designed, and maintained accordingly (see the later section on SISs).

Quench Systems Quench systems are used for essentially all types of reactive chemicals. A quench system involves the addition of flooding quantities of water or other quenching medium to the reactive material; the quenching medium might be a subcooled material such as liquid nitrogen or dry ice in special applications.

The means by which a quench system works depends on the nature of the reactive material; e.g., for water-reactive materials, a quench system will destroy the material in a last-resort situation and generally form less-hazardous products, and will at the same time absorb some of the heat of reaction. Most quench systems are designed to both cool down and dilute a material that may be reacting uncontrollably; the quenching medium may also actually interfere with the chemical reaction or deactivate a catalyst.

Dump Systems For an inhibitor injection or quench system, the inhibitor or quenching medium is transferred from an external supply to the reactive material; in a dump system, the reactive material is transferred from the storage/handling facility to a safer location that is the same size or, more commonly, larger than the normal capacity of the facility. This allows depressurizing and deinventory of the reacting mass from the facility in an out-of-control situation, such as an incipient runaway reaction.

Depressuring Systems A last-resort depressurizing system can be added to a reactive system to vent off excessive pressure buildup in a tank vessel in a controlled manner before reaching the relief valve or rupture disk set pressure. Such a depressurizing system typically consists of a remotely actuated vent valve connected to the vapor space of the vessel, with the venting discharge directed to a scrubber or other treatment system of adequate capacity. The system can be designed to be actuated either manually, by a control room or field operator, or by detection of high pressure and/or high temperature in the vessel.

Reactive Hazard Reviews and Process Hazard Analyses Reactive hazards should be evaluated using reviews on all new processes and on all existing processes on a periodic basis. Reviews should include

1. Review of process chemistry, including reactions, side reactions, heat of reaction, potential pressure buildup, and characteristics of intermediate streams
2. Review of reactive chemicals test data for evidence of flammability characteristics, exotherms, shock sensitivity, and other evidence of instability
3. Review of planned operation of process, especially the possibility of upsets, modes of failure, unexpected delays, redundancy of equipment and instrumentation, critical instruments and controls, and worst-credible-case scenarios

These reviews can be either in addition to or combined with periodic process hazard analyses (PHAs) by using methods such as what-if analysis and HAZOP studies. The latter should consciously focus on identifying scenarios in which intended reactions could get out of control and unintended reactions could be initiated. One means of accomplishing this as part of a HAZOP study has been to include *chemical reaction* as one of the parameters to be investigated for each study node. Johnson and Unwin (2003) describe other PHA-related approaches for studying chemical reactivity hazards.

Worst-Case Thinking At every point in the operation, the process designer should conceive of the worst possible combination of circumstances that could realistically exist, such as loss of cooling water, power failure, wrong combination or amount of reactants, wrong valve position, plugged lines, instrument failure, loss of compressed air, air leakage, loss of agitation, deadheaded pumps, and raw material impurities. An engineering evaluation should then be made of the worst-case consequences, with the goal that the plant will be safe even if the worst case occurs. The previous discussion of calculating the maximum adiabatic temperature rise, then considering what might happen if it is realized, is an example of this type of analysis. A hazard and operability (HAZOP) study could be used to help identify abnormal situations and worst-case consequences.

Reactivity Testing Many of the data needed for the design of facilities with reactivity hazards involve the determination of thermal stability and of

1. The temperature at which an exothermic reaction starts
2. The rate of reaction as a function of temperature
3. Heat generated per unit mass of material

In many cases, data on the increase of pressure during a reaction are also required, especially for vent sizing, and on the composition of the product gases.

The term *onset temperature* T_{onset} is used in two contexts:

1. In a testing context, it refers to the first detection of exothermic activity on the thermogram. The differential scanning calorimeter (DSC) has a scan rate of 10°C/min, whereas the accelerating rate calorimeter (ARC®) has a sensitivity of 0.02°C/min. Consequently, the temperature at which thermal activity is detected by the DSC can be as much as 50°C different from ARC data.

2. The second context is the process reactor. There is a potential for a runaway if the net heat gain of the system exceeds its total heat loss capability. A self-heating rate of 3°C/day is not unusual for a monomer storage tank in the early stages of a runaway. This corresponds to 0.00208°C/min, which is 10 percent of the ARC's detection limit.

Sources of Reactivity Data Several important sources of reactivity data are described in the following paragraphs.

Calculations Potential energy that can be released by a chemical system can often be predicted by thermodynamic calculations. If there is little energy, the reaction still may be hazardous if gaseous products are produced. Kinetic data are usually not available in this way. Thermodynamic calculations should be backed up by actual tests.

Differential Scanning Calorimetry Sample and inert reference materials are heated in such a way that the temperatures are always equal. Onset-of-reaction temperatures reported by the DSC are higher than the true onset temperatures, so the test is mainly a screening test.

Differential Thermal Analysis (DTA) A sample and inert reference material are heated at a controlled rate in a single heating block. This test is basically qualitative and can be used for identifying exothermic reactions. Like the DSC, it is also a screening test. Reported temperatures are not reliable enough to be able to make quantitative conclusions. If an exothermic reaction is observed, it is advisable to conduct tests in the ARC.

Mixing Cell Calorimetry (MCC) The MCC provides information regarding the instantaneous temperature rise resulting from the mixing of two compounds. Together, DSC and MCC provide a reliable overview of the thermal events that may occur in a process.

Accelerating Rate Calorimetry (ARC) This equipment determines the self-heating rate of a chemical under near-adiabatic conditions. It usually gives a conservative estimate of the conditions for, and consequences of, a runaway reaction. Pressure and rate data from the ARC may sometimes be used for pressure vessel emergency relief design. Activation energy, heat of reaction, and approximate reaction order can usually be determined. For multiphase reactions, agitation can be provided. Nonstirred ARC runs may give answers that do not adequately duplicate plant results when there are reactants that may settle out or that require mixing for the reaction to be carried out (DeHaven and Dietsche, "Catalyst Explosion: A Case History," *Plant/Oper. Prog.*, April 1990).

Vent Sizing Package (VSP2™) The VSP is an extension of ARC technology. The VSP2 is a bench-scale apparatus for characterizing runaway chemical reactions. It makes possible the sizing of pressure relief systems with less engineering expertise than is required with the ARC or other methods.

Advanced Reactive System Screening Tool (ARSST™) The ARSST measures sample temperature and pressure within a sample containment vessel. The ARSST determines the potential for runaway reactions and measures the rate of temperature and pressure rise (for gassy reactions) to allow determinations of the energy and gas release rates. This information can be combined with simplified methods to assess reactor safety system relief vent requirements.

Shock Sensitivity Shock-sensitive materials react exothermically when subjected to a pressure pulse. Materials that do not show an exotherm on a DSC or DTA are presumed not to be shock-sensitive. Testing methods include

- *Drop weight test* A weight is dropped on a sample in a metal cup. The test measures the susceptibility of a chemical to decompose explosively when subjected to impact. This test should be applied to any materials known or suspected to contain unstable atomic groupings.
- *Confinement cap test* Detonatability of a material is determined by using a blasting cap.
- *Adiabatic compression test* High pressure is applied rapidly to a liquid in a U-shaped metal tube. Bubbles of hot compressed gas are driven into the liquid and may cause explosive decomposition of the liquid. This test is intended to simulate water hammer and sloshing effects in transportation, such as humping of railway tank cars. It is very severe and gives worst-case results.

Obtaining test data for designing a facility with significant reactivity hazards requires familiarity with a range of test equipment and a significant amount of experience in the interpretation of test results.

TOXICITY

Introduction Many natural and artificial substances are toxic to humans (and animals). Liquids and solids can be ingested, or exposure can be through the skin, eyes, or other external passages to the body. Where these substances are gaseous or volatile, toxic effects can result from inhalation. As a result of accidents and tests, it has been discovered that some of these substances are more toxic than others. Quantification of the degree of hazard has become important in devising appropriate measures for containing these substances.

Several chemical companies have established toxicology laboratories to develop quantitative information concerning the toxicity of raw materials, products, by-products, and waste materials. They include Dow, Du Pont, Eastman Kodak, and Union Carbide (Fawcett and Wood, *Safety and Accident Prevention in Chemical Operations,* 2d ed., pp. 262 and 281, 1982). Also, the Chemical Industry Institute of Toxicology was established in 1975 to provide toxicological hazards services, and The Netherlands Organization of Applied Scientific Research [TNO] has performed similar toxicological research in the Netherlands (The Institution of Chemical Engineers, *Chlorine Toxicity Monograph,* p. 34, 1989).

The present "Process Safety Management" standard of the Occupational Safety and Health Act requires "toxicity information" and "a qualitative evaluation of a range of the possible safety and health effects of failure of controls on employees in the workplace" [U.S. Department of Labor, *Occupational Safety and Health Standards,* 29 CFR 1910.119(d)(1)(i), (e)(3)(vii), and (f)(1)(iii)(A), 1992]. Similarly, the "Risk Management Programs for Chemical Accidental Release Prevention" standard of the Environmental Protection Agency's Clean Air Act Amendments [U.S. Environmental Protection Agency, *Risk Management Programs for Chemical Accidental Release Prevention,* 40 CFR 68.15(b)(3), 15(c), 24(c)(7), and 26(b)(1), 1993] requires toxicity information, a qualitative evaluation of a range of the possible safety and health effects of failure of controls on public health and the environment, and analysis of the off-site consequences of the worst-case release scenario and the other more likely significant accidental release scenarios.

To perform safety and health evaluations, quantitative knowledge of the effects of exposure to toxic materials would be needed. Some of the available data for inhalation toxicity (quantitative), skin-absorption toxicity (qualitative), and ingestion (quantitative) of hazardous materials are presented in Table I (American Industrial Hygiene Association, *Emergency Response Planning Guidelines and Workplace Environmental Exposure Level,* No. AEAH05-559, 2005; National Institute for Occupational Safety and Health, *Pocket Guide to Chemical Hazards,* 1994; American Conference of Governmental Industrial Hygienists, *Threshold Limit Values for Chemical Substances and Physical Agents,* 2001; National Institute for Occupational Safety and Health, *Registry of Toxic Effects of Chemical Substances,* 1983). To facilitate use of these data, several types of graphical and "probit" equation methods are available (Griffiths, "The Use of Probit Expressions in the Assessment of Acute Population Impact of Toxic Releases," *Journal of Loss Prevention in the Process Industries,* 4: 49, January 1991; Prugh, "Quantitative Evaluation of Inhalation-Toxicity Hazards," 29th Annual Loss Prevention Symposium, 1995).

The scope of this section is limited to dangerous and life-threatening exposures of the public to toxic materials (primarily gases and vapors) and non-life-threatening exposures of employees to toxic materials. Data concerning life-threatening concentrations and doses of many toxic gases, vapors, and liquids are available (National Institute for Occupational Safety and Health, *Registry of Toxic Effects of Chemical Substances,* 1983).

Inhalation Toxicity: The Haber Equation In 1924, Fritz Haber reported on his analysis of the results of animal inhalation tests on chemical warfare agents. He discovered that the product of gas or vapor concentration and duration of exposure was nearly constant for a given physiological effect. This relationship has been termed the *Haber law* (Haber, *Funf Vortrage aus den Jahren 1920–1923,* Springer-Verlag, 1924; Fleming et al., *Modern Occupational Medicine,* p. 78, 1960):

$$Ct = K = D \tag{23-22}$$

When the concentration C is expressed in parts per million (ppm) and the duration of exposure t is expressed in minutes, the values of the constant K and the dose D are in units of ppm-minutes.

It is now recognized that Haber's law does not apply for long exposures to low concentrations. Apparently, there are metabolic processes in the human body (and in animals) that can (for many toxic materials) result in biotransformation or detoxification, elimination, or excretion of toxic materials, or can repair damaged cells or tissues (Elkins, *The Chemistry of Industrial Toxicology,* 2d ed., p. 242, 1959; U.S. Federal Emergency Management Agency, *Handbook of Chemical Hazard Analysis Procedures,* p. 6-7, 1989). It is likely that the absorption process functions in proportion to the square root of the duration of exposure (Perry, *Chemical Engineers' Handbook,* 4th ed., p. 14-13 and Figs. 14-7, 14-9, and 14-21, 1963).

Dosage Equation The Haber law apparently applies to short exposures (less than 30 min) (The Institution of Chemical Engineers, *Chlorine Toxicity Monograph,* Table 5, Rat Group Codes U, N, E, X, and Z, and Mouse Group Codes N and R, 1989), but does not apply for long exposures to toxic vapors and gases. Eisenberg and others (Eisenberg et al., *Vulnerability Model—A Simulation System for Assessing Damage Resulting from Marine Spills,* U.S. Coast Guard Report CG-D-136-75, pp. 77, 83–89, and 257–267, 1975); also, 3d International Symposium on Loss Prevention in the Process Industries, p. 15/1158, and Proceedings, p. 190, 1980) attempted to modify the dose equation to fit the data over a useful range of interest, between 5 min and 2 h (Lees, *Loss Prevention in the Process Industries,* pp. 206–209, 527, 594, 599, 651–653, and 661, 1980). They found that the following equation could be used:

$$C^n t = K \tag{23-23}$$

Eisenberg found that a value of 2.75 for n was appropriate for the chlorine and ammonia data which were available.

In the 20 years since Eisenberg's report, many inhalation toxicity tests have been conducted, and many of the earlier data have been reexamined, with the result that values for n ranging from 0.6 to 4.9 have been applied to the above dose equation. It appears (Griffiths, "The Use of Probit Expressions in the Assessment of Acute Population Impact of Toxic Releases," *Journal of Loss Prevention in the Process Industries,* 4, p. 49, 1991) that the value of n may be related to the degree of breathing rate stimulation (high value of n) or repression (low values of n), and the value of n apparently increases with increasing exposure times (decreasing slope of ordinate C versus abscissa t). A value of 1.0 is frequently used by investigators if there are few data.

Probit Equation The probit equation has been used in an attempt to quantitatively correlate hazardous material concentration, duration of exposure, and probability of effect/injury, for several types of exposures. The objective of such use is to transform the typical sigmoidal (S-shaped) relationship between cause and effect to a straight-line relationship (Mannan, *Lees' Loss Prevention in the Process Industries,* 3d ed., p. 9/68, 2005).

Probit equations have the following form (Mannan, *Lees' Loss Prevention in the Process Industries,* 3d ed., Table 9.29, 2005):

$$Y = k_1 + k_2 \ln V \tag{23-24}$$

where Y = probit value
V = value of "intensity of causative factor which harms the vulnerable resource"
k_1 = constant that is intercept of Y versus V line (where value of V is 1.0 and ln V is 0).
k_2 = constant that is slope of Y versus ln V line.

The following table can be used to convert from probit values to probability percentages (Mannan, *Lees' Loss Prevention in the Process Industries,* 3d ed., Table 9.29, 2005):

Percentage	Probit value P
0.001 (1 in 10^5)	0.73
0.01 (1 in 10^4)	1.28
0.1 (1 in 10^3)	1.90
1 [1%] (1 in 100)	2.67
3 [3%] (3 in 100)	3.12
10 [10%] (1 of 10)	3.72
30 [30%] (3 of 10)	4.48
50 [50%] (5 of 10)	5.00

For the inhalation hazards of toxic vapors and gases, the function V has the form

$$V = C^n t \qquad (23\text{-}25)$$

where C = concentration by volume, ppm
t = duration of exposure, min
n = exponent that expresses difference between dosage and dose

The term *dosage* typically refers to an environmental hazard and is the product of the concentration of toxic gas or vapor at a particular point and the duration of the hazardous environment at that point. Thus, the *dosage* can be expressed as an average concentration multiplied by an average duration, or

$$D = Ct \qquad \text{ppm·min} \qquad (23\text{-}26)$$

Dose typically refers to the amount of toxic material actually retained and is sometimes referred to as the *toxic load*. Thus, the dose can be expressed as the product of a concentration term and a duration-of-exposure term, by either of the following relationships:

$$\text{TL} = V = C^n t \qquad \text{ppm}^n\text{·min} \qquad (23\text{-}27)$$

$$\text{TL} = V = Ct^{1/n} \qquad \text{ppm·min}^{1/n} \qquad (23\text{-}28)$$

Ingestion Toxicity Data are available for the *acute* (single-dose) ingestion/oral toxicity of many toxic materials (National Institute for Occupational Safety and Health, *Registry of Toxic Effects of Chemical Substances,* 1983; Lewis, *Sax's Dangerous Properties of Industrial Materials*, 9th ed., 1996). However, very few data are available for prolonged ingestion or periodic doses of toxic materials. It is likely that metabolic processes would operate to increase the total *burden* required for toxic effects for such chronic exposures, except for some materials (such as mercury and lead) which apparently can accumulate in the body.

The primary route for ingestion of toxic materials (especially dusts, mists, and vapors) is by the swallowing of mucus and saliva that has absorbed these materials during breathing. Cilia in the nose and esophagus (windpipe) sweep foreign materials that have been embedded or absorbed by these fluids toward the pharynx, where the contaminated fluid is swallowed (Guyton, *Textbook of Medical Physiology*, 3d ed., pp. 555, 556, 880, and 894, 1966).

Skin-Contact Toxicity Data for acute (short-term) exposures of the skin to corrosive and toxic liquids, solids, and gases are extremely limited, particularly where the consequences are severe or fatal injury, and the available data may not be useful, from an engineering standpoint. For example, the skin toxicity of hydrogen peroxide to rats is stated as 4060 mg/kg, but the skin area and duration of exposure are not stated. Thus, it is not possible (with the available data) to estimate the relationship among percent of body surface exposed to a corrosive material, the concentration of the corrosive material, the duration of exposure (before removal of the corrosive material), and the severity of the effect.

Somewhat in contrast, there are considerable data concerning the relatively long-term effects of exposure to toxic materials, where there are irritation, tumorigenic, reproductive, or mutation consequences (National Institute for Occupational Safety and Health, *Registry of Toxic Effects of Chemical Substances,* 1983). In the absence of better skin-contact data, it might be appropriate to use parenteral or subcutaneous injection data for a worst-case exposure (e.g., through a cut in the skin). However, use of intravenous or intraperitoneal data might overstate the skin exposure toxicity of a material.

The U.S. Department of Transportation and others have developed guidance for the corrosivity of chemical substances (U.S. Department of Transportation, *Shippers—General Requirements for Shipments and Packagings*, 49 CFR 173.136, *Definitions*, and 49 CFR 173.137, *Assignment of Packing Group*, 1998); American Society for Testing and Materials, *Practice for Laboratory Immersion Corrosion Testing of Metals*, G-31, 2002; Organization for Economic Cooperation and Development, *Guideline for Testing of Chemicals—Acute Dermal Irritation/Corrosion*, no. 404, 1992). The following definitions apply for Class 8 corrosive materials:

Packing Group 1—Great Danger: Full thickness destruction of human skin (exposure time, 3 min or less; observation time, 60 min).

Packing Group 2—Medium Danger: Full thickness destruction of human skin (exposure time, 3 to 60 min; observation time, 14 days).

Packing Group 3—Minor Danger: Full thickness destruction of human skin (exposure time, 1 to 4 h; observation time, 14 days).

Examples of assignments to packing groups are shown in the table at the bottom of the page.

Another effect of skin-contact toxicity is dermatitis. This can be caused by "physical" agents, such as detergents and solvents that remove the natural oils from the skin and thereby render the skin susceptible to materials that ordinarily do not affect the skin (National Safety Council, *Fundamentals of Industrial Hygiene*, 3d ed., p. 23, 1988). Dermatitis also can be caused by dessicants and water-reactive chemicals that remove moisture from the skin, generating heat and causing burns. Other causes are oxidizers; protein precipitants; allergic or anaphylactic proteins; friction, pressure, and trauma; thermal and electromagnetic radiation; biological agents; and plant poisons. Dermatitis can be prevented or controlled by containment of skin-contact hazards and use of tools to avoid contact (engineering controls) or by the wearing of protective clothing, including gloves and eye and face protection, and good personal hygiene, including hand and face washing (administrative controls) (National Safety Council, *Fundamentals of Industrial Hygiene*, 3d ed., pp. 106, 108, 467, 469, and 471, 1988).

Compilation of Data Table 23-16 presents inhalation toxicity data for the following criteria:

The emergency response planning guidelines (ERPG) concentrations for the following types of effects, for 1-h exposures (American Industrial Hygiene Association, *Emergency Response Planning Guidelines and Workplace Environmental Exposure Level*, no. AEAH05-559, 2005):

ERPG-1 Mild, transient health effects, without objectionable odor
ERPG-2 No irreversible or action-impairing effects
ERPG-3 No life-threatening effects

The immediately dangerous to life and health (IDLH) concentrations (National Institute for Occupational Safety and Health, *Pocket Guide to Chemical Hazards,* 1994), for 30-min exposures.

Values for workplace environmental exposure levels (WEELs) for many materials not listed in Table 23-16 can be obtained from the American Industrial Hygiene Association, at *www.aiha.org*.

The threshold limit values (TLVs) or time-weighted averages (TWAs) for 8-h exposures of workers (American Conference of Governmental Industrial Hygienists, *Threshold Limit Values for Chemical Substances and Physical Agents,* 2001).

Group	Acids	Alcohols	Oxides	Hydroxides	Halogens	Anhydrides
1	Hydrofluoric acid; selenic acid	Alkyl phenols			Fluorine; chlorine; bromine	
2	Hydrochloric acid; acetic acid; nitric acid	Pentol	Phosphorus pentoxide	Potassium hydroxide; sodium hydroxide		Acetic anhydride
3	Phosphoric acid		Calcium oxide	Ammonium hydroxide		Propionic anhydride

TABLE 23-16 ERPG Values and Other Toxicity Values for Toxic Materials

For materials that are listed in the USEPA 40 CFR 68.130 list and in the NJTCPA Group A and B lists (as of January 1, 2005; refer to current issues for changes).

Definitions:

ERPG-1 The maximum airborne concentration below which it is believed nearly all individuals could be exposed for up to 1 h without experiencing other than mild, transient adverse health effects or without perceiving a clearly defined objectionable odor.

ERPG-2 The maximum airborne concentration below which it is believed nearly all individuals could be exposed for up to 1 h without experiencing or developing irreversible or other serious health effects or symptoms that could impair an individual's ability to take protective action.

ERPG-3 The maximum airborne concentration below which it is believed nearly all individuals could be exposed for up to 1 h without experiencing or developing life-threatening health effects.

IDLH Immediately dangerous to life and health, for 1-h exposures. Where no IDLH data are available, the 50% lethal concentration is shown, as LC50: ppm/time.

WEEL-8 Workplace environmental exposure level, for 8-h time-weighted average (TWA).

WEEL-C Workplace environmnetal exposure level, as a ceiling (not to be exceeded) value.

TLV-TWA Threshold limit value, time-weighted average for 8-h exposures, with ceiling concentrations shown as C, and with skin absorption hazard as S. The OSHA permissible exposure limit (PEL) is the lower of the TWA or the ceiling limit.

Oral LD50 data are recorded where they are available (— indicates a toxicity listing but no oral toxicity data). Where the material is a gas at normal temperatures and pressure (25°C and 1 atm), the atmospheric-pressure boiling point is given.

Material	CAS no.	ERPG-1	ERPG-2	ERPG-3	IDLH	WEEL-8	WEEL-C	TLV-TWA	Oral, mg/kg
Acetaldehyde	75-07-0	10 ppm	200 ppm	1000 ppm	Ca (2000 ppm)	—	—	0.25 ppm C	1930
Acrolein	107-02-8	0.1 ppm	0.5 ppm	3 ppm	2 ppm	—	—	0.1 ppm C	7
Acrylonitrile	107-13-1	10 ppm	35 ppm	75 ppm	LC50: 576/4 h	—	—	2 ppm	27
Ammonia	7664-41-7	25 ppm	150 ppm	750 ppm	300 ppm	—	—	25 ppm	350
Arsine	7784-42-1	—	0.5 ppm	1.5 ppm	Ca (3 ppm)	—	—	0.05 ppm	−62.5°C
Boron trifluoride	7637-07-2	2 mg/m³	30 mg/m³	100 mg/m³	25 ppm	—	—	1 ppm C	−99.8°C
Bromine	7726-95-6	0.1 ppm	0.5 ppm	5 ppm	3 ppm	—	—	0.1 ppm	14
Carbon disulfide	75-15-0	1 ppm	50 ppm	500 ppm	500 ppm	—	—	10 ppm	2125
Carbon monoxide	630-08-0	200 ppm	350 ppm	500 ppm	1200 ppm	—	—	25 ppm	−191.5°C
Chlorine	7782-50-5	1 ppm	3 ppm	20 ppm	10 ppm	—	—	0.5 ppm	−34.1°C
Chlorine dioxide	10049-04-4	—	0.5 ppm	3 ppm	5 ppm	—	—	0.1 ppm	10.9°C
Chloroform	67-66-3	—	50 ppm	5000 ppm	Ca (500 ppm)	—	—	10 ppm	36
Chloropicrin	76-06-2	0.1 ppm	0.3 ppm	1.5 ppm	2 ppm	—	—	0.1 ppm	250
Chloroprene	126-99-8	—	—	—	Ca (300 ppm)	—	—	10 ppm	260
Dichlorosilane	4109-96-0	—	—	—	LC50: 215/	—	—	—	—
Diethylamine	109-89-7	—	—	—	200 ppm	—	—	5 ppm	540
Dimethylamine	124-40-3	0.6 ppm	100 ppm	350 ppm	500 ppm	—	—	5 ppm	240
Dimethylhydrazine	57-14-7	—	—	—	Ca (15 ppm)	—	—	0.01 ppm	122
Epichlorohydrin	106-89-8	2 ppm	20 ppm	100 ppm	Ca (75 ppm)	—	—	0.5 ppm	90
Ethylamine	75-04-7	—	—	—	600 ppm	—	—	5 ppm	400
Ethylenediamine	107-15-3	—	—	—	1000 ppm	—	—	10 ppm	470
Ethyleneimine	151-56-4	—	—	—	Ca (100 ppm)	—	—	0.5 ppm	15
Ethylene oxide	75-21-8	NA	50 ppm	500 ppm	Ca (800 ppm)	—	—	1 ppm	72
Ethylmercaptan	75-08-1	—	—	—	500 ppm	—	—	0.5 ppm	1960
Fluorine	7782-41-4	0.5 ppm	5 ppm	20 ppm	25 ppm	—	—	1 ppm	−188.3°C
Formaldehyde	50-00-0	1 ppm	10 ppm	25 ppm	Ca (20 ppm)	—	—	0.3 ppm C	42
Furan	110-00-9	—	—	—	LC50: 43/1 h	Minimize	—	—	—
Hydrazine	302-01-2	0.5 ppm	5 ppm	30 ppm	Ca (50 ppm)	—	—	0.01 ppm	49
Hydrogen bromide	10035-10-6	—	—	—	30 ppm	—	—	3 ppm C	−66.7°C
Hydrogen chloride	7647-01-0	3 ppm	20 ppm	150 ppm	50 ppm	—	—	5 ppm C	−85.0°C
Hydrogen cyanide	74-90-8	NA	10 ppm	25 ppm	50 ppm	—	—	4.7 ppm C	2
Hydrogen fluoride	7664-39-3	2 ppm	20 ppm	50 ppm	30 ppm	—	—	3 ppm C	−19.5°C
Hydrogen selenide	7783-07-5	NA	0.2 ppm	2 ppm	1 ppm	—	—	0.05 ppm	−41°C
Hydrogen sulfide	7783-06-4	0.1 ppm	30 ppm	100 ppm	100 ppm	—	—	10 ppm	−60°C
Iron pentacarbonyl	13463-40-6	—	—	—	LC50: 870/10 min	—	—	0.1 ppm	12
Isopropylamine	75-31-0	—	—	—	750 ppm	—	—	5 ppm	820
Ketene	143-50-0	—	—	—	5 ppm	—	—	0.5 ppm	1300
Methacrylonitrile	126-98-7	—	—	—	LC50: 36/4 h	—	—	1 ppm	15
Methylamine	74-89-5	10 ppm	100 ppm	500 ppm	100 ppm	—	—	5 ppm	−6.3°C
Methyl bromide	74-83-9	NA	50 ppm	200 ppm	Ca (250 ppm)	—	—	1 ppm	214
Methyl chloride	74-87-3	NA	400 ppm	1000 ppm	Ca (2000 ppm)	—	—	50 ppm	−24°C
Methylhydrazine	60-34-4	—	—	—	Ca (20 ppm)	—	—	0.01 ppm	22
Methyl iodide	74-88-4	25 ppm	50 ppm	125 ppm	Ca (100 ppm)	—	—	2 ppm	150
Methyl isocyanate	624-83-9	0.025 ppm	0.25 ppm	1.5 ppm	3 ppm	—	—	0.02 ppm	69
Methyl mercaptan	74-93-1	0.005 ppm	25 ppm	100 ppm	150 ppm	—	—	0.5 ppm	6.0°C
Methyl chlorosilane	75-79-6	0.5ppm	3 ppm	15 ppm	LC50: 29/2 h	1 ppm	1 ppm	—	1000
Methylvinylketone	78-94-4	—	—	—	LC50: 2.4/4 h	—	—	0.2 ppm	31
Nickel carbonyl	13463-39-3	—	—	—	Ca (2 ppm)	—	—	0.05 ppm	43°C
Nitric acid (white fuming)	7697-37-2	1 ppm	6 ppm	78 ppm	25 ppm	—	—	2 ppm	430
Nitric oxide	10102-43-9	—	—	—	LC50: 315/15 min	—	—	25 ppm	−151.8°C
Nitrogen dioxide	10102-44-0	1 ppm	15 ppm	30 ppm	20 ppm	—	—	3 ppm	20.8°C
Oleum	8014-95-7	2 mg/m³	10 mg/m³	30 mg/m³	LC50: 347/1 h	—	—	—	—
Ozone	10028-15-6	—	—	—	5 ppm	—	—	0.05 ppm	−112°C
Perchloryl fluoride	7616-94-6	—	—	—	100 ppm	—	—	3 ppm	−47°C
Phosgene	75-44-5	NA	0.2 ppm	1 ppm	2 ppm	—	—	0.1 ppm	7.6°C
Phosphine	7803-51-2	NA	0.5 ppm	5 ppm	50 ppm	—	—	0.3 ppm	−87°C
Phosphorus oxychloride	10025-87-3	—	—	—	LC50: 48/4 h	—	—	0.1 ppm	380
Phosphorus trichloride	7719-12-2	0.5 ppm	3 ppm	15 ppm	25 ppm	—	—	0.2 ppm	550
Propylene oxide	75-56-9	50 ppm	250 ppm	750 ppm	Ca (400 ppm)	—	—	2 ppm	380
Stibine	7803-52-3	ID	0.5 ppm	1.5 ppm	5 ppm	—	—	0.1 ppm	−18°C

TABLE 23-16 ERPG Values and Other Toxicity Values for Toxic Materials *(Concluded)*

Material	CAS no.	ERPG-1	ERPG-2	ERPG-3	IDLH	WEEL-8	WEEL-C	TLV-TWA	Oral, mg/kg
Sulfur dioxide	7446-09-5	0.3 ppm	3 ppm	15 ppm	100 ppm	—	—	2 ppm	−10°C
Sulfuric acid	7664-93-9	2 mg/m^3	10 mg/m^3	30 mg/m^3	15 mg/m^3	—	—	0.25 ppm	2140
Sulfur trioxide	7446-11-9	2 mg/m^3	10 mg/m^3	30 mg/m^3	LC50: >9/6 h	—	—	—	—
Tetranitromethane	509-14-8	—	—	—	4 ppm	—	—	0.005 ppm	130
Thionyl chloride	7719-09-7	0.2 ppm	2 ppm	10 ppm	LC50: 500/1 h	—	—	1 ppm	—
Titanium tetrachloride	7550-45-0	5 mg/m^3	20 mg/m^3	100 mg/m^3	LC50: 13/2 h	0.5 mg/m^3	—	—	—
Toluene di-isocyanate	584-84-9	0.01 ppm	0.15 ppm	0.6 ppm	Ca (2.5 ppm)	—	—	0.005 ppm	5800
Trimethylamine	75-50-3	—	—	—	LC50: 3500/4 h	1 ppm	—	5 ppm	2.9°C
Vinyl acetate	108-05-4	5 ppm	75 ppm	500 ppm	LC50: 1550/4 h	—	—	10 ppm	1613
Vinyl trichlorosilane	75-94-5	—	—	—	LC50: 500/4 h	1 ppm	1 ppm	—	1.280

The 50 percent lethal doses of ingested toxic materials that could cause fatal injury (National Institute for Occupational Safety and Health, *Registry of Toxic Effects of Chemical Substances,* 1983).

Where data for the above categories could not be found in the available literature, but the material was listed in the USEPA or NJTCPA standards, the LC50 value (op. cit.) was entered in the IDLH column.

Additional data concerning relatively long-term exposures of the public to toxic chemicals are presented in Table 23-17.

Safeguards against Toxicity Hazards Certainly the best protection against toxicity hazards is complete containment of hazardous materials within processing equipment.

Where complete containment is impractical, exhaust ventilation (preferably to a scrubber) can limit or eliminate exposure to toxic materials. The exhaust ventilation rate (velocity or volumetric rate) may be calculable for volatile liquids from spill size and vapor pressure (U.S. Environmental Protection Agency, *Risk Management Program Guidance for Offsite Consequence Analysis*, Appendix D, Equation D-1, 1999), but tests to determine concentrations in air usually would be needed for dusty processes and fugitive releases of gases.

If containment and exhaust ventilation are not considered adequate, cartridge respirators or self-contained breathing apparatus can provide protection against inhalation (and, in some cases, ingestion) of toxic materials. In 1994, the Occupational Safety and Health Standards were amended to require that "the employer shall assess the workplace to determine if hazards are present, or are likely to be present, which necessitate the use of personal protective equipment (PPE). If such hazards are present, or likely to be present, the employer shall select, and have each affected employee use, the types of PPE that will protect the affected employee from the hazards identified in the hazard assessment" [U.S. Department of Labor, *Occupational Safety and Health Standards,* 29 CFR 1910.132(d), 1999]. This hazard assessment would aid in determining the type of breathing protection that is appropriate for the toxicity hazard. Guidelines for appropriate use of breathing protection are given in this Standard (U.S. Department of Labor, *Occupational Safety and Health Standards*, 29 CFR 1910.134 and Appendices A, B, and C to 1910.134, and Appendix B to Subpart I, 1999). OSHA has not yet provided official Assigned Protection Factors, in Table I of this Standard, but manufacturers do provide these factors. As examples, a full-face cartridge respirator typically has a protection factor of 50 × PEL, and a self-contained breathing apparatus typically has a protection factor of 10,000 × PEL.

Conclusion Toxicity data are available for many thousands of solid, liquid, and gaseous chemicals and other materials. The data for inhalation toxicity provide guidance for concentration and duration limits, for protection of the public, chemical plant employees, and emergency response personnel. Similar data for ingestion and skin contact with toxic materials are not as readily available. Investigation into toxic effects is continuing, so that toxic materials can be handled safely.

OTHER HAZARDS

Hazards of Vacuum

Introduction Storage tanks and many other equipment items often have a relatively low resistance to the damage that can be caused by internal vacuum. The low vacuum rating for such equipment and the amount of damage that can result are often surprising and potentially costly lessons learned by plant engineers and operators.

Equipment Limitations A robust internal pressure rating for a piece of equipment is no guarantee that it will withstand an appreciable vacuum. Industry loss experience includes failures of vessels with design pressure ratings in excess of 25 psig (Sanders, "Victims of Vacuum," *Proceedings of the 27th Annual Loss Prevention Symposium*, AIChE, 1993). Low-pressure storage tanks are particularly fragile. For example, an atmospheric fixed-roof storage tank may only withstand a vacuum of 2.5 mbar (0.036 psi or 1 in water) (British Petroleum, *Hazards of Trapped Pressure and Vacuum*, 2003).

Jacketed vessels can be particularly vulnerable to internal vacuum, since the operating pressure of the heat-transfer medium in the jacket adds to the differential pressure that would otherwise exist between the atmosphere and the vessel interior.

While many pressure vessels may withstand a significant vacuum, design calculations are required to confirm this. Unless specifically rated for vacuum service, equipment should be assumed to be subject to damage by vacuum. When equipment is procured, consideration should be given to including the vacuum rating in the pressure vessel calculations and code stamp. In many instances, the additional cost of doing so will be an insignificant fraction of the total procurement cost for the vessel (see Protective Measures for Equipment).

Consequences of Vacuum Damage Vessels, tank trucks, or railcars can be dimpled by partial collapse or, more significantly, crushed like used drink cans. Fortunately, equipment damaged by underpressurization does not fail explosively, as might occur with overpressurized equipment. Nevertheless, loss of containment of equipment contents is a real risk, due to damage to the vessel or to the piping connected to the vessel. Significant releases of toxic, flammable, or otherwise hazardous materials can result, with severe consequences.

Alternatively, vacuum within equipment could lead to ingress of air into inerted or fuel-rich systems, posing a fire or explosion hazard within the equipment.

The potential for "knock-on" effects resulting from equipment damage should be considered.

Common Causes of Equipment Underpressurization Equipment can be exposed to excessive vacuum due to an unanticipated mechanism creating a vacuum and/or the failure or inadequate design of protective systems provided to mitigate the hazard.

A common scenario involves the pumping, draining, or siphoning of liquid from a tank that has no, or an inadequate, venting capacity and thus cannot allow the entry of air at a rate sufficient to backfill behind the dropping liquid level.

Similarly, vacuums can be created when a blower, fan, compressor, or jet ejector removes gases from equipment. The magnitude of the vacuum attainable will be governed by the performance characteristics of the device. Other mechanisms for generating a vacuum, which have been demonstrated by industry experience, include the following.

- *Condensation of vapors or cooling of hot gases.* For example, steam is commonly used to clean vessels and, less frequently, to create an inert atmosphere inside of equipment. Steam condensing inside a closed vessel can create a significant vacuum, and vessels (e.g., railcars) have collapsed when all vessel inlets were

TABLE 23-17 Emergency and Continuous Exposure Guidance Levels for Selected Airborne Contaminants

Committee on Toxicology National Research Council

Vol.	Material (concentrations in ppm, except for mg/m³)	Emergency exposure limits		Continuous exposure limit
		60-min	24-h	90-day
1	Acetone	8,500	1,000	200
1	Acrolein	0.05	0.01	0.01
4	Aluminum oxide	15*	N/L	N/L
IV	Ammonia	10	N/L	N/L
1	Arsine	1.0	0.1	0.01
3	Bromotrifluoromethane	25,000	N/L	100
1	Carbon disulfide	200	100	50
VI	Carbon monoxide	25	N/L	N/L
4	Carbon monoxide	400	50	20
VIII	Chlorine	0.5	N/L	N/L
2	Chlorine	3	0.5	0.1
2	Chlorine trifluoride	1	N/L	N/L
1	Chloroform	100	30	1
2	Dichlorodifluoromethane	10,000	1,000	100
2	Dichlorofluoromethane	100	3	1
2	Dichlorotetrafluoroethane	10,000	1,000	100
V	Dimethyl hydrazine	15	N/L	N/L
5	Dimethyl hydrazine	0.24	0.01	N/L
2	Ethanolamine	50	3	0.5
4	Ethylene glycol	40	20	4
1	Fluorine	15	10	7.5
V	Hydrazine	5	N/L	N/L
5	Hydrazine	0.12	0.005	N/L
II	Hydrogen chloride	2	N/L	N/L
III	Hydrogen fluoride	4	N/L	N/L
4	Hydrogen sulfide	N/L	10	1
2	Isopropyl alcohol	400	200	1
1	Mercury vapor	N/L	0.2*	0.01*
1	Methane	N/L	5,000	5,000
4	Methanol	200	10	N/L
V	Methyl hydrazine	15	N/L	N/L
5	Methyl hydrazine	0.24	0.01	N/L
4	Nitrogen dioxide	1	0.04	N/L
I	Nitrogen oxides	2	N/L	N/L
4	Nitrous oxide	10,000	N/L	N/L
1	Ozone	1	0.1	0.02
2	Phosgene	0.2	0.02	0.01
1	Sulfuric acid	5*	2*	1*
2	Trichlorofluoromethane	1,500	500	100
2	Trichlorotrifluoroethane	1,500	500	100
2	Sodium hydroxide	2*	N/L	N/L
2	Sulfur dioxide	10	5	1
2	Vinylidene chloride	N/L	10	0.15
2	Xylene	200	100	50

N/L = not listed; no guidance is given.
*Concentration in milligrams per cubic meter.

closed immediately after steam cleaning. The rapid addition of cool liquid to a vessel containing a hot, volatile liquid can markedly reduce the vapor pressure of the liquid. The sudden cooling of a storage tank by a thunderstorm can create a vacuum when gases in the vessel head space cool and/or vapors of volatile liquids condense. The American Petroleum Institute (API, *Venting Atmospheric and Low-Pressure Storage Tanks*, Standard 2000, Washington, 1998) provides guidance for in-breathing requirements as a function of tank capacity to protect against this latter scenario.

- *Absorption of a gas in a liquid.* Vessels have collapsed when ammonia vapor from the head space dissolved in water within the vessel (Lees, *Loss Prevention in the Process Industries*, 2d ed., Butterworths, London, 1996). A similar potential should be considered for HCl and water.
- *Chemical reactions that remove gases from the head space.* The corrosion of the interior of a steel vessel, especially if the vessel is newly fabricated or has been chemically cleaned, can consume and remove a significant quantity of the oxygen from the vessel atmosphere. Other chemical reactions (e.g., ammonia reacting with hydrogen chloride to form ammonium chloride) can reduce the amount of gas or vapor in the vessel.

Prudent design requires that equipment be protected from credible underpressurization scenarios. Equipment damage can result when such protections are omitted, improperly sized, incorrectly designed or installed, or inadequately maintained. Common failures include the following.

- *Failure to consider appropriate challenges when determining the required relief capacity* (e.g., maximum rates of liquid withdrawal or cooling of vessel contents). Credible contingencies (e.g., thunderstorm cooling a vessel during steam-out) should be considered.
- *Inadequate capacity, or failure, of vessel blanketing systems.* Inert gas supplies are often piped to vessels to maintain a reduced-oxygen atmosphere during liquid withdrawal. Coincident high demand for inert elsewhere, closure of a valve, or depletion of the supply could result in the failure to prevent a vacuum. A common means of initially inerting a vessel is to fill the vessel with liquid, then drain the liquid while allowing the blanketing system to backfill the head space with inert gas. Unless the blanketing system is sized to accommodate the drainage rate (which may exceed the normal process demand), there is a risk of collapsing the vessel.
- *Operating errors.* Many vessel collapses have resulted from closing or failing to open a valve in a vent line. For this reason, valves in vacuum relief lines should be avoided, and they may be prohibited by some design codes.
- *Maintenance errors.* One common error is the failure to remove an isolation blind in a vent line when returning a vessel to service. Even a thin sheet of plastic placed over an open nozzle may be sufficient to allow a vessel-damaging vacuum to be produced (BP, *Hazards of Trapped Pressure and Vacuum*, 2003).
- *Inappropriate modifications.* In one incident, a hose was connected to a vent line that was provided for both pressure and vacuum protection. The hose was submerged into a drum of liquid in an attempt to scrub vapors emitted from the vent. Only a few inches of submergence were required to ensure that the vent was effectively blocked the next time a vacuum was pulled on the vessel (Lees, *Loss Prevention in the Process Industries*, 2d ed., Butterworths, London, 1996).
- *Failure of vacuum control loop.* Control failures can either initiate events (e.g., increase the speed of an exhauster) or disable protections (e.g., reduce the rate of supply of inert gas to a vessel).
- *Plugging of vent lines or devices.* Process materials can migrate into and occlude vent systems when they polymerize, crystallize, condense, or solidify. Monomers requiring an inhibitor to prevent polymerization can evaporate from a tank and then condense in the vent line, free of the inhibitor. Waxes and other high-melting-point materials can solidify upon cooling in the vent system, dusts can accumulate, and water vapor can condense to form liquid seals in low points of vent lines or freeze in the winter. Such scenarios are a particular problem in cases where flame arrestors, screens, and other devices introduce small apertures in the vent flow path. Plugging of vent lines by animal or insect nests is not uncommon.
- *Inadequate or incorrect maintenance.* Mechanical devices such as vacuum breakers and flame arrestors require routine maintenance attention to ensure that they provide their intended protective function. Incorrect maintenance (e.g., changing the vacuum breaker set pressure) could defeat the intended protection.

Lees (*Loss Prevention in the Process Industries*, 2d ed., Butterworths, London, 1996), BP (*Hazards of Trapped Pressure and Vacuum*, 2003), and Kletz (*What Went Wrong?—Case Histories of Process Plant Disaster*, Gulf Publishing Company, 1989) include additional case histories providing valuable lessons about how equipment failures and human errors can combine to inflict vacuum damage.

Protective Measures for Equipment If equipment is subject to experiencing a vacuum, the inherently safer alternative would be to design the equipment to withstand a full vacuum. While this may not be economically feasible for large storage tanks, the incremental cost for smaller vessels may not be prohibitive, particularly when traded off against the capital and continued operating and maintenance costs of some alternatives (e.g., protective instrumentation systems). The incremental fabrication cost of providing a suitable vacuum rating can be less than 10 percent for vessels of up to 3000-gal nominal capacity and having a 15-psig pressure rating (Wintner, "Check the Vacuum Rating of Your Tanks," *Chemical Engineering*, pp. 157–159, February 1991).

Careful process hazards analysis may show that a particular vessel need not be designed to withstand a full vacuum (e.g., if the maximum attainable vacuum is limited to the performance characteristics of an exhauster). Whatever the vacuum rating, rated vessels must be periodically inspected to ensure that internal or external corrosion has not diminished the vessel strength.

Reliable protections against excessive vacuum should be provided whenever equipment cannot withstand the vacuums that can credibly be achieved. In some low-risk situations, protections may consist of administrative controls implemented by adequately trained personnel. Where the risk of damage is higher or where design standards or codes require, engineered protections should be implemented.

Where process, safety, and environmental considerations permit, vacuum protection may be provided by properly sized ever-open vents. Alternatively, active protective devices and systems are required. Vacuum breaker valves designed to open and admit air at a predetermined vacuum in the vessel are commonly used on storage tanks, but may not be suitable for some applications involving flammable liquids. Inert gas blanketing systems may be used if adequate capacity and reliability can be ensured. Where the source of the vacuum can be deenergized or isolated, suitably reliable safety instrumented systems (e.g, interlocks) can be provided.

API (*Venting Atmospheric and Low-Pressure Storage Tanks*, Standard 2000, Washington, 1998) provides guidance for vacuum protection of low-pressure storage tanks. Where vacuum relief devices are provided, they should communicate directly with the vapor space in the vessel and should be installed so that they cannot be sealed off by the liquid contents in the vessel. Valves should be avoided in the inlets or outlets of vacuum relief devices unless the valves are reliably carsealed or locked open, or excess relief capacity is provided (e.g., via multiple-way valves).

Hazards of Inerts

Introduction The use of inert gases to displace oxygen from equipment atmospheres in order to prevent combustion and, perhaps, consequent explosions has been described in the subsection "Flammability." Other applications for inerting exist, including preventing (1) corrosion or other deterioration of out-of-service equipment, (2) degradation of oxygen-sensitive products, or (3) exothermic reactions with air- or water-reactive materials. While the risk of personnel asphyxiation in an oxygen-deficient environment is the most frequently recognized concern, other hazards such as toxicity, temperature and pressure extremes, and chemical incompatibilities also need to be considered.

Sources of Inerts The most commonly used inert gases are N_2 and CO_2, but other gases and vapors such as argon (Ar), helium (He), steam, and exhaust gases from combustion devices are also used. The choice of the most appropriate inert for a given application must be based upon factors such as cost, availability, reliability of supply, effectiveness, and compatibility with process streams (Cunliff, "Avoiding Explosions by Means of Inerting Systems," IChemE Symposium Series no. 148, 2001; Grossel and Zalosh, *Guidelines for Safe Handling of Powders and Bulk Solids*, CCPS-AIChE, 2004).

Traditionally, inerts have been obtained from sources such as high-pressure gas cylinders or tube trailers or through evaporation of cryogenic liquids from bulk tanks. Other sources of inerts include (NFPA 69, *Standard on Explosion Prevention Systems*, National Fire Protection Association, 2002; FM Global, Loss Prevention Data Sheet 7-59, *Inerting and Purging of Tanks, Process Vessels, and Equipment*, 2000)

- On-site cryogenic air separation plants
- Gas generators burning or catalytically oxidizing a hydrocarbon to produce an oxygen-deficient product gas
- Nitrogen produced by the air oxidation of ammonia
- Nitrogen produced by removal of oxygen from air using pressure swing adsorption (PSA) or membrane separation units

Inert gas streams generated on site should be carefully monitored to ensure detection of an excessively high O_2 concentration in the product gas in the event of equipment failure or operational upset (e.g., due to a too high air-to-fuel ratio in a combustion generator or the failure of a membrane in a membrane separator). Consideration should be given to monitoring other indicators of problems in the inert generator (e.g., monitoring for low differential pressure across the membrane as an indication of the failure of a membrane separator).

Asphyxiation and Toxicity Hazards An asphyxiant is a chemical (either a gas or a vapor) that can cause death or unconsciousness by suffocation (BP, *Hazards of Nitrogen and Catalyst Handling*, 2003). A simple asphyxiant is a chemical, such as N_2, He, or Ar, whose effects are caused by the displacement of O_2 in air, reducing the O_2 concentration below its normal value of approximately 21 vol %. The physiological effects of oxygen concentration reduction by simple asphyxiants are illustrated in Table 23-18 (BP, *Hazards of Nitrogen and Catalyst Handling*, 2003).

The physiological processes leading to death from hypoxia (i.e., insufficient supply of oxygen to the body tissues) are described by Air Products (Air Products, *Dangers of Oxygen-Deficient Atmospheres*, Safetygram 17, 1998). At very low oxygen concentrations, loss of consciousness occurs within about 10 s of the first breath, followed by death within 2 to 4 min. A person exposed to an oxygen-deficient environment may not recognize the warning signs and may not be able to reason or take protective action before unconsciousness occurs. Victims removed from an O_2-deficient atmosphere require resuscitation through the administration of O_2 to prevent death [U.S. Chemical Safety and Hazard Investigation Board (CSB), *Hazards of Nitrogen Asphyxiation*, Safety Bulletin no. 2003-10-B, 2003].

Physical exertion increases oxygen demand and may result in oxygen deficiency symptoms at higher oxygen concentrations (CGA, *Oxygen-Deficient Atmospheres*, Publication SB-2, 2001), and individuals in poor health may be less tolerant of reduced oxygen concentrations. The guidance in Table 23-18 assumes a sea-level location and should be applied cautiously for facilities at significant altitudes; however, OSHA's Respiratory Protection Standard accepts 19.5 vol % as a safe O_2 concentration up to an altitude of 8000 ft (OSHA, 29 CFR 1910.134, Respiratory Protection Standard, 1998).

In its safety bulletin on the hazards of nitrogen asphyxiation, CSB identified 80 nitrogen asphyxiation deaths and 50 injuries occurring in 85 incidents between 1992 and 2002 (CSB, *Hazards of Nitrogen Asphyxiation*, Safety Bulletin no. 2003-10-B, 2003).

A chemical asphyxiant works by interfering with the body's ability to absorb or transport O_2 to the tissues. A relevant example of a chemical asphyxiant is CO, which can be present in inert gas streams produced

TABLE 23-18 Physiological Effects of Reduced O_2 Atmospheres

O_2 (vol %)	Effects
23.5	Maximum "safe level" (23 vol % is often the high-level alarm of most O_2 detectors)
21	Typical O_2 concentration in air
19.5	Minimum "safe level" (19 vol % is often the low-level alarm of most O_2 detectors)
15–19	First sign of hypoxia. Decreased ability to work strenuously. May induce early symptoms in persons with coronary, pulmonary, or circulatory problems.
12–14	Respiration increases with exertion; pulse up; impaired muscular coordination, perception, and judgment
10–12	Respiration further increases in rate and depth, poor judgment, lips blue
8–10	Mental failure, fainting, unconsciousness, ashen face, blueness of lips, nausea, vomiting, inability to move freely
6–8	6 min, 50% probability of death; 8 min, 100% probability of death
4–6	Coma in 40 s, convulsions, respiration ceases, death

TABLE 23-19 Physiological Effects of Exposure to CO_2

CO_2 (vol %)	Effects
1	Slight increase in breathing rate.
2	Breathing rate increases to 50% above normal. Prolonged exposure can cause headache and tiredness.
3	Breathing increases to twice the normal rate and becomes labored. Weak narcotic effect. Impaired hearing, headache, increase in blood pressure and pulse rate.
4–5	Breathing increases to approximately four times the normal rate, symptoms of intoxication become evident, and slight choking may be felt.
5–10	Characteristic sharp odor noticeable. Very labored breathing, headache, visual impairment, and ringing in the ears. Judgment may be impaired, followed within minutes by loss of consciousness.
50–100	Unconsciousness occurs more rapidly above 10 vol % level. Prolonged exposure to high concentrations may eventually result in death from asphyxiation.

by combustion. Exposure to CO concentrations of approximately 1000 and 13,000 ppm can cause, respectively, loss of consciousness after 1 h and unconsciousness and danger of death after 1 to 3 min (Meidl, *Explosive and Toxic Hazardous Materials*, Table 28, p. 293, Glencoe Press, 1970).

Note that CO_2 acts as neither a simple asphyxiant (like N_2) nor a chemical asphyxiant (like CO). The normal concentration of CO_2 in air is approximately 300 ppm (0.03 vol %). Table 23-19 (Air Products, *Carbon Dioxide*, Safetygram 18, 1998) illustrates that exposure to air diluted by 5 vol % CO_2 (yielding an oxygen concentration of 21×0.95, or approximately 20 vol %) prompts physiological effects that are more severe than those inferred from Table 23-18 for dilution by the same amount of nitrogen.

Injuries and fatalities from asphyxiation are often associated with personnel entry into inerted equipment or enclosures. Guidance on safe procedures for confined space access are provided by OSHA (OSHA, 29 CFR 1910.146, *Confined Space Entry Standard*, 2000), the American National Standards Institute (ANSI, Z117.1, *Safety Requirements for Confined Spaces*, 2003), Hodson (Hodson, "Safe Entry into Confined Spaces," *Handbook of Chemical Health and Safety*, American Chemical Society, 2001), and BP (BP, *Hazards of Nitrogen and Catalyst Handling*, 2003). OSHA has established 19.5 vol % as the minimum safe oxygen concentration for confined space entry without supplemental oxygen supply (see Table 23-18). Note that OSHA imposes a safe upper limit on O_2 concentration of 23.5 vol % to protect against the enhanced flammability hazards associated with O_2-enriched atmospheres.

Physical Hazards A variety of physical hazards are presented by the various inerts in common usage.

High temperature The high-temperature off-gases from combustion-based sources of inerts typically must be quenched before use. Water scrubbing, in addition to reducing the temperature, can remove soot and sulfur compounds (which could react with moisture to form corrosive acids) present in the off-gas. The humidity of the resultant gas stream may make it unsuitable for inerting applications where moisture cannot be tolerated.

Use of steam as an inert requires that equipment be maintained at an elevated temperature to limit condensation that would lower the inert concentration. FM Global (FM Global, Loss Prevention Data Sheet 7-59, *Inerting and Purging of Tanks, Process Vessels, and Equipment*, 2000) recommends a minimum temperature of 160°F. The Compressed Gas Association (CGA, *Safe Handling of Compressed Gases in Containers*, Publication P-1, 2000) cautions against the use of steam in (1) systems where brittle materials (such as cast iron) may be stressed by thermal expansion, (2) systems with close clearances where high temperatures may cause permanent warping or maladjustment, and (3) systems where pipe coatings or plastic materials may be damaged by high temperatures. Protection for personnel to prevent thermal burns from equipment may be required.

In addition, some equipment or equipment supports may not have the strength to support a significant load of condensate, and provisions must be made for removal of condensate from the inerted equipment.

Low temperature The atmospheric boiling points for N_2, CO_2, He, and Ar are −196, −79, −269, and −186°C, respectively. The potential for cryogenic burns must be addressed in operating and maintenance procedures and in specifying personal protective equipment requirements.

Cryogenic temperatures can cause embrittlement of some materials of construction (e.g., carbon steel) and must be considered in the design of inert gas delivery systems. Controls should be provided to ensure that operational upsets do not allow the migration of cryogenic liquids to piping or equipment not designed to withstand such low temperatures.

The potential for the condensation and fractional distillation of air on the outside of equipment containing cryogenic liquids with boiling points less than that of O_2 must be considered. For example, because N_2 boils at a lower temperature than O_2 (−196 versus −183°C), air can condense on the outside of liquid N_2-bearing piping. The liquid that drops off of the piping will be enriched in O_2 and can pose an enhanced fire or explosion risk in the vicinity of the equipment.

High pressure Cryogenic liquids produce large volumes of gas upon evaporation (for example, 1 volume of liquid N_2 produces 694 volumes of gas at 20°C) (Air Products, *Safe Handling of Cryogenic Liquids*, Safetygram 16, 1999). Containers such as transport and storage vessels must be provided with overpressure relief to address this hazard. An additional concern is the hydrostatic pressure that can be produced if cryogenic liquids are trapped in a liquid-full system. Absent a vapor space to allow liquid expansion, extremely high pressures can be produced; accordingly, pressure relief devices must be installed in sections of equipment where cryogenic liquids might become trapped between closed valves.

Given the large liquid-to-gas expansion ratio, consideration should be given to limiting the quantity of cryogenic liquid stored inside tight enclosures or buildings that could become pressurized. The asphyxiation hazard associated with inert gases was addressed previously.

Portable containers of high-pressure inert gases can operate at pressures of thousands of pounds per square inch. Suitable precautions are required to protect containers and associated regulators and piping from damage. Refer to CGA (CGA, *Safe Handling of Compressed Gases in Containers*, Publication P-1, 2000; CGA, *Precautions for Connecting Compressed Gas Containers to Systems*, Publication SB-10, 2003) and Air Products (Air Products, *Handling, Storage, and Use of Compressed Gas Cylinders*, Safetygram 10, 2000) for guidance.

Air Products (Air Products, *Product Migration of Liquefied Compressed Gases in Manifolded Systems*, Safetygram 38, 2003) provides precautionary guidance with respect to manifolding of liquid-containing cylinders. A temperature difference of only a few degrees between cylinders can cause gas from the warmer cylinder to migrate through the manifold to the cooler cylinder, where it could condense and potentially fill the cylinder. A liquid-filled cylinder could rupture if it was subsequently valved closed.

Static electricity The use of high-pressure CO_2 for inerting poses a concern for potential static electricity hazards. CO_2 converts directly to a solid if the liquid is depressurized below 61 psig (Air Products, *Carbon Dioxide*, Safetygram 18, 1998). Consequently, discharge of liquid CO_2 produces CO_2 "snow" that, when moving at a high velocity, can generate static electric charge. Incendive sparks (5 to 15 mJ at 10 to 20 kV) have been reported (Urben, *Bretherick's Handbook of Reactive Chemical Hazards*, 6th ed., Butterworth-Heinemann Ltd., 1999).

Chemical Incompatibility Hazards While N_2 and CO_2 may act as inerts with respect to many combustion reactions, they are far from being chemically inert. Only the noble gases (e.g., Ar and He) can, for practical purposes, be regarded as true inerts. Frank (Frank, "Inerting for Explosion Prevention," *Proceedings of the 38th Annual Loss Prevention Symposium*, AIChE, 2004) lists a number of incompatibilities for N_2, CO_2, and CO (which can be present in gas streams from combustion-based inert gas generators). Notable incompatibilities for N_2 are lithium metal and titanium metal (which is reported to burn in N_2). CO_2 is incompatible with many metals (e.g., aluminum and the alkali metals), bases, and amines, and it forms carbonic acid in water, which can corrode some materials. CO is a strong reducing agent and is incompatible with oxidizers, potassium, sodium, some aluminum compounds, and certain metal oxides. Trace metals and residual organic compounds may contaminate gas streams from combustion-based inert gas generators, posing a variety of potential incompatibility and product quality concerns.

Certain polymerization inhibitors added to stabilize monomers require a small concentration of dissolved O_2 to be effective (NFPA 69, *Standard on Explosion Prevention Systems,* National Fire Protection Association, 2002). For example, methyl acrylate and ethyl acrylate are commonly stabilized with hydroquinone monomethyl ether. Industry guidance recommends a minimum concentration of 5 vol % O_2 in the atmosphere above the acrylate to prevent polymerization (Intercompany Committee for the Safety and Handling of Acrylic Monomers, *Acrylate Esters, A Summary of Safety and Handling*, 3d ed., 2002).

INHERENTLY SAFER DESIGN AND OTHER PRINCIPLES

INHERENTLY SAFER AND MORE USER-FRIENDLY DESIGN

Introduction For many years the usual procedure in plant design was to identify the hazards, by one of the systematic techniques described later or by waiting until an accident occurred, and then add protective equipment to control them or to protect people from their consequences. This protective equipment is often complex and expensive and requires regular testing and maintenance. It often interferes with the smooth operation of the plant and is sometimes bypassed. Gradually the industry came to realize that, whenever possible, we should design user-friendly plants that can withstand human error and equipment failure without serious effects on safety (and output and efficiency). When we handle flammable, explosive, toxic, or corrosive materials, we can tolerate only very low failure rates, of people and equipment, rates which it may be impossible or impracticable to achieve consistently for long periods.

The most effective way of designing user-friendly plants is to avoid, when possible, large inventories of hazardous materials in process or storage. "What you don't have, can't leak." This sounds obvious, but until the explosion at Flixborough in 1974, little systematic thought was given to ways of reducing inventories. The industry simply designed a plant and accepted whatever inventory the design required, confident it could be kept under control. Flixborough weakened that confidence, and 10 years later Bhopal almost destroyed it. Plants in which we avoid a hazard, by reducing inventories or avoiding hazardous reactions, are usually called inherently safer.

The principle ways of designing inherently safer plants, and other ways of making plants user-friendly, are summarized below, with examples (Kletz, *Process Plants: A Handbook for Inherently Safer Design*, Taylor & Francis, 1998).

Intensification or Minimization One approach is to use so little hazardous material that it does not matter if it all leaks out. For example, at Bhopal methyl isocyanate (MIC), the material that leaked and killed over 2000 people, was an intermediate that was convenient but not essential to store. Within a few years many companies had reduced their stocks of MIC and other hazardous intermediates.

Intensification is the preferred route to inherently safer design as the plants, being smaller, are also cheaper (Bell, *Loss Prevention in the Process Industries*, Institution of Chemical Engineers Symposium Series no. 34, 1971 p. 50).

Substitution If intensification is not possible, then an alternative is to use a safer material in place of a hazardous one. Thus it is possible to replace flammable solvents, refrigerants, and heat-transfer media by nonflammable or less flammable (high-boiling) ones, hazardous products by safer ones, processes that use hazardous raw materials or intermediates by processes that do not. As an example of the latter, the product manufactured at Bhopal (carbaryl) was made from three raw materials. Methyl isocyanate is formed as an intermediate. It is possible to react the same raw materials in a different order so that a different and less hazardous intermediate is formed.

Attenuation or Moderation Another alternative to intensification is attenuation, or using a hazardous material under the least hazardous conditions. Thus large quantities of liquefied chlorine, ammonia, and petroleum gas can be stored as refrigerated liquids at atmospheric pressure instead of under pressure at ambient temperature. (Leaks from the refrigeration equipment should also be considered, so there is probably no net gain in refrigerating quantities less than a few hundred tons.) Dyestuffs that form explosive dusts can be handled as slurries.

Limitation of Effects of Failures Effects of failures can be limited by equipment design or change in reaction conditions, rather than by adding protective equipment. For example:

- Heating media such as steam or hot oil should not be hotter than the temperature at which the materials being heated are liable to ignite spontaneously or react uncontrollably.
- Spiral-wound gaskets are safer than fiber gaskets because if the bolts work loose or are not tightened correctly, the leak rate is much lower.
- Tubular reactors are safer than pot reactors as the inventory is usually lower and a leak can be stopped by closing a valve.
- Vapor-phase reactors are safer than liquid-phase ones as the mass flow rate through a hole of a given size is much less. (This is also an example of attenuation.)
- A small, deep diked area around a storage tank is safer than a large shallow one as the evaporation rate is lower and the area of any fire is smaller.
- Changing the order of operations, reducing the temperature, or changing another parameter can prevent many runaway reactions.
- Reduce the frequency of hazardous operations such as sampling or maintenance. We should consider the optimum balance between reliability and maintenance.

Simplification Simpler plants are friendlier than complex ones as they provide fewer opportunities for error and less equipment which can go wrong. Some of the reasons for complication in plant design are

- The need to control hazards. If we can intensify or carry out one of the other actions already discussed, we need less added protective equipment and plants will therefore be simpler.
- A desire for flexibility. Multistream plants with numerous crossovers and valves, so that any item can be used on any stream, have numerous leakage points, and errors in valve settings are easily made.
- Lavish provision of installed spares with the accompanying isolation and changeover valves.
- Continued adherence to traditional rules or practices that are no longer necessary.
- Design procedures that result in a failure to identify hazards until late in design. By this time it is impossible to avoid the hazard, and all we can do is to add complex equipment to control it.

Knock-on Effects Plants should be designed so that those incidents that do occur do not produce knock-on or domino effects. This can be done, e.g., by

- Providing firebreaks, about 15 m wide, between sections, similar to firebreaks in a forest, to restrict the spread of fire.
- Locating equipment that is liable to leak out-of-doors so that leaks of flammable gases and vapors are dispersed by natural ventilation. Indoors a few tens of kilograms are sufficient for an explosion that can destroy the building. Outdoors a few tons are necessary for serious damage. A roof over a piece of equipment, such as a compressor, is acceptable, but walls should be avoided. If leaks of toxic gases are liable to occur, it may be safer to locate the plant indoors, unless leaks will disperse before they reach members of the public or employees on other units.

- Constructing storage tanks so that the roof-wall weld will fail before the base-wall weld, thus preventing spillage of the contents. In general, in designing equipment we should consider the way in which it is most likely to fail and, when possible, locate or design the equipment so as to minimize the consequences.

Making Incorrect Assembly Impossible Plants should be designed so that incorrect assembly is difficult or impossible. For example, compressor valves should be designed so that inlet and exit valves cannot be interchanged; hose connections of different types or sizes should be used for compressed air and nitrogen.

Making Status Clear It should be possible to see at a glance if equipment has been assembled or installed incorrectly or whether it is in the open or shut position. For example:

- Check valves should be marked so that installation the wrong way round is obvious. It should not be necessary to look for a faint arrow hardly visible beneath the dirt.
- Gate valves with rising spindles are friendlier than valves with nonrising spindles, as it is easy to see whether they are open or shut. Ball valves and cocks are friendly if the handles cannot be replaced in the wrong position.
- Figure 8 plates are friendlier than slip plates (blinds) as their position is apparent at a glance. If slip plates are used, their projecting tags should be readily visible, even when the line is insulated. In addition, spectacle plates are easier to fit than slip plates, if the piping is rigid, and are always available on the job. It is not necessary to search for one, as with slip plates.

Tolerance Whenever possible, equipment should tolerate poor installation or operation without failure. Expansion loops in pipework are more tolerant of poor installation than are expansion joints (bellows). Fixed pipes, or articulated arms, if flexibility is necessary, are friendlier than hoses. For most applications, metal is friendlier than glass or plastic.

Bolted joints are friendlier than quick-release couplings. The former are usually dismantled by a fitter after issue of a permit-to-work. One worker prepares the equipment and another opens it up; the issue of the permit provides an opportunity to check that the correct precautions have been taken. In addition, if the joints are unbolted correctly, any trapped pressure is immediately apparent and the joint can be remade or the pressure allowed to blow off. In contrast, many accidents have occurred because operators opened up equipment that was under pressure, without independent consideration of the hazards, using quick-release couplings. There are, however, designs of quick-release coupling which give the operator a second chance.

Low Leak Rate If friendly equipment does leak, it does so at a low rate which is easy to stop or control. Examples already mentioned are spiral-wound gaskets, tubular reactors, and vapor-phase reactors.

Ease of Control Processes with a flat response to change are obviously friendlier than those with a steep response. Processes in which a rise of temperature decreases the rate of reaction are friendlier than those with a positive temperature coefficient, but this is a difficult ideal to achieve in the chemical industry. However, there are a few examples of processes in which a rise in temperature reduces the rate of reaction. For example, in the manufacture of peroxides, water is removed by a dehydrating agent. If magnesium sulfate is used as the agent, a rise in temperature causes release of water by the agent, diluting the reactants and stopping the reaction (Gerrison and van't Land, *I&EC Process Design*, **24,** 1985, p. 893).

Software In some programmable electronic systems (PESs), errors are much easier to detect and correct than in others. Accidentally pressing the wrong key should never produce serious consequences. If we press the delete key on our computers, sometimes we are asked if we really want to do so; but stocks and currency have been accidentally sold or bought because someone pressed the wrong key. If we use the term *software* in the wider sense to cover all procedures, as distinct from hardware or equipment, some software is much friendlier than others. For example, if many types of gaskets or nuts and bolts are stocked, sooner or later the wrong type will be installed. It is better, and cheaper in the long run, to keep the number of types stocked to a minimum even though more expensive types than are strictly necessary are used for some applications.

Actions Needed for the Design of Inherently Safer and User-Friendly Plants

1. Designers need to be made aware that there is scope for improving the friendliness of the plants they design.

2. To achieve many of the changes suggested above, it is necessary to carry out much more critical examination and systematic consideration of alternatives during the early stages of design than has been customary in most companies. Two studies are suggested, one at the conceptual or business analysis stage when the process is being chosen and another at the flow sheet stage. For the latter the usual hazard and operability study (HAZOP) questions may be suitable but with one difference. In a normal HAZOP study on a line diagram, if we are discussing "more of temperature," say, we assume that it is undesirable and look for ways of preventing it. In a HAZOP of a flow sheet, we should ask if "more of temperature" would be better. For the conceptual study, different questions are needed.

Many companies will say that they do consider alternatives during the early stages of plant design. However, what is lacking in many companies is a formal, systematic structured procedure of the HAZOP type.

When a new plant is needed, it is usually wanted as soon as possible, and so there is no time to consider and develop inherently safer designs (or other innovations). When we are designing a new plant, we are conscious of all the improvements we could have made if we had had more time. These possible improvements should be noted and work on their feasibility started, ready for the plant after next. Unless we do so, we will never innovate and will ultimately lose to those who do.

3. To achieve the more detailed improvements suggested above, it may be necessary to add a few questions to those asked during a normal HAZOP. For example, what types of valve, gasket, blind, etc. will be used?

INCIDENT INVESTIGATION AND HUMAN ERROR

Although most companies investigate accidents (and many investigate dangerous incidents in which no one was injured), these investigations are often superficial and we fail to learn all the lessons for which we have paid the high price of an accident. The collection of evidence is usually adequate, but often only superficial conclusions are drawn from it. Identifying the causes of an accident is like peeling an onion. The outer layers deal with the immediate technical causes and triggering events while the inner layers deal with ways of avoiding the hazard and with the underlying weaknesses in the management system (Kletz, *Learning from Accidents*, 3d ed., Gulf Professional, 2001).

Dealing with the immediate technical causes of a leak, e.g., will prevent another leak for the same reason. If we can use so little of the hazardous material that leaks do not matter, or a safer material instead, as discussed above, we prevent all significant leaks of this hazardous material. If we can improve the management system or improve our designs, we may be able to prevent many more accidents.

Other points to watch when you are drawing conclusions from the facts are as follows:

1. Avoid the temptation to list causes we can do little or nothing about. For example, a source of ignition should not be listed as the primary cause of a fire or explosion as leaks of flammable gases are liable to ignite even though we remove known sources of ignition. The cause is whatever led to the formation of a flammable mixture of gas or vapor and air. (Removal of known sources of ignition should, however, be included in the recommendations.)

Similarly, human error should not be listed as a cause. See item 7 below.

2. Do not produce a long list of recommendations without any indication of the relative contributions they will make to the reduction of risk or without any comparison of costs and benefits. Resources are not unlimited, and the more we spend on reducing one hazard, the less there is left to spend on reducing others.

3. A named person should be made responsible for carrying out each agreed recommendation, and a completion date agreed with him or her. The report should be brought forward at this time; otherwise, nothing will happen except a repeat of the accident.

4. Avoid the temptation to overreact after an accident and install an excessive amount of protective equipment or complex procedures that

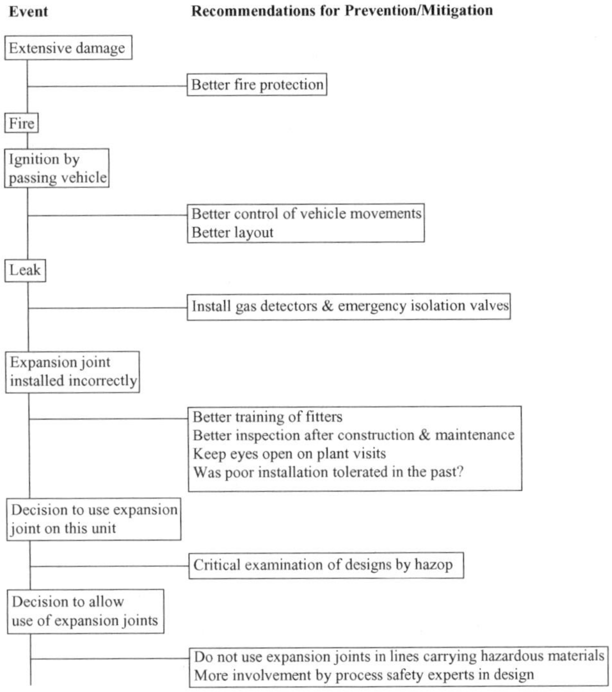

FIG. 23-17 An example of the many ways by which an accident could have been prevented.

are unlikely to be followed after a few years have elapsed. Sometimes an accident occurs because the protective equipment available was not used, but nevertheless the report recommends installation of more protective equipment; or an accident occurs because complex procedures were not followed, and the report recommends extra procedures. It would be better to find out why the original equipment was not used or the original procedures were not followed.

5. Remember that few, if any, accidents have a single cause. In most cases many people had an opportunity to prevent it, from the chemist who developed the process to the operator who closed the wrong valve. Figure 23-17 shows by example the opportunities that were available to prevent a fire or minimize the consequences of an apparently simple incident: an expansion joint (bellows) was incorrectly installed in a pipeline so that it was distorted. After some months it leaked, and a passing vehicle ignited the escaping vapor. Damage was extensive as the surrounding equipment had not been fire-protected to save on costs.

The fitter who installed the expansion joint incorrectly could have prevented the fire. So could the person who was responsible for her or his training and supervision; so could the designers if they had not specified an expansion joint, had carried out a HAZOP, or had consulted experts; so could the author of the company's design standards, and those responsible for the training of designers; those responsible for inspection of workmanship, and anyone who kept his or her eyes open when walking round the plant.

6. When you are reading an accident report, look for the things that are not said. For example, a gland leak on a liquefied flammable gas pump caught fire and caused considerable damage. The report drew attention to the congested layout, the amount of redundant equipment in the area, the fact that a gearbox casing had been made of aluminum, which melted, and several other unsatisfactory features. It did not stress that there had been a number of gland leaks on this pump over the years, that reliable glands are available for liquefied gases at ambient temperatures, and therefore there was no need to have tolerated a leaky pump on this duty.

7. At one time most accidents were said to be due to human error, and in a sense they all are. If someone, designer, manager, operator, or maintenance worker had done something differently, the accident would not have occurred. However, the term *human error* is not very helpful as different types of error require quite different actions to prevent their happening again. The following classification of errors is recommended as it helps us see the type of action needed to prevent a repeat (Kletz, *An Engineer's View of Human Error*, 3d ed., Institution of Chemical Engineers, United Kingdom, and Taylor and Francis, 2001).

a. Some errors, called mistakes, are due to poor training or instructions—someone did not know what to do. It is a management responsibility to provide good training and instructions and avoid instructions that are designed to protect the writer rather than help the reader. However many instructions we write, problems will arise that are not covered, and so people, particularly operators, should be trained in flexibility, i.e., the ability to diagnose and handle unforeseen situations. If the instructions are hard to follow, can the job be simplified?

b. Some accidents, called violations or noncompliance, occur because someone knew what to do but made a decision not to do it. We should, if possible, simplify the job (if the correct method is difficult, an incorrect one will be used); explain the reasons for the instructions; carry out checks from time to time to see that instructions are being followed; and not turn a blind eye if they are not. While some violations make the job easier, many are made by people who think they have found a better way of doing the job. If instructions are incorrect, a violation can prevent an accident.

The methods of behavioral science can be used to reduce violations. Specially trained members of the workforce keep their eyes open and tactfully draw the attention of fellow workers to violations such as failures to wear protective clothing. These techniques can reduce the incidence of everyday accidents, but they have little or no influence on process safety.

c. Some accidents—mismatches—occur because the job is beyond the physical or mental ability of the person asked to do it, sometimes beyond anyone's ability. We should improve the plant design or method of working.

d. The fourth category is the commonest, a momentary slip or lapse of attention. This happens to everyone from time to time and cannot be prevented by telling people to be more careful, telling them to keep their minds on the job, or better training. In fact, slips and lapses of attention occur only when we are well trained. We put ourselves on autopilot and carry out the task without continually monitoring our progress, though we may check it from time to time. These errors are more likely to occur when we are distracted or stressed. To avoid slips and lapses of attention, all we can do is to change the plant design or method of working so as to remove opportunities for error (or minimize the consequences or provide opportunities for recovery). We should, whenever possible, design user-friendly plants (see above) that can withstand errors (and equipment failures) without suffering serious effects on safety (and output and efficiency). It is more effective to change the behavior of equipment than to try to change the behavior of people.

When an accident report says that an accident was due to human error, the writer usually means an error by an operator or other front line worker. But designers and managers also make errors, not slips or lapses of attention as they usually have time to check their work, as well as mistakes or, less often, violations.

INSTITUTIONAL MEMORY

Most accidents occur not because we do not know how to prevent them but because we do not use the information that is available. The recommendations made after an accident are forgotten when the people involved have left the plant; the procedures they introduced are allowed to lapse, the equipment they installed is no longer used, and the accident happens again. The following actions can prevent or reduce this loss of information.

- Include a note on the reason why in every instruction, code, and standard as well as accounts of accidents that would not have occurred if the instruction, code, or standard had been followed.
- Describe old accidents as well as recent ones in safety bulletins and newsletters, and discuss them at safety meetings.
- Follow up at regular intervals (e.g., during audits) to see that the recommendations made after accidents are being followed, in design as well as operations.
- Make sure that recommendations for changes in design are acceptable to the design organization.
- Remember that the first step down the road to an accident is taken when someone turns a blind eye to a missing blind (or other feature).
- On each unit keep a memory book, a folder of reports on past accidents, which is compulsory reading for new recruits and which others dip into from time to time. It should include relevant reports from other companies but should not include cuts and bruises.
- Never remove equipment before you know why it was installed. Never abandon a procedure before you know why it was adopted.
- When people are moving to other jobs in the company or leaving it, make sure that the remaining employees at all levels have adequate knowledge and experience.
- Include important accidents of the past in the training of undergraduates and company employees. The training should start with accounts of accidents that demonstrate the need for codes, standards, or instructions. Ask audience members to say what *they think* should be done to prevent similar accidents from happening again. More will be remembered after a discussion than after a lecture, and audience members are more likely to be convinced when they have worked out for themselves the actions that should be taken (Kletz, *Lessons from Disaster—How Organizations Have No Memory and Accidents Recur*, Institution of Chemical Engineers, Rugby, United Kingdom, 1993, Chap. 10). Suitable accidents for discussion can be found in books of accident case histories, but local accidents have the greatest impact.
- There are many databases of accidents as well as books of case histories, but they have been little used. We need better retrieval systems so that we can find, more easily than at present, details of past accidents, in our own and other companies, and the recommendations made afterward.

PROCESS SAFETY ANALYSIS

HAZARD ANALYSIS

General References: Alaimo, *Handbook of Chemical Health and Safety*, Oxford University Press, New York, 2001; *Chemical Exposure Index*, 2d ed., AIChE, New York, 1994; Council of the European Union, Directive 96/82/EC, *Prevention of Major Accidents*, December 9, 1996; CPQRA, *Guidelines for Chemical Process Quantitative Risk Analysis*, 2d ed., CCPS-AIChE, New York, 2000; Crawley, Preston, and Tyler, *HAZOP: Guide to Best Practice*, IChemE, Rugby, Warwickshire, U.K., 2002; Crowl and Louvar, *Chemical Process Safety: Fundamentals with Applications*, 2d ed., Prentice-Hall, Englewood Cliffs, N.J., 2002; Dowell, "Managing the PHA Team," *Process Safety Progress* **13**, no. 1, January 1994; *Dow Fire and Explosion Index*, AIChE, New York, January 1994; Harris, *Patty's Industrial Hygiene*, 5th ed., Wiley, New York, 2000; *Guidelines for Design Solutions for Process Equipment Failures*, AIChE, New York, 1997; *HEP Guidelines for Hazard Evaluation Procedures*, 2d ed., AIChE, New York, 1992; Johnson, Rudy, and Unwin, *Essential Practices for Managing Chemical Reactivity Hazards*, AIChE, New York, 2003; Knowlton, *Hazard and Operability Studies*, Chemetics International Co., Ltd., Vancouver, British Columbia, February 1989; *LOPA Layer of Protection Analysis*, AIChE, New York, 2001; Mannan, *Lees' Loss Prevention in the Process Industries*, Butterworth-Heinemann, London, 2005; Mosley, Ness, and Hendershot, "Screen Reactive Chemical Hazards Early in Process Development," *Chemical Engineering Progress*, 96(1), pp. 51–65; NOAA, *Chemical Reactivity Worksheet*, Version 1.6, U.S. National Oceanic and Atmospheric Administration, 2004 (*http://response.restoration.noaa.gov/chemaids/react.html*); Skelton, *Process Safety Analysis: An Introduction*, Gulf Publishing, Houston, Tex., 1997; *SVA Guidelines for Analyzing and Managing the Security Vulnerabilities of Fixed Chemical Sites*, AIChE, New York, 2003; Tweeddale, *Managing Risk and Reliability of Process Plants*. Gulf Professional, Houston, Tex., 2003; USEPA, "Risk Management Programs for Chemical Accidental Release Prevention Requirements," 40 CFR 68; U.S. OSHA, "Process Safety Management of Highly Hazardous Chemicals, Explosives and Blasting Agents," 29 CFR 1910.119; Wells, *Hazard Identification and Risk Assessment*, IChemE, Rugby, Warwickshire, U.K., 1996.

Introduction The meaning of *hazard* is often confused with *risk*. *Hazard* is defined as the inherent potential of a material or activity to harm people, property, or the environment. Hazard does not have a probability component.

There are differences in terminology on the meaning of *risk* in the published literature that can lead to confusion. *Risk* has been defined in various ways (CPQRA, 2000, pp. 5, 6). In this edition of the handbook, *risk* is defined as "a measure of human injury, environmental damage, or economic loss in terms of both the incident likelihood and magnitude of the injury, damage, or loss." *Risk* implies a probability of something occurring.

Definitions of Terms Following are some definitions that are useful in understanding the components of hazards and risk.

Accident A specific combination of events or circumstances that leads to an undesirable consequence.

Acute hazard The potential for injury or damage to occur as a result of a short-duration exposure to the effects of an accident.

Cause-consequence A procedure using diagrams to illustrate the causes and consequences of a particular scenario. They are not widely used because, even for simple systems, displaying all causes and outcomes leads to very complex diagrams.

Chemical Exposure Index (CEI) A method of rating the relative potential of acute health hazard to people from possible chemical release incidents, developed by Dow Chemical Company.

Chronic hazard The potential for injury or damage to occur as a result of prolonged exposure to an undesirable condition.

Consequence The direct, undesirable result of an accident, usually measured in health and safety effects, environmental impacts, loss of property, or business costs.

Consequence analysis Once hazards and specific incident scenarios through which those hazards might impact people, the environment, or property have been identified, methods exist for analyzing their consequences (size of vapor cloud, blast damage radius, overpressure expected, etc.). This is independent of frequency or probability.

Domino effect An incident which starts in one piece of equipment and affects other nearby items, such as vessels containing hazardous materials, by thermal blast or fragment impact. This can lead to escalation of consequences or frequency of occurrence. This is also known as a *knock-on effect*.

Event An occurrence involving equipment performance or human action or an occurrence external to the system that causes system upset. An event is associated with an incident, either as a cause or a contributing cause of the incident, or as a response to an initiating event.

Event tree A graphical logic model that identifies and potentially quantifies possible outcomes following an initiating event.

Failure mode and effect analysis (FMEA) A hazard identification technique in which all known failure modes of components or features of a system are considered in turn and undesired outcomes are noted. It is often used in combination with hazard and operability (HAZOP) studies or fault tree analysis.

Fault tree A graphical logic model for representing the combinations of various system states which lead to a particular outcome, known as the *top event*.

Fire and Explosion Index (F&EI) A hazard index developed by Dow Chemical Company used to rank fire and explosion hazards in a chemical process.

Hazard An inherent physical or chemical characteristic that has the potential for causing harm to people, the environment, or property.

HAZOP (HAZard and OPerability study) A formal hazard identification and evaluation procedure based on the application of guide words to identify possible deviations from the intended operation of the process.

Incident The loss of containment of material or energy, e.g., a leak of a flammable and toxic gas.

Interaction matrix A two-dimensional matrix listing all components of interest on the x and y axes and recording the consequences of mixing of these components for each combination of the components. It is useful for identifying chemical reaction hazards and incompatibilities.

Layer-of-protection analysis (LOPA) A method, based on event tree analysis, of evaluating the effectiveness of independent protection layers in reducing the likelihood or severity of an undesired event.

Process hazard analysis (PHA) Any of a number of techniques for understanding and managing the risk of a chemical process or plant. Examples of PHA techniques include HAZOP, checklists, what-if methods, fault tree analysis, event tree analysis, and others.

Process Hazard Analysis Regulations In the United States, the OSHA rule for Process Safety Management (PSM) of Highly Toxic Hazardous Chemicals, 29 CFR 1910.119, part (e), requires an initial PHA and an update every 5 years for processes that handle listed chemicals or contain over 10,000 lb (4356 kg) of flammable material. The PHA must be done by a team, must include employees such as operators and mechanics, and must have at least one person skilled in the methodology employed. Suggested methodologies from Process Safety Management are listed in Table 23-20. The PHA must consider hazards listed in the PSM Rule, part (e), including information from previous incidents with potential for catastrophic consequences, engineering and administrative controls and consequences of their failure, facility siting, and human factors. Consequences of failure of controls must be considered. The key to good PHA documentation is to do it right away while the information is fresh. Periodic follow-up of the PHA study is needed by management and safety professionals to confirm that all recommendations have been addressed. The PHA must be reviewed as part of the management of change procedures when the facility is modified in any way. (Dowell, 1994, pp. 30–34.)

As required by the Clean Air Act Amendments of 1990, the USEPA mandates a Risk Management Program (RMP) for listed substances (40 CFR 68). RMP requires (1) a hazard assessment that details the potential effects of an accidental release, an accident history of the last 5 years, and an evaluation of worst-case and alternative accidental releases; (2) a prevention program that includes safety precautions and maintenance, monitoring, and employee training measures; and (3) an emergency response program that spells out emergency health care, employee training measures, and procedures for informing the public and response agencies, should an accident occur.

Most countries also have regulations analogous to the U.S. regulations. For example, the European Union issued the "Seveso II" Directive in 1996 (replacing the original 1982 directive) which requires all member states to implement regulations for the control of major accident hazards. Also, in addition to the U.S. government requirements, many state and local governments have implemented regulations requiring process hazard analysis and risk management.

TABLE 23-20 Process Hazard Analysis Methods Listed in the OSHA Process Safety Management Rule

- What-if
- Checklist
- What-if/checklist
- Hazard and operability study (HAZOP)
- Failure mode and effect analysis (FMEA)
- Fault tree analysis (FTA)
- An appropriate equivalent methodology

SOURCE: Dowell, 1994, pp. 30–34.

Hazard Identification and Analysis Tools The hazard and risk assessment tools used vary with the stage of the project from the early design stage to plant operations. Many techniques are available. In the following discussion, they will be categorized as hazard identification and analysis tools, hazard ranking methods, and logic model methods. Reviews done early in projects often result in easier, more effective changes.

Safety, Health, Environmental, and Loss Prevention Reviews Most chemical companies have specific internal protocols defining these reviews, which may have different names or descriptions in different organizations. In most organizations, these reviews are conducted at various stages in the process life cycle, from initial process conceptualization, through laboratory development, scale-up, plant design, start-up, operation, modification, and shutdown. The scope and focus of the review will be different at different stages in development, with reviews early in process development focusing on major hazards and strategies for managing the risks. As the process and plant become clearly defined, the reviews will focus more on details of the design and operation. The purpose of the reviews is to have an independent (from the development, design, or operation team) evaluation of the process and layout from safety, industrial hygiene, environmental, and loss prevention points of view. It is often desirable to combine these reviews to improve the efficiency of the use of time for the reviewers (*HEP* Chaps. 4.1, 4.4, 6.1, 6.4, 13).

Checklists Checklists are simple means of applying experience to designs or situations to ensure that the features appearing in the list are not overlooked. Checklists tend to be general and may not be appropriate to a specific situation. They may not handle adequately the novel design or unusual process. (*HEP,* Chaps. 4.2, 6.2, 16, 20). The CCPS *Design Solutions for Process Equipment Failures* (1997) provides a useful set of checklists for common chemical processing equipment.

What-if At each process step, what-if questions are formulated and answered to evaluate the effects of component failures or procedural errors. This technique relies on the experience level of the questioner. What-if methods are often used in conjunction with checklists (*HEP* Chaps. 4.5, 4.6, 6.5, 6.6, 12, 20).

Failure Mode and Effect Analysis (FMEA) This is a systematic study of the causes of failures and their effects. All causes or modes of failure are considered for each element of a system, and then all possible outcomes or effects are recorded (*HEP* Chaps. 4.8, 6.8, 19).

Reactive Chemistry Reviews The process chemistry is reviewed for evidence of exotherms, shock sensitivity, and other instability, with emphasis on possible exothermic reactions. The purpose of this review is to prevent unexpected and uncontrolled chemical reactions. Reviewers should be knowledgeable people in the field of reactive chemicals and include people from loss prevention, manufacturing, and research. The CCPS *Essential Practices for Managing Chemical Reactivity Hazards* provides a useful protocol for identifying chemical reactivity hazards (Johnson et al., 2003). A series of questions about the chemical handling operations and the materials are used to determine if there are possible reactivity hazards. Figure 23-18 summarizes the CCPS protocol for identifying reaction hazards.

Interaction Matrix (Compatibility Chart) An interaction matrix is a tool for identifying and understanding potential hazards, including reaction hazards, from combinations of materials (*HEP,* 1992, Chaps. 3.3, 11.3). The most common representation is a two-dimensional matrix, listing all components of interest (including, e.g., chemicals; materials of construction; potential contaminants; environmental contaminants such as air, rust, or water; utilities). The consequences of mixing the materials for each row-column intersection are identified. Figure 23-19 is an example of an interaction matrix. Johnson et al. (2003) describe the use of the interaction matrix, and Mosley et al. (2000) provide a specific example. The U.S. National Oceanic and Atmospheric Administration (NOAA) has developed a computer tool, the *Chemical Reactivity Worksheet*, which can generate an interaction matrix for materials in the program's database (NOAA, 2004).

Industrial Hygiene Reviews These reviews evaluate the potential of a process to cause harm to the health of people. The review

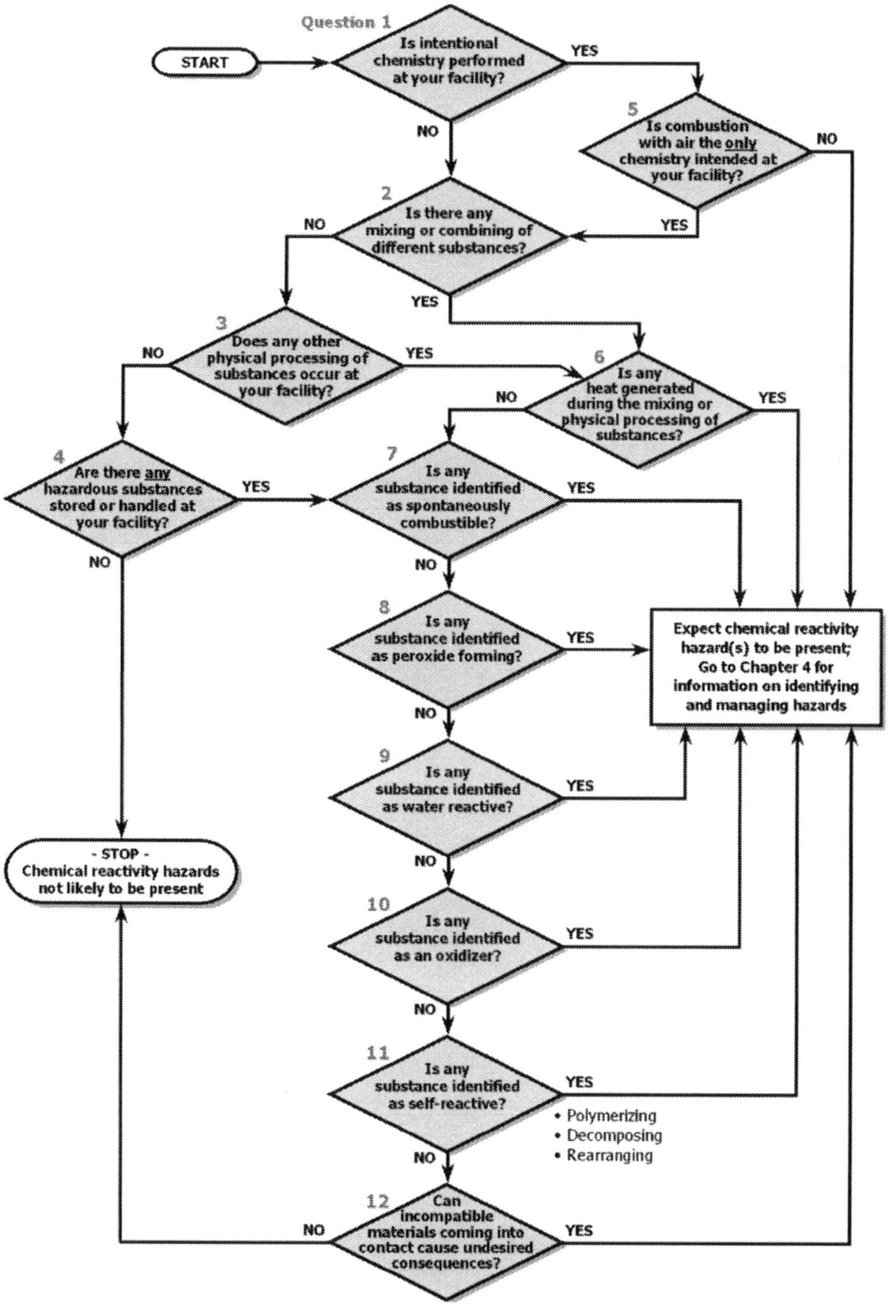

FIG. 23-18 CCPS preliminary screening for chemical reactivity hazards. [*From Johnson et al. (2003); copyright AIChE; reproduced with permission.*]

normally deals with chronic effects of exposure to chemicals and other harmful agents (e.g., noise, heat, repetitive motion) in the workplace. Chapter 44 of *Patty's Industrial Hygiene* reviews industrial hygiene survey procedures.

Facilities Reviews There are many kinds of facilities reviews that are useful in detecting and preventing process safety problems. They include prestart-up reviews (before the plant operates), new-plant reviews (the plant has started, but is still new), reviews of existing plants (safety, technology, and operations audits and reviews), management reviews, critical instrument reviews, and hazardous materials transportation reviews.

HAZOP (*HEP*, 1992, Chaps. 4.7, 6.7, 14, 18; Knowlton, 1989; Lees, 2005; CPQRA, 2000, pp. 583–587). HAZOP is a formal hazard identification and evaluation procedure designed to identify hazards to people, process plants, and the environment. The techniques aim to stimulate, in a systematic way, the imagination of designers and people

SUBSTANCE

	Substance	1	2	3	4	5	6
1	Hydrochloric Acid (35%)	1					
2	Sulfuric Acid (90%)	R[1]	2				
3	Acetic Acid	?[8]	R[2]	3			
4	Ethanol	NR[3]	R[4]	NR[15]	4		
5	Ethylenediamine	R[5]	R[6]	R[7]	NR[12]	5	
6	Water	R[9]	R[10]	NR[11]	NR[13]	R[14]	6

Legend:

R	Reactive under the stated scenario - **incompatible**
NR	Non-Reactive under the stated scenario - **compatible**
?	unknown – assume incompatible until further information is obtained

Footnotes/Information Sources:

1 Unlikely to be compatible – USCG chart NVC 4-75 indicates a hazard with non-oxidizing acids plus sulfuric acid. Heat of mixing may be significant.
2 Unlikely to be compatible – the P-168 chart indicates that gas and heat are formed; USCG chart NVC 4-75 indicates a hazard when combining sulfuric and organic acids.
3 Primary alcohols do not react with aq. HCl at ambient temperature.
4 Heat of solution followed by reaction to form ethyl hydrogen sulfate.
5 Lab experiment 980001 (50/50 mix) resulted in a significant heat of neutralization.
6 Lab experiment 980002 (50/50 mix by volume) resulted in a XXX C adiabatic temperature rise.
7 Organic acids and amines are generally incompatible due to acid/base neutralization heat.
8 The P-168 and USCG charts indicate no hazard; most likely compatible, but lab testing should be performed.
9 Heat of mixing may be a concern in some circumstances. The maximum adiabatic temperature rise is XX C (see XYZ Encyclopedia of Chemical Technology).
10 Heat of mixing may be a concern in some circumstances. The maximum adiabatic temperature rise is XX C (see XYZ Encyclopedia of Chemical Technology). Violent reaction with splattering if water is added to the acid.
11 Lab experiment 98005 showed that mixing acetic acid and water is endothermic at room temperature.
12 Lab experiments 98003 and 98008 indicate that the materials do not generate heat or gases when mixed nor when heated to 100 C. Although the USCG chart NVC 4-75 indicates that some alcohols and amines are incompatible, ethylene diamine has been found to be compatible with many alcohols; see Appendix of USCG Guide.
13 Plant experience has shown materials to be compatible.
14 Milidly exothermic hydrate formation.
15 Very slow, nearly thermoneutral, equilibrium-limited esterification at ambient temperature.

FIG. 23-19 Hypothetical compatibility chart. (*Copyright ASTM International. Reprinted with permission.*)

who operate plants or equipment to identify potential hazards. HAZOP studies assume that a hazard or operating problem can arise when there is a deviation from the design or operating intention. Actions to correct identified hazard or operational scenarios can then be taken before a real incident occurs. The primary goal in performing a HAZOP study is to identify, not analyze or quantify, the hazards in a process. The end product of a HAZOP is a list of concerns and recommendations for prevention of problems, not an analysis of the occurrence, frequency, overall effects, and the definite solution. A HAZOP study is most cost-effective when done during plant design—it is easier and cheaper to change a design than to modify an existing plant. However, HAZOP is a valuable process hazard analysis tool at any stage in the life cycle of a plant.

These studies make use of the combined experience and training of a group of knowledgeable people in a structured setting. Some key HAZOP terms are as follows.

Intention How the part or process is expected to operate.

Guide words Simple words used to qualify the intention in order to guide and stimulate creative thinking and so discover deviations. Table 23-21 describes commonly used guide words.

Deviations Departures from the intention discovered by systematic application of guide words.

Causes Reasons that deviations might occur.

Consequences Results of deviations if they occur.

Safeguards Prevention, mitigation, and control features which already exist in the plant, or which are already incorporated in a new design.

Actions Prevention, mitigation, and control features which do not currently exist and are recommended by the HAZOP team. Actions may also include recommendations for additional study if the HAZOP team does not have sufficient information, or time to understand a concern sufficiently to make a specific recommendation.

TABLE 23-21 Some HAZOP Guide Words Used in Conjunction with Process Parameters

Guide word	Meanings	Comments
No, Not, None	Complete negation of design intentions	No part of intention is achieved and nothing else occurs
More	Quantitative increases	Quantities and relevant physical properties such as flow rates, heat, pressure
Less	Quantitative decreases of any relevant physical parameters	Same as above
As well as	Qualitative increase	All design and operating intentions are achieved as well as some additional activity
Part of	A qualitative decrease	Some parts of the intention are achieved, others are not
Reverse	Logical opposite of intention	Activities such as reverse flow or chemical reaction, or poison instead of antidote
Other than	Complete substitution	No part of intention is achieved; something quite different happens

SOURCE: Knowlton, 1989.

The HAZOP study is not complete until response to actions has been documented. Initial HAZOP planning should establish the management follow-up procedure that will be used.

The guide words are used in conjunction with the process intentions to generate possible deviations from the intended operation (see Table 23-21). Some examples of deviations that might be generated in the course of a HAZOP study include

- No flow
- Reverse flow
- Less flow
- Increased temperature
- Decreased temperature
- Composition change
- Sampling
- Corrosion/erosion

The HAZOP team then determines the specific causes of each deviation; e.g., no flow of a particular material in a specified pipe might include causes such as a manual valve improperly closed, pump stops, pipe plugged with solids, etc. The HAZOP team then determines the consequences of the deviation for each cause and qualitatively decides the magnitude of hazard. The team identifies any existing safeguards in the plant or design and qualitatively judges whether they are adequate. If the team determines that additional safeguards are required, it may recommend specific actions. The team may determine that the issue requires greater study than can be accommodated in the time frame of a HAZOP meeting and recommend more extensive evaluation to determine if further action is needed, and what that action should be.

Many HAZOP studies incorporate a qualitative evaluation of risk to assist the team in evaluating the adequacy of existing safeguards, and the need for additional safeguards. This involves constructing a risk matrix, such as the one shown in Fig. 23-20. The team determines, based on its knowledge of the plant, experience, and engineering judgment, which of the several consequence and likelihood categories in the risk matrix best describe the particular deviation-cause-consequence sequence under consideration. Scenarios with high consequence and high frequency represent a large risk, those with low likelihood and consequence are of low risk. An organization can use the matrix to establish guidelines for which of the boxes in the risk matrix require action.

HAZOP studies may be made on batch as well as continuous processes. For a continuous process, the working document is usually a set of flow sheets or piping and instrument diagrams (P&IDs). Batch processes have another dimension: time. Time is usually not significant with a continuous process that is operating smoothly, although start-up and shutdown must also be considered, when the continuous process will resemble a batch process. For batch processes, the working documents consist not only of the flow sheets or P&IDs but also

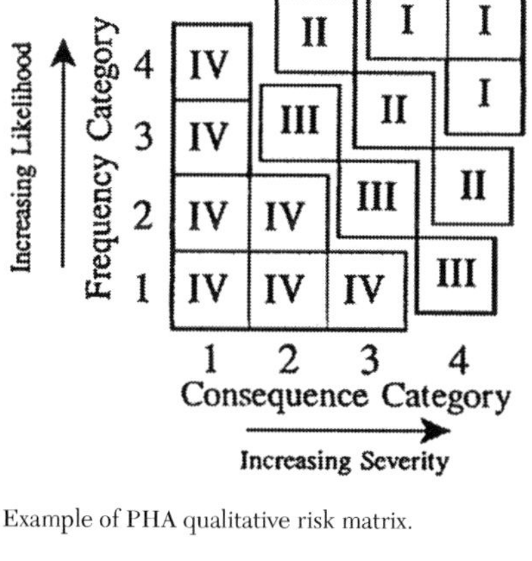

FIG. 23-20 Example of PHA qualitative risk matrix.

TABLE 23-22 HAZOP Guide Words Associated with Time

Guide word	Meaning
No time	Step(s) missed
More time	Step does not occur when it should
Less time	Step occurs before previous step is finished
Wrong time	Flow or other activity occurs when it should not

SOURCE: Knowlton, 1989.

the operating procedures. One method to incorporate this fourth dimension is to use guide words associated with time, such as those described in Table 23-22.

HAZOP studies involve team members, at least some of whom have had experience in the plant design to be studied. These team members apply their expertise to achieve the aims of HAZOP. There are four overall aims to which any HAZOP study should be addressed:

1. Identify as many deviations as possible from the way the design is expected to work, their causes, and problems associated with these deviations.
2. Decide whether action is required, and identify ways in which the problem can be solved.
3. Identify cases in which a decision cannot be made immediately, and decide what information or action is required.
4. Ensure that required actions are followed through.

The team leader is a key to the success of a HAZOP study and should have adequate training for the job. Proper planning is important to success. The leader is actually a facilitator (a discussion leader and one who keeps the meetings on track) whose facilitating skills are just as important as technical knowledge. The leader outlines the boundaries of the study and ensures that the design intention is clearly understood. The leader applies guide words and encourages the team to discuss causes, consequences, and possible remedial actions for each deviation. Prolonged discussions of how a problem may be solved should be avoided. Ideally, the team leader should be accompanied by a scribe or recorder, freeing the leader for full-time facilitating. The scribe should take notes in detail for full recording of as much of the meeting as is necessary to capture the intent of actions and recommendations.

Many companies offer computer tools to help facilitate and document HAZOP studies.

Team size is important. Having fewer than three contributing members, excluding the secretary and leader, will probably reduce team effectiveness. A team size of five to eight, including the leader and scribe, is probably optimum. The time required for HAZOP studies is significant. It has been estimated that each line or *node* (a node is usually a line or an item of equipment) may require in the range of 30 min for an experienced team, although the time may vary widely depending on the complexity of the system. It should be recognized that the time required for HAZOP studies may not really be additional time for the project as a whole, particularly if started early enough in the design, and may actually save time on the project. It also should make possible smoother start-ups and make the process or plant safer and easier to operate, which will more than pay back the cost of the HAZOP study during the life of the plant.

Hazard Ranking Methods Hazard ranking methods (*HEP*, 1992, Chaps. 4.3, 6.3) allow the analyst to compare the hazards of several processes, plants, or activities. They can be used to compare alternative chemical process routes, plant designs, plant siting options, or other design choices. Hazard ranking methods can also be useful for prioritizing facilities for additional risk management studies. They generally produce a numerical score for the process being evaluated. The scores generally do not have units and are only meaningful relative to each other in the context of the hazard index being used. Some of the more commonly used hazard ranking methods are briefly discussed.

NFPA Standard System for Identification of Health, Flammability, Reactivity, and Related Hazards (NFPA 704, 2001) This printed material is not the complete and official position of the National Fire Protection Association on the referenced subject, which is represented only by the standard in its entirety.

TABLE 23-23 NFPA 704 System for Identification of Hazards

Degree of hazard	Health hazard color code: blue	Flammability hazard color code: red	Instability hazard color code: yellow
4	Materials that, under emergency conditions, can be lethal	Materials that will rapidly or completely vaporize at atmospheric pressure and normal ambient temperature or that are readily dispersed in air and will burn readily.	Materials that in themselves are readily capable of detonation or explosive decomposition or explosive reaction at normal temperatures and pressures.
3	Materials that, under emergency conditions, can cause serious or permanent injury	Liquids and solids that can be ignited under almost all ambient temperature conditions. Materials in this degree produce hazardous atmospheres with air under almost all ambient temperatures or, though unaffected by ambient temperatures, are readily ignited under almost all conditions.	Materials that in themselves are capable of detonation or explosive decomposition or explosive reaction, but that require a strong initiating source or that must be heated under confinement before initiation.
2	Materials that, under emergency conditions, can cause temporary incapacitation or residual injury	Materials that must be moderately heated or exposed to relatively high ambient temperatures before ignition can occur. Materials in this degree would not, under normal conditions, form hazardous atmospheres with air, but under high ambient temperatures or under moderate heating could release vapor in sufficient quantities to produce hazardous atmospheres with air.	Materials that readily undergo violent chemical change at elevated temperatures and pressures.
1	Materials that, under emergency conditions, can cause significant irritation	Materials that must be preheated before ignition can occur. Materials in this degree require considerable preheating, under all ambient temperature conditions, before ignition and combustion can occur.	Materials that in themselves are normally stable, but that can become unstable at elevated temperatures and pressures.
0	Materials that, under emergency conditions, would offer no hazard beyond that of ordinary combustible materials	Materials that will not burn under typical fire conditions, including intrinsically noncombustible materials such as concrete, stone, and sand.	Materials that in themselves are normally stable, even under fire conditions.

SOURCE: Reprinted with permission from NFPA 704, *Standard System for the Identification of the Fire Hazards of Materials,* National Fire Protection Association, Quincy, Mass., 2001. This printed material is not the complete and official position of the National Fire Protection Association on the referenced subject, which is represented only by the standard in its entirety.

This is a brief summary of NFPA 704 which addresses hazards that may be caused by short-term exposure to a material during handling under conditions of fire, spill, or similar emergencies. This standard provides a simple, easily recognized and understood system of markings. The objective is to provide on-the-spot identification of hazardous materials. The markings provide a general idea of the hazards of a material and the severity of these hazards as they relate to handling, fire protection, exposure, and control. This standard is not applicable to transportation or to use by the general public. It is also not applicable to chronic exposure. For a full description of this standard, refer to NFPA 704.

The system identifies the hazards of a material in four principal categories: health, flammability, reactivity, and unusual hazards such as reactivity with water. The degree of severity of health, flammability, and reactivity is indicated by a numerical rating that ranges from 0 (no hazard) to 4 (severe hazard). Table 23-23 describes the characteristics associated with the various ratings; for a detailed description of the degree of severity ratings, see NFPA 704. The information is presented in a square-on-point (diamond) field of numerical ratings, e.g., as shown in Figs. 23-21 through 23-23. Information is presented as follows:

- Health rating in blue at nine o'clock
- Flammability rating in red at twelve o'clock
- Reactivity hazard rating in yellow at three o'clock
- Unusual hazards at six o'clock

Materials that demonstrate unusual reactivity with water are identified as W, and materials that possess oxidizing properties shall be identified by the letters OX. Other special hazard symbols may be used to identify radioactive hazards, corrosive hazards, substances that are toxic to fish, and so on. The use of this system provides a standard method of identifying the relative degree of hazard that is contained in various tanks, vessels, and pipelines.

Fire and Explosion Index (F&EI) (*Dow Fire and Explosion Index Hazard Classification Guide,* 1994; Mannan, 2005, pp. 8/13–8/22.)

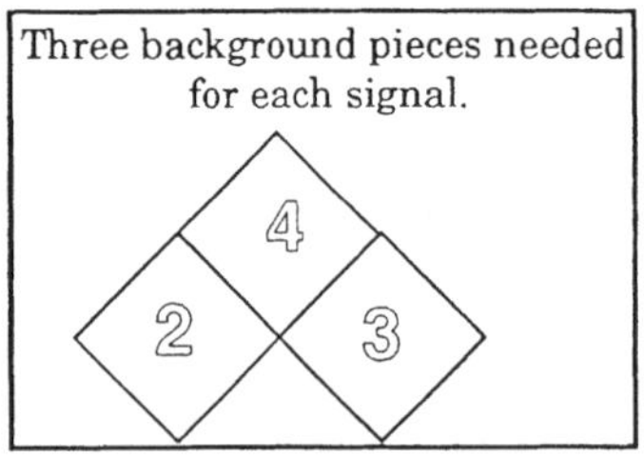

FIG. 23-21 NFPA 704 hazard ratings for use where specified color background is used with numerals of contrasting colors. (*NFPA 704, 2001.*)

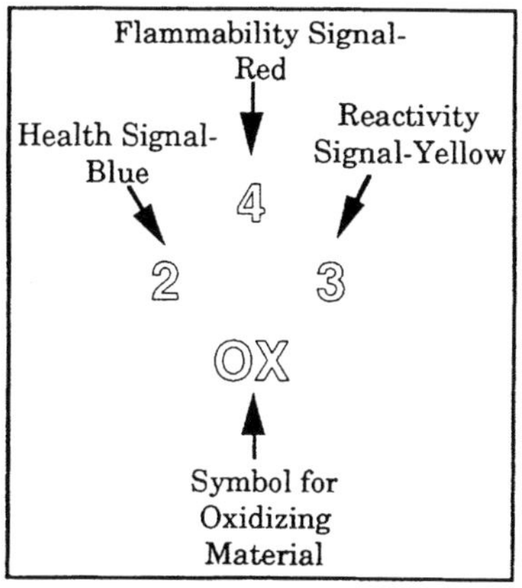

FIG. 23-22 For use where white background is used. (*NFPA 704, 2001.*)

The F&EI is used to rate the potential of hazard from fires and explosions. Its purpose is to quantify damage from an incident. It identifies equipment that could contribute to an incident and ways to mitigate possible incidents. It is a way to communicate to management the quantitative hazard potential. The F&EI measures realistic maximum loss potential under adverse operating conditions. It is based on quantifiable data. It is designed for flammable, combustible, and reactive materials that are stored, handled, or processed. It does not address frequency (risk) except indirectly, nor does it address specific hazards to people except indirectly. The goals of the F&EI are to raise awareness of loss potential and identify ways to reduce potential severity and potential dollar loss in a cost-effective manner. The index number has significance as a comparison and in calculations to estimate the *maximum probable property damage* (MPPD). It also provides a method for measuring the effect of outage (plant being shut down) on the business. It is easy for users to get credible results with a small amount of training.

Chemical Exposure Index (CEI) (*Chemical Exposure Index,* 1994; Mannan, 2005, pp. 8/22–8/26.) The CEI provides a method of rating the relative potential of acute health hazard to people from possible chemical release incidents. It may be used for prioritizing initial process hazard analysis and establishing the degree of further analysis needed. The CEI also may be used as part of the site review process. The system provides a method of ranking one risk relative to another. It is not intended to define a particular containment system as safe or unsafe, but provides a way of comparing toxic hazards. It deals with acute, not chronic, releases. Flammability and explosion hazards are not included in this index. To develop a CEI, information needs include

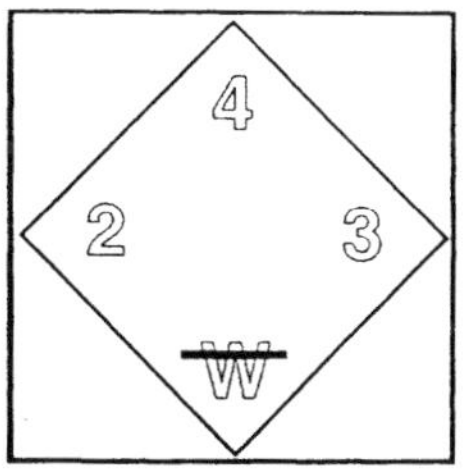

FIG. 23-23 For use where white background is used or for signs or placards. (*NFPA 704, 2001.*)

- An accurate plot plan of the plant and surrounding area
- A simplified process flow sheet showing containment vessels, major piping, and quantity of chemicals
- Physical, chemical, and toxicological properties of the chemicals
- Process variables such as pressures, temperatures, and quantities of materials

Substance Hazard Index The Substance Hazard Index (SHI) was developed for use by U.S. OSHA for ranking material hazards during the development of the OSHA Process Safety Management regulations. SHI is defined as "the equilibrium vapor concentration (EVC) of a material at 20°C divided by an acute toxicity concentration." SHI is a measure of the ability of a released material to be transported through the air and impact people. The EVC is defined as the vapor pressure of the material at 20°C × 10^6/760. Different organizations using the SHI may use different toxicity measures, for example, ERPG-3, IDLH, or AEGL.

Consequence-Based Ranking Systems Release consequence modeling can be used to rank potential chemical hazards. For example, the USEPA's RMP regulations require consequence modeling for a predefined worst-case scenario—release of the entire contents of the largest container of a material in 10 min. EPA provides lookup tables and software (RMPComp) to assist in estimating the hazard distances for materials covered by the RMP regulations.

Logic Model Methods The following tools are most commonly used in quantitative risk analysis, but can also be useful qualitatively to understand the combinations of events which can cause an accident. The logic models can also be useful in understanding how protective systems impact various potential accident scenarios. These methods will be thoroughly discussed in the "Risk Analysis" subsection. Also, hazard identification and evaluation tools discussed in this section are valuable precursors to a quantitative risk analysis (QRA). Generally a QRA quantifies the risk of hazard scenarios which have been identified by using tools such as those discussed above.

- Fault tree analysis
- Event tree analysis
- Cause-consequence diagram
- Layer-of-protection analysis (LOPA)

RISK ANALYSIS

GENERAL REFERENCES: *Guidelines for Chemical Process Quantitative Risk Analysis,* 2d ed., CCPS-AIChE, New York, 2000. Arendt, "Management of Quantitative Risk Assessment in the Chemical Process Industry," *Plant Operations Progress,* vol. 9, no. 4, AIChE, New York, October 1990. CMA, "A Manager's Guide to Quantitative Risk Assessment," Chemical Manufacturers' Association, December 1989. EFCE, "Risk Analysis in the Process Industries," European Federation of Chemical Engineering, Publication Series no. 45, 1985. Lees, *Loss Prevention in the Process Industries* 2d ed., Butterworths, Boston, 1996. World Bank, *Manual of Industrial Hazard Assessment Techniques,* Office of Environmental and Scientific Affairs, World Bank, Washington, D.C., 1985. *Guidelines for Chemical Transportation Risk Analysis,* CCPS-AIChE, New York, 1995.

FREQUENCY ESTIMATION REFERENCES: *Guidelines for Process Equipment Reliability Data,* CCPS-AIChE, New York, 1989. Billington and Allan, *Reliability Evaluation of Engineering Systems: Concepts and Techniques,* Plenum Press, New York, 1983. Fussell, Powers, and Bennetts, "Fault Trees: A State of the Art Discussion," *IEEE Transactions on Reliability,* 1974. Roberts, N. H. et al., *Fault Tree Handbook,* NUREG-0492, Washington, D.C. Swain and Guttmann, *Handbook of Human Reliability Analysis with Emphasis on Nuclear Power Plant Applications,* NUREG/CR-1278, USNRC, Washington, D.C., 1983.

CONSEQUENCE ESTIMATION REFERENCES: *Guidelines for Use of Vapor Cloud Dispersion Models,* CCPS-AIChE, New York, 1987. TNO, *Methods for the Calculation of the Physical Effects of the Escape of Dangerous Materials: Liquids and Gases* ("The Yellow Book"), Apeldoorn, The Netherlands, 1979. *Understanding Explosions,* CCPS-AIChE, New York, 2003. *Guidelines for Evaluating the Characteristics of Vapor Cloud Explosions,* CCPS-AIChE, New York, 1994. *Guidelines for Use of Vapor Cloud Dispersion Models,* 2d ed., CCPS-AIChE, New York, 1996. *Guidelines for Consequence Analysis of Chemical Releases,* CCPS-AIChE, New York, 1999. Technical Report no. 55, *Techniques for Assessing Industrial Hazards,* World Bank, Washington, 1988.

RISK ESTIMATION REFERENCES: Health and Safety Executive, *Canvey—An Investigation of Potential Hazards from the Operations in the Canvey*

Island/Thurrock Area, HMSO, London, 1978. Rasmussen, *Reactor Safety Study: An Assessment of Accident Risk in U.S. Commercial Nuclear Power Plants,* WASH-1400 NUREG 75/014, Washington, D.C., 1975. Rijnmond Public Authority, *A Risk Analysis of 6 Potentially Hazardous Industrial Objects in the Rijnmond Area—A Pilot Study,* D. Reidel, Boston, 1982. Considine, *The Assessment of Individual and Societal Risks,* SRD Report R-310, Safety and Reliability Directorate, UKAEA, Warrington, 1984. Baybutt, "Uncertainty in Risk Analysis," Conference on Mathematics in Major Accident Risk Assessment, University of Oxford, U.K., 1986.

RISK CRITERIA AND RISK DECISION MAKING: Ale, "The Implementation of an External Safety Policy in the Netherlands," *Intl. Conference on Hazard Identification and Risk Analysis, Human Factors and Reliability in Process Safety,* AIChE, New York, pp. 173–183, 1992; Gibson, "Hazard Analysis and Risk Criteria," *Chemical Engineering Progress,* Nov. 1980, pp. 46–50; Gibson, "The Use of Risk Criteria in the Chemical Industry," *Trans. of IChemE 71,* Pt. B, pp. 117–123, 1993; Health and Safety Executive, *Reducing Risks Protecting People, HSE's Decision Making Process,* HSE Books, London, 2001; Helmers and Schaller, "Calculated Process Risks and Health Management," *Plant Operations Progress,* 1, no. 3, pp. 190–194, 1982; Renshaw, "A Major Accident Prevention Program," *Plant Operations Progress,* 9, no. 3, pp. 194–197, 1990; Pikaar and Seaman, *A Review of Risk Control,* Zoetermeer, Ministerie VROM, Netherlands, 1995; *Tools for Making Acute Risk Decisions with Chemical Process Safety Implications,* CCPS-AIChE, New York, 1995; Hamm and Schwartz, "Issues and Strategies in Risk Decision Making," *Intl. Process Safety Management Conference and Workshop,* San Francisco, AIChE, New York, pp. 351–371, 1993.

Introduction The previous sections dealt with techniques for qualitative hazard analysis only. This section addresses the quantitative methodologies available to analyze and estimate risk, which is a function of both the consequences of an incident and its frequency. The application of these methodologies in most instances is not trivial. A significant allocation of resources is necessary. Therefore, a selection process or risk prioritization process is advised before considering a risk analysis study.

Important definitions are as follows.

Accepted risk The risk is considered tolerable for a given activity by those responsible for managing or regulating the operation of a hazardous facility. The term *acceptable risk* has often been used, but this inevitably raises the question, "Acceptable to whom?" Tweeddale (*Managing Risk and Reliability of Process Plants,* Gulf Professional, Houston, Tex, 2003, p. 70) suggests that *accepted risk* is a better term because it makes it clear that the risk has been accepted by those responsible for the decisions on how to build, operate, and regulate the facility.

Event sequence A specific, unplanned sequence of events composed of initiating events and intermediate events that may lead to an incident.

Frequency The rate at which observed or predicted events occur.

Incident outcome The physical outcome of an incident; e.g., a leak of a flammable and toxic gas could result in a jet fire, a vapor cloud explosion, a vapor cloud fire, or a toxic cloud.

Probability The likelihood of the occurrence of events, the values of which range from 0 to 1.

Probability analysis Way to evaluate the likelihood of an event occurring. By using failure rate data for equipment, piping, instruments, and fault tree techniques, the frequency (number of events per unit time) can be quantitatively estimated.

Probit model A mathematical model of dosage and response in which the dependent variable (response) is a probit number that is related through a statistical function directly to a probability.

Quantitative risk assessment (QRA) The systematic development of numerical estimates of the expected frequency and consequence of potential accidents associated with a facility or an operation. Using consequence and probability analyses and other factors such as population density and expected weather conditions, QRA estimates the fatality rate for a given set of events.

Risk A measure of economic loss or human injury in terms of both incident likelihood (frequency) and the magnitude of the loss or injury (consequence).

Risk analysis The development of an estimate of risk based on engineering evaluation and mathematical techniques for combining estimates of incident consequences and frequencies. Incidents in the context of the discussion in this chapter are acute events which involve loss of containment of material or energy.

Risk assessment The process by which results of a risk analysis are used to make decisions, either through a relative ranking of risk reduction strategies or through comparison with risk targets. The terms *risk analysis* and *risk assessment* are often used interchangeably in the literature.

A typical hazard identification process, such as a hazard and operability (HAZOP) study, is sometimes used as a starting point for selection of potential major risks for risk analysis. Other selection or screening processes can also be applied. However major risks are chosen, a HAZOP study is a good starting point to develop information for the risk analysis study. A major risk may qualify for risk analysis if the magnitude of the incident is potentially quite large (high potential consequence) or if the frequency of a severe event is judged to be high (high potential frequency) or both. A flowchart which describes a possible process for risk analysis is shown in Fig. 23-24.

The components of a risk analysis involve the estimation of the frequency of an event, an estimation of the consequences, and the selection and generation of the estimate of risk itself.

A risk analysis can have a variety of potential goals:

1. To screen or bracket a number of risks in order to prioritize them for possible future study
2. To estimate risk to employees
3. To estimate risk to the public
4. To estimate financial risk
5. To evaluate a range of risk reduction measures
6. To meet legal or regulatory requirements
7. To assist in emergency planning

The scope of a study required to satisfy these goals will be dependent upon the extent of the risk, the depth of the study required, and the level of resources available (mathematical models and tools and skilled people to perform the study and any internal or external constraints).

A risk analysis can be applied to fixed facilities or transportation movements, although much of the attention today still centers on the former. In a fixed-facility risk analysis, QRA can aid risk management decisions with respect to

1. Chemical processes
2. Process equipment
3. Operating procedures
4. Chemical inventories
5. Storage conditions

In a transportation risk analysis (TRA), the risk parameters are more extensive, but more restrictive in some ways. Examples of risk parameters that could be considered are

1. Alternate modes of transport
2. Routes
3. Travel restrictions
4. Shipment size
5. Shipping conditions (e.g., pressure, temperature)
6. Container design
7. Unit size (e.g., bulk versus drums)

The objective of a risk analysis is to reduce the level of risk wherever practical. Much of the benefit of a risk analysis comes from the discipline which it imposes and the detailed understanding of the major contributors of the risk that follows. There is general agreement that if risks can be identified and analyzed, then measures for risk reduction can be effectively selected.

The expertise required in carrying out a risk analysis is substantial. Although various software programs are available to calculate the frequency of events or their consequences, or even risk estimates, engineering judgment and experience are still very much needed to produce meaningful results. And although professional courses are available in this subject area, there is a significant learning curve required not only for engineers to become practiced risk analysts, but also for management to be able to understand and interpret the results. For these reasons, it may be useful to utilize a consultant organization in this field.

The analysis of a risk—that is, its estimation—leads to the assessment of that risk and the decision-making processes of selecting the

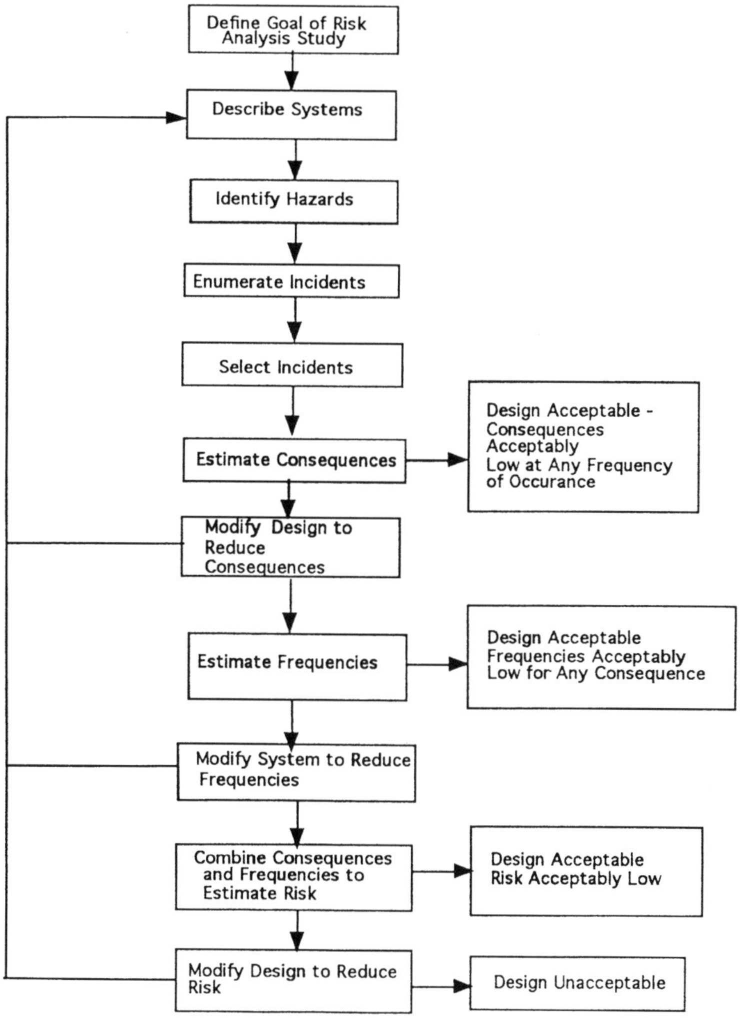

FIG. 23-24 One version of a risk analysis process. (*CCPS-AIChE, 1989, p. 13, by permission.*)

appropriate level of risk reduction. In most studies this is an iterative process of risk analysis and risk assessment until the risk is reduced to some specified level. The question of how safe is safe enough has to be addressed either implicitly or explicitly in the decision-making process. This subject is discussed in greater detail later on in this section.

Frequency Estimation There are two primary sources for estimates of incident frequencies. These are historical records and the application of fault tree analysis and related techniques, and they are not necessarily applied independently. Specific historical data can sometimes be usefully applied as a check on frequency estimates of various subevents of a fault tree, for example.

The use of historical data provides the most straightforward approach to the generation of incident frequency estimates but is subject to the applicability and the adequacy of the records. Care should be exercised in extracting data from long periods of the historical record over which design or operating standards or measurement criteria may have changed.

An estimate of the total population from which the incident information has been obtained is important and may be difficult to obtain.

Fault tree analysis and other related event frequency estimation techniques, such as event tree analysis, play a crucial role in the risk analysis process. Fault trees are logic diagrams that depict how components and systems can fail. The undesired event becomes the top event and subsequent subevents, and eventually basic causes, are then developed and connected through logic gates. The fault tree is completed when all basic causes, including equipment failures and human errors, form the base of the tree. There are general rules for construction, which have been developed by practitioners, but no specific rules for events or gates to use. The construction of a fault tree is still more of an art than a science. Although a number of attempts have been made to automate the construction of fault trees from process flow diagrams or piping instrumentation diagrams, these attempts have been largely unsuccessful [P. K. Andow, "Difficulties in Fault Tree Synthesis for Process Plant," *IEEE Transactions on Reliability* **R-29**(1): 2, 1980].

Once the fault tree is constructed, quantitative failure rate and probability data must be obtained for all basic causes. A number of equipment failure rate databases are available for general use. However, specific equipment failure rate data is generally lacking and, therefore, data estimation and reduction techniques must be applied to generic databases to help compensate for this shortcoming. Accuracy and applicability of data will always be a concern, but useful results from quantifying fault trees can generally be obtained by experienced practitioners.

Human error probabilities can also be estimated using methodologies and techniques originally developed in the nuclear industry. A number of different models are available (Swain, "Comparative Evaluation of Methods for Human Reliability Analysis," GRS Project RS 688, 1988). This estimation process should be done with great care, as many factors can affect the reliability of the estimates. Methodologies using expert opinion to obtain failure rate and probability estimates have also been used where there is sparse or inappropriate data.

In some instances, plant-specific information relating to frequencies of subevents (e.g., a release from a relief device) can be compared against results derived from the quantitative fault tree analysis, starting with basic component failure rate data.

An example of a fault tree logic diagram using AND and OR gate logic is shown in Fig. 23-25.

The logical structure of a fault tree can be described in terms of boolean algebraic equations. Some specific prerequisites to the application of this methodology are as follows.

- Equipment states are binary (working or failed).
- Transition from one state to another is instantaneous.
- Component failures are statistically independent.
- The failure rate and repair rate are consistent for each equipment item.
- After repair, the component is returned to the working state.

Minimal cut set analysis is a mathematical technique for developing and providing probability estimates for the combinations of basic component failures and/or human error probabilities, which are necessary and sufficient to result in the occurrence of the top event.

A number of software programs are available to perform these calculations, given the basic failure data and fault tree logic diagram (AIChE-CCPS, 2000). Other less well known approaches to quantifying fault tree event frequencies are being practiced, which result in gate-by-gate calculations using discrete-state, continuous-time, Markov models (Doelp et al., "Quantitative Fault Tree Analysis, Gate-by-Gate Method," *Plant Operations Progress* **4**(3): 227–238, 1984).

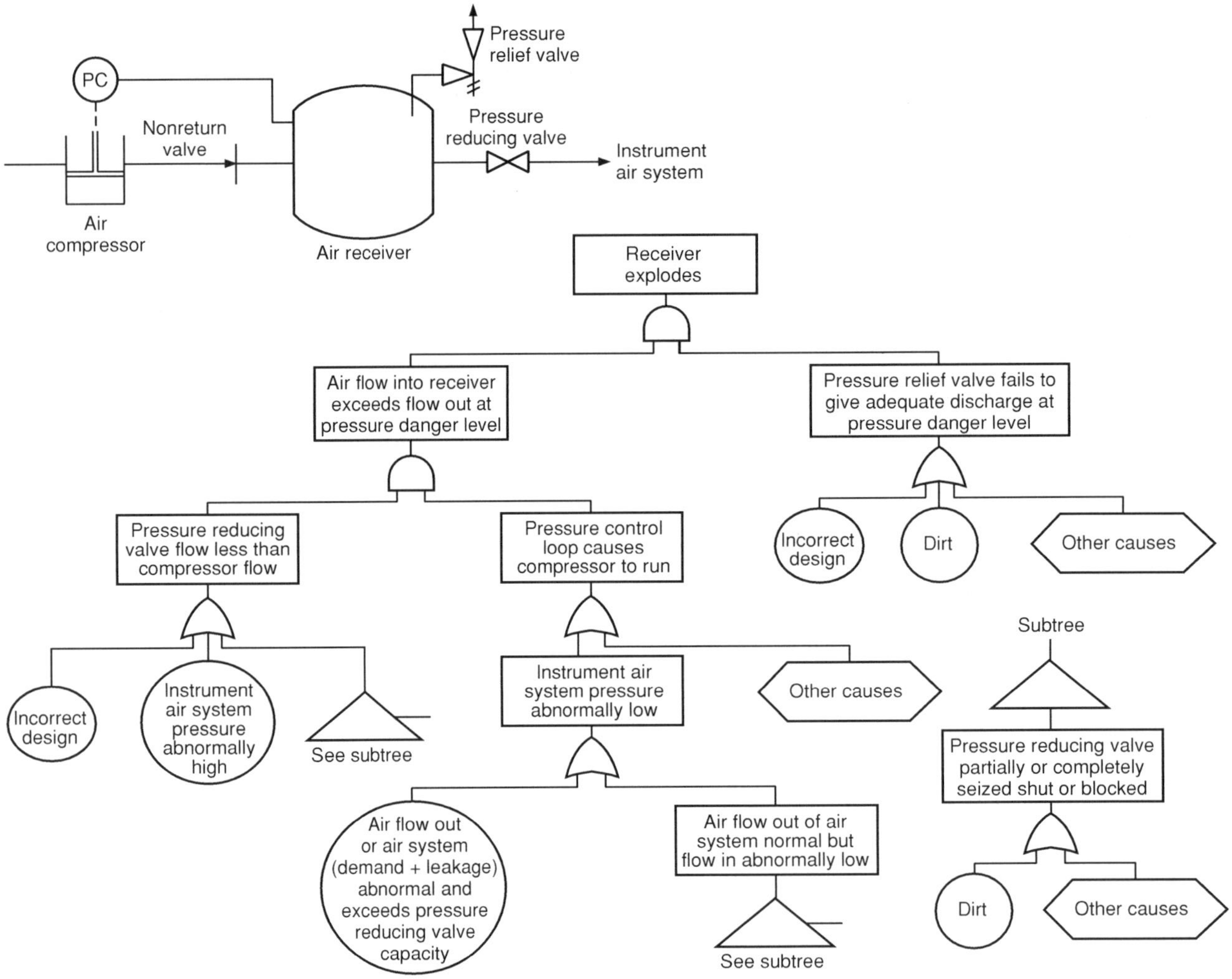

FIG. 23-25 Process drawing and fault tree for explosion of an air receiver. (*From Lees, 1980, pp. 200, 201, by permission.*)

Identification and quantitative estimation of common-cause failures are general problems in fault tree analysis. Boolean approaches are generally better suited to mathematically handle common-cause failures. The basic assumption is that failures are completely independent events, but in reality dependencies will exist and these are categorized as common cause failures (CCFs). Both qualitative and quantitative techniques can be applied to identify and assess CCFs. An excellent overview of CCF is available (AIChE-CCPS, 2000).

Event tree analysis is another useful frequency estimation technique used in risk analysis. It is a bottom-up logic diagram, which starts with an identifiable event. Branches are then generated, which lead to specific chronologically based outcomes with defined probabilities. Event tree analysis can provide a logic bridge from the top event, such as a flammable release into specific incident outcomes (e.g., no ignition, flash fire, or vapor cloud explosion). Probabilities for each limb in the event tree diagram are assigned and, when multiplied by the starting frequency, produce frequencies at each node point for all the various incident outcome states. The probabilities for all of the limbs at any given level of the event tree must sum to 1.0. Event trees are generally very helpful toward the generation of a final risk estimate. Figure 23-26 shows an event tree for a release of LPG (AIChE-CCPS, 2000).

Layers-of-protection analysis (LOPA) is a semiquantitative methodology for analyzing and assessing risk. It is typically applied after a qualitative hazards analysis has been completed, which provides the LOPA team with a listing of hazard scenarios with associated safeguards for consideration. LOPA uses simplified methods to characterize the process risk based on the frequency of occurrence and consequence severity of potential hazard scenarios. The process risk is compared to the owner/operator risk criteria. When the process risk exceeds the risk criteria, protection layers are identified that reduce the process risk to the risk criteria.

The protection layers are safeguards that are designed and managed to achieve seven core attributes: independence, functionality, integrity, reliability, auditability, access security, and management of change. Protection layers may include inherently safer design, control, supervisory, prevention, mitigation, limitation, barriers, and emergency response systems, depending on the owner/operator's risk management strategy.

In general, risk reduction is accomplished by implementing one or more protective layers, which reduce the frequency and/or consequence of the hazard scenario. LOPA provides specific criteria and restrictions for the evaluation of protection layers, eliminating the subjectivity of qualitative methods at substantially less cost than fully quantitative techniques. LOPA is a rational, defensible methodology that allows a rapid, cost-effective means for identifying the protection layers that lower the frequency and/or the consequence of specific hazard scenarios.

Consequence Estimation Given that an incident (release of material or energy) has been defined, the consequences can be estimated. The general logic diagram in Fig. 23-27 illustrates these calculations for the release of a volatile hazardous substance.

For any specific incident there will be an infinite number of incident outcome cases that can be considered. There is also a wide degree of consequence models which can be applied. It is important, therefore, to understand the objective of the study to limit the number of incident outcome cases to those which satisfy that objective. An example of variables which can be considered is as follows.

- Quality, magnitude, and duration of the release
- Dispersion parameters (wind speed, wind direction, weather stability)
- Ignition probability, ignition sources/location, ignition strength (flammable releases)
- Energy levels contributing to explosive effects (flammable releases)
- Impact of release on people, property, or environment (thermal radiation, projectiles, shock-wave overpressure, toxic dosage)

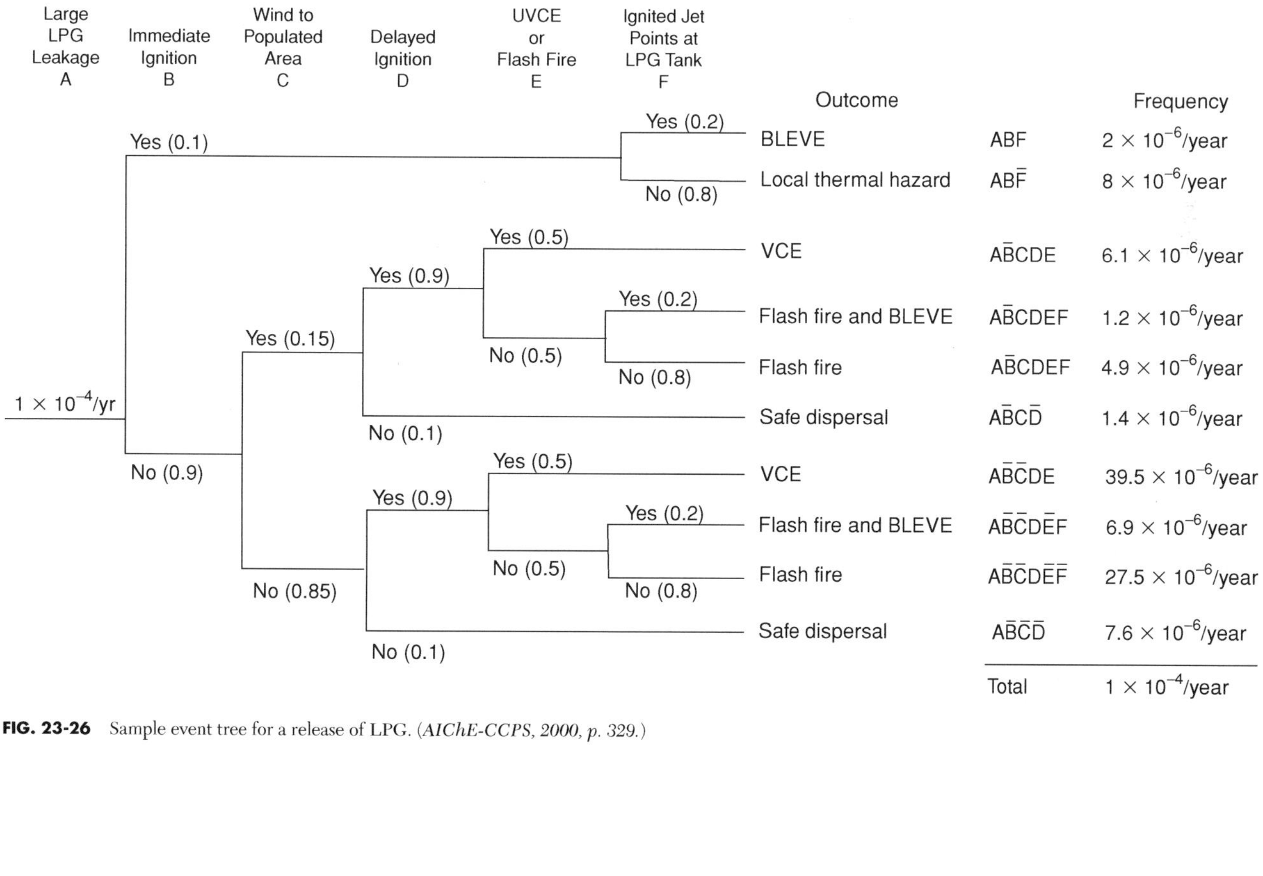

FIG. 23-26 Sample event tree for a release of LPG. (*AIChE-CCPS, 2000, p. 329.*)

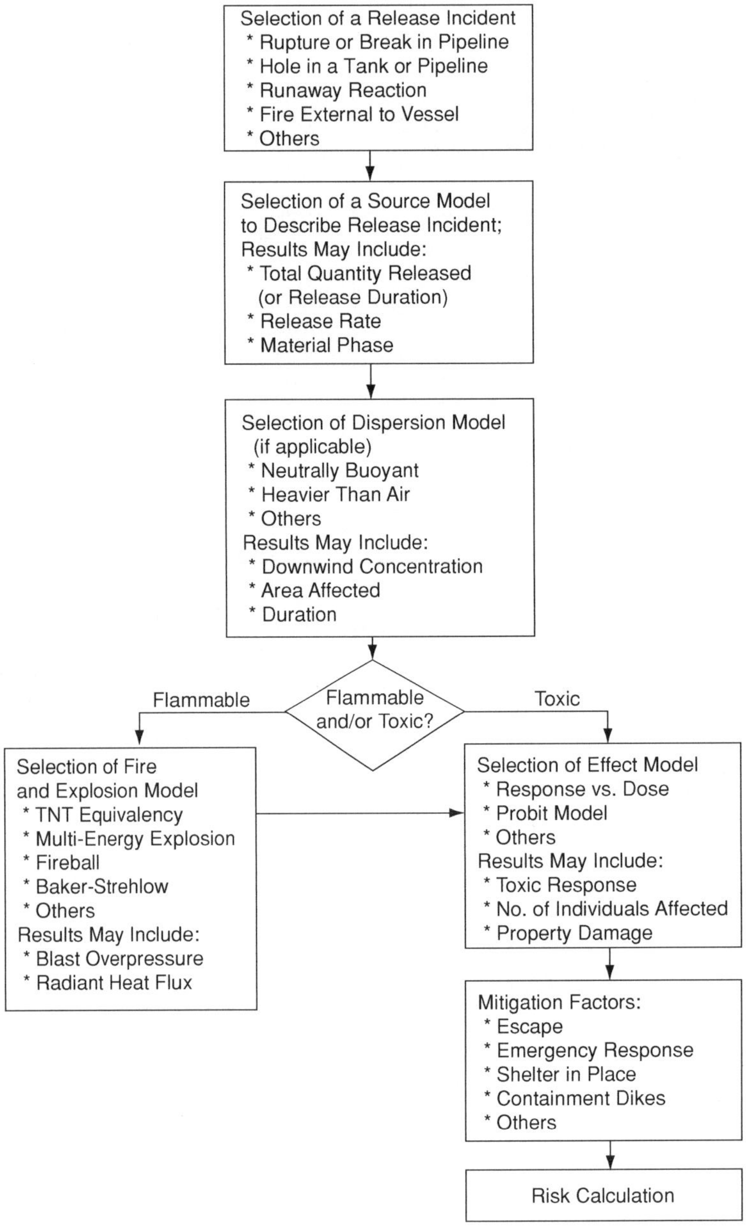

FIG. 23-27 Overall logic diagram for consequence analysis of volatile hazardous substances. (*CCPS-AIChE, 1989, p. 60.*)

- Mitigation effects (safe havens, evacuation, daytime/nighttime populations)

Probit models have been found generally useful to describe the effects of incident outcome cases on people or property for more complex risk analyses. At the other end of the scale, the estimation of a distance within which the population would be exposed to a concentration of ERPG-2 or higher may be sufficient to describe the impact of a simple risk analysis.

Portions or all of the more complex calculation processes, using specific consequence models, have been incorporated into a few commercially available software packages (AIChE-CCPS, 2000). These programs should be used by risk analysts with extensive engineering experience, as significant judgment will still be required.

The output of these calculation processes is one or more pairs of an incident or incident outcome case frequency and its effect (consequence or impact).

Risk Estimation There are a number of risk measures which can be estimated. The specific risk measures chosen are generally related to the study objective and depth of study, and any preferences or requirements established by the decision makers. Generally, risk measures can be broken down into three categories: risk indices, individual risk measures, and societal risk measures.

Risk indices are usually single-number estimates, which may be used to compare one risk with another or used in an absolute sense compared to a specific target. For risks to employees the *fatal accident rate* (FAR) is a commonly applied measure. The FAR is a single-number index, which is the expected number of fatalities from a specific event based on 10^8 exposure hours. For workers in a chemical plant the FAR would be calculated for a specific event as

$$\text{FAR} = \frac{10^8}{8760} E P \tau \qquad (23\text{-}29)$$

where FAR = expected number of fatalities from a specific event based on 10^8 exposure hours
E = frequency of event, yr^{-1}
P = probability of being killed by event at a specific location
τ = fraction of time spent at specific location

However, the worker in a chemical plant could be exposed to other potential events that might result in a fatality. In this case the overall FAR for the worker would be

$$\text{FAR} = \frac{10^8}{8760} \sum_i E_i \sum_j P_{ij} \tau_j \qquad (23\text{-}30)$$

Each subscript i is a specific event and each subscript j is a specific location.

References are available which provide FAR estimates for various occupations, modes of transportation, and other activities (Kletz, "The Risk Equations—What Risk Should We Run?," *New Scientist,* May 12, pp. 320–325, 1977).

Figure 23-28 is an example of an individual risk contour plot, which shows the expected frequency of an event causing a specified level of harm at a specified location, regardless of whether anyone is present at that location to suffer that level of harm.

The total individual risk at each point is equal to the sum of the individual risks at that point from all incident outcome cases.

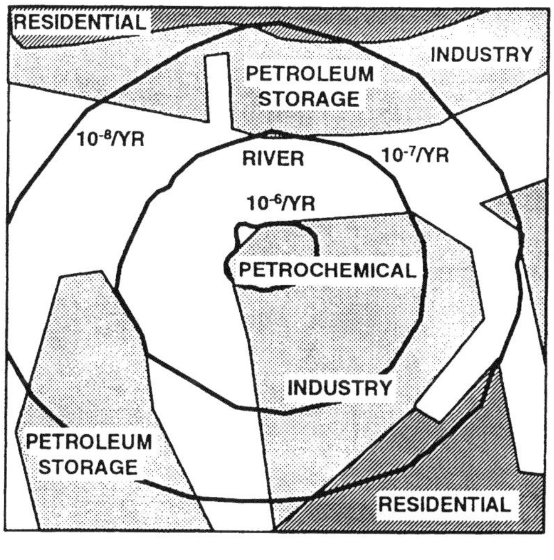

FIG. 23-28 Example of an individual risk contour plot. (*CCPS-AIChE, 1989, p. 269.*)

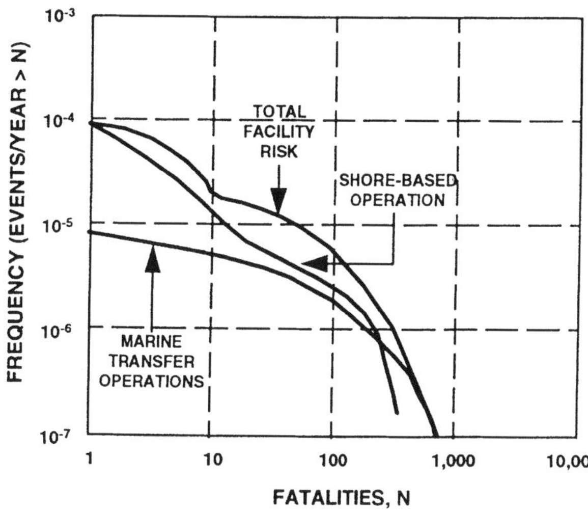

FIG. 23-29 Example of a societal risk F-N curve. (*AIChE-CCPS, 1989, p. 4.4.*)

$$\text{IR}_{x,y} = \sum_{i=1}^{n} \text{IR}_{x,y,i} \qquad (23\text{-}31)$$

where $\text{IR}_{x,y}$ = total individual risk of fatality at geographical location x,y
$\text{IR}_{x,y,i}$ = individual risk of fatality at geographical location x,y from incident outcome case i
n = total number of incident outcome cases

A common form of societal risk measure is an F-N curve, which is normally presented as a cumulative distribution plot of frequency F versus number of fatalities N. An example of this type of measure is shown in Fig. 23-29.

Any individual point on the curve is obtained by summing the frequencies of all events resulting in that number of fatalities or greater. The slope of the curve and the maximum number of fatalities are two key indicators of the degree of risk.

Risk Criteria Once a risk estimate is prepared, alternatives to reduce the risk can be determined; but one always faces the challenge of how low a risk level is low enough. Different countries in the world have established numerical criteria [ALE (1992) and HSE (2001)] primarily as a tool to address the Seveso regulations. In addition, over the years various chemical companies have established numerical targets [Gibson (1980), Helmers (1982), and Renshaw (1990)]. A good survey paper on the use of risk guidelines by both governments and operating companies is provided by Pikaar and Seaman (1995).

Risk Decision Making Risk criteria represent the first step in risk decision making. Efforts must be made to reduce the risk at least to a "tolerable" risk level, which may be dictated by a government or an operating company. Once the tolerable level is reached, then additional risk reductions should still be evaluated. A number of criteria also provide guidance on a level of risk that requires no further risk reduction. In between these two levels is a range of risk where risk reduction options need to be evaluated. How far to continue reducing risk below the tolerable level is up to the individual company and is a function of costs and benefits (degree of risk reduction). At some point additional risk reductions have little benefit. Hamm and Schwartz (1993) summarize some strategies for consideration, and CCPS-AIChE (1995) provides an introduction to a number of decision analysis tools which could be applied.

DISCHARGE RATES FROM PUNCTURED LINES AND VESSELS

Nomenclature

a	Coefficient in nonequilibrium compressibility factor, Eq. (23-74)
A	Cross-sectional area perpendicular to flow, m^2
C_D	Discharge coefficient
C_{DG}	Discharge coefficient for gas flow
C_{DL}	Discharge coefficient for liquid flow
C_p	Heat capacity at constant pressure, J/(kg·K)
C_v	Heat capacity at constant volume, J/(kg·K)
D	Pipe diameter, m
D_T	Tank diameter, m
f	Fanning friction factor
F_I	Pipe inclination factor defined by Eq. (23-43)
g	Gravitational acceleration, m^2
G	Mass flux, kg/(m^2·s)
H	Specific enthalpy, J/kg
H_{GL}	Heat of vaporization ($H_G - H_L$) at saturation, J/kg
K	Slip velocity ratio, u_G/u_L
K_e	Number of velocity heads for fittings, expansions, contractions, and bends
L	Length of pipe, m
N	$4fL/D + K_e$, pipe resistance or nonequilibrium compressibility factor
P	Pressure, Pa = N/m^2
Q	Heat-transfer rate, Wkg
R	Gas constant, J/(kgmol·K)
S	Specific entropy, J/(kg·K)
t	Time, s
T	Temperature, K
u	Velocity, m/s
v	Specific volume, m^3/kg
v_{GL}	Specific volume difference, $v_G - v_L$, m^3/kg
w	Mass discharge rate, kg/s
x	Vapor quality, kg vapor/kg mixture or mole fraction of liquid components
y	Mole fraction of vapor components
X_m	Lockhart Martinelli parameter
z	Vertical distance, m

Greek

α	Volume fraction vapor, m^3 vapor/m^3 mixture
ε	Dimensionless specific volume, v/v_0
γ	Heat capacity ratio, C_p/C_v
η	Pressure ratio, P/P_0
φ	Two-phase multiplier, pressure drop for two-phase flow divided by pressure drop for single-phase flow
μ	Single- or two-phase viscosity, Pa · s
θ	Overall inclination angle of pipe to horizontal from source to break, deg
ρ	Density, kg/m^3
ω	Slope of dimensionless specific volume to reciprocal dimensionless pressure, defined by Eqs. (23-61a) and (23-61b)

Subscripts

a	Ambient
b	Backpressure
bub	Bubble point, pressure and saturation temperature when first vapor bubbles appear
c	η_{ch}/η_s, ratio of choked pressure to saturation pressure
ch	Choked
C	Condensable components (the contaminant)
d	Discharge
dew	Dew point, pressure and saturation temperature when first (or last) liquid occurs
diff	Differential form
e	Equivalent, for two-phase specific volume with slip
g,G	Gas or vapor
H	Homogeneous flow (slip velocity ratio of unity)
int	Integral form
I	Inert or padding gas component
L	Liquid
max	Maximum
N	Nonequilibrium
ori	Orifice
p	Pipe flow
s	Saturation (or bubble point)
sol	Solids
S	Constant specific entropy
v	Vapor pressure
0	Initial stagnation conditions
1	Point at which backpressure from pipe is felt after entrance from tank
2	Plane at vena contracta or at pipe puncture
*	Dimensionless

GENERAL REFERENCES: Bragg, "Effect of Compressibility on the Discharge Coefficient of Orifices and Convergent Nozzles," *J. Mech. Eng. Sci.*, **2**: 35–44, 1960. Cheremisinoff and Gupta, eds., *Handbook of Fluids in Motion*, Ann Arbor Science, Ann Arbor, Mich., 1983. Chisholm, *Two-Phase Flow in Pipelines and Heat Exchangers*, George Godwin, New York, in association with the Institution of Chemical Engineers, 1983. Daubert and Danner, "Physical and Thermodynamic Properties of Pure Chemicals, Data Compilation," DIPPR (Design Inst. of Phys. Props. Res.) AIChE, New York, 1989. Diener and Schmidt, "Sizing of Throttling Device for Gas/Liquid Two-Phase Flow, Part 1: Safety Valves, Part 2: Control Valves, Orifices, and Nozzles," *J. Hazardous Mat.* **23**(4): 335–344, December 2004; **24**(1): 29–37, March 2005. Ely and Huber, "NIST Thermophysical Properties of Hydrocarbon Mixtures Database (STRAPP Program)," Version 1.0 Users' Guide, U.S. Dept. of Commerce, National Institute of Standards and Technology (NIST), Gaithersburg, Md., Feb. 1990. Fisher et al., *Emergency Relief System Design Using DIERS Technology*, AIChE, New York, 1992. Graham, "The Flow of Air-Water Mixtures Through Nozzles," National Engineering Laboratories (NEL) Report no. 308, East Kilbride, Glasgow, 1967. Jobson, "On the Flow of Compressible Fluids Through Orifices," *Proc. Instn. Mech. Engrs.* **169**(37): 767–776, 1955. "The Two-Phase Critical Flow of One-Component Mixtures in Nozzles, Orifices, and Short Tubes," *Trans. ASME J. Heat Transfer*, **93**(5): 179–N87, 1976. Lee and Sommerfeld, "Maximum Leakage Times Through Puncture Holes for Process Vessels of Various Shapes," *J. Hazardous Mat.* **38**(1) 27–40, July 1994. Lee and Sommerfeld, "Safe Drainage or Leakage Considerations and Geometry in the Design of Process Vessels," *Trans. IChemE* **72**, part B: 88–89, May 1994. Leung, "A Generalized Correlation for One-Component Homogeneous Equilibrium Flashing Choked Flow," *AIChE J.* **32**(10): 1743–1746, 1986. Leung, "Similarity Between Flashing and Non-flashing Two-Phase Flow," *AIChE J.* **36**(5): 797, 1990. Leung, "Size Safety Relief Valves for Flashing Liquids," *Chem. Eng. Prog.* **88**(2): 70–75, February 1992. Leung, "Two-Phase Flow Discharge in Nozzles and Pipes—A Unified Approach," *J. Loss Prevention Process Ind.* **3**(27): 27–32, January 1990. Leung and Ciolek, "Flashing Flow Discharge of Initially Subcooled Liquid in Pipes," *ASME Trans. J. Fluids Eng.* **116**(3), September 1994. Leung and Epstein, "Flashing Two-Phase Flow Including the Effects of Noncondensable Gases," *ASME Trans. J. Heat Transfer* **113**(1): 269, February 1991. Leung and Epstein, "A Generalized Correlation for Two-Phase Non-flashing Homogeneous Choked Flow," *ASME J. Heat Transfer* **112**(2), May 1990. Leung and Grolmes, "The Discharge of Two-Phase Flashing Flow in a Horizontal Duct," *AIChE J.* **33**(3): 524–527, 1987; also errata, **34**(6): 1030, 1988. Levenspiel, "The Discharge of Gases from a Reservoir Through a Pipe," *AIChE J.* **23**(3): 402–403, 1977. Lockhart and Martinelli, "Proposed Correlation of Data for Isothermal Two-Phase, Two-Component Flow in Pipes," *Chem. Eng. Prog.* **45**(1): 39–48, January 1949. Simpson, Rooney, and Grattan, "Two-Phase Flow Through Gate Valves and Orifice Plates," Int. Conf. on the Physical Modeling of Multi-phase Flow, Coventry, U.K., April 19–20, 1983. Sozzi and Sutherland, "Critical Flow of Saturated and Subcooled Water at High Pressure," General Electric Co. Report no. NEDO-13418, July 1975; also *ASME Symp. on Non-equilibrium Two-Phase Flows*, 1975. Tangren, Dodge, and Siefert, "Compressibility Effects in Two-Phase Flow," *J. Applied Phys.* **20:** 637–645, 1949. Uchida and Narai, "Discharge of Saturated Water Through Pipes and Orifices," *Proc. 3d Intl. Heat Transfer Conf., ASME*, Chicago, **5**: 1–12, 1966. Van den Akker, Snoey, and Spoelstra, "Discharges of Pressurized Liquefied Gases Through Apertures and Pipes," *I. Chem. E. Symposium Ser.* (London) **80:** E23–35, 1983. Watson, Vaughan, and McFarlane, "Two-Phase Pressure Drop with a Sharp-Edged Orifice," National Engineering Laboratories (NEL) Report no. 290, East Kilbride, Glasgow, 1967. Woodward, "Discharge Rates Through Holes in Process Vessels and Piping," in Fthenakis, ed., *Prevention and Control of Accidental Releases of Hazardous Gases*, Van Nostrand Reinhold, New York, 1993, pp. 94–159. Woodward and Mudan, "Liquid and Gas Discharge Rates Through Holes in Process Vessels," *J. Loss Prevention*: **4**(3): 161–165, 1991.

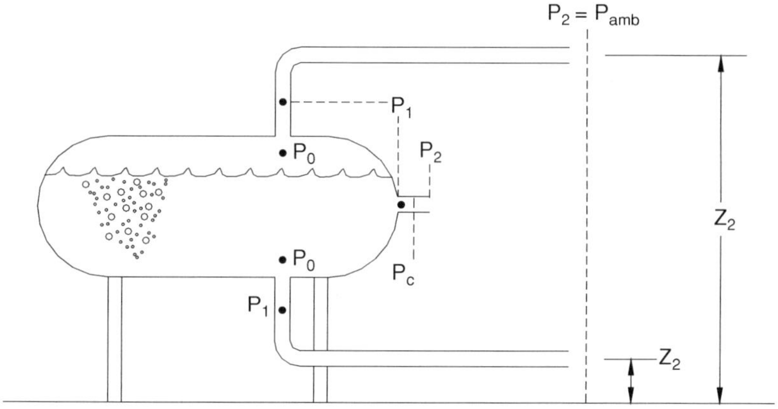

FIG. 23-30 Types of accidental discharge and pressure evaluation points.

Overview Modeling the consequences of accidental releases of hazardous materials begins with the calculation of discharge rates. In the most general case, the discharged material is made up of a volatile flashing liquid and vapor along with noncondensable gases and solid particles. Most of the model development is given in terms of two-phase, vapor + liquid flow, but it can be readily extended to three phases with vapor, liquid, plus suspended solids flow. The solids contribute primarily to the density and heat capacity of the mixture, contributing energy to the flashing and to the subsequent dispersion.

Types of Discharge Hazardous accidental releases can occur from vessels, pipelines, reactors, and distillation columns. The more common accidental events are breaks in a vessel or its associated piping. Figure 23-30 illustrates this. For a vessel, the piping attachments can be from a dip leg or from the bottom or top of the vessel. A puncture in the vapor space of the vessel or a break in a top-attachment pipe, initially at least, discharges vapor plus the padding gas used to maintain vessel pressure. This padding gas can be air for an atmospheric vessel. The discharged vapor can cool upon expansion and condense liquids when the pressure in the jet drops to the dew point pressure P_{dew}.

A puncture in the liquid space of the vessel or a break in the bottom-attachment or dip-leg pipe, initially, at least, discharges liquid plus any solids present without any noncondensable components. The liquid can begin to flash when the pressure drops to the bubble point pressure P_{bub}. If the liquid is extremely volatile, it could totally evaporate when the pressure drops below the dew point, producing vapor plus solids. The initial mass vapor fraction x_0 is zero as is the initial volume fraction α_0.

For a puncture, break, or pressure relief valve (PRV) opening from a reactor or distillation column, there may be no clear-cut level distinguishing the liquid and vapor phases. That is, the system is initially mixed. In this case, noncondensable gases, condensable vapors, and liquid plus solids are initially discharged. The value of α_0 is nonzero and less than unity, reflecting the contributions of the gases and vapors.

For a blow-down calculation, the conditions change. With a tank vapor space release, as the pressure decreases, the liquid could reach its bubble point and begin to flash. Vapor bubbles in the liquid generate a liquid swell, so a frothy, two-phase interface rises and could drive out all the vapor (along with noncondensables) and begin two-phase flashing flow of liquid without noncondensables. Thus, the discharge calculation would also change from a vapor discharge to a two-phase flashing liquid discharge.

The energy and momentum balance equations are drawn across planes at points 0, 1, and 2, as illustrated for a general case in Fig. 23-31.

Energy Balance Method for Orifice Discharge Solutions can be found for the discharge rate by solving the energy balance and/or the momentum balance. The energy balance solution is quite simple and general, but is sensitive to inaccuracies in physical properties correlations. The equations below apply to orifice flow. A separate momentum balances solution is applied to pipe flow to find the pressure losses. The balances are written across an orifice from a plane 0 inside the tank at stagnant conditions (i.e., far enough away from the orifice that the velocity inside the tank is negligible) to a plane at the backpressure P_b at point 2.

$$H_0 = H_2 + \frac{1}{2}(G^2 v_e^2)_2 + Q \qquad (23\text{-}32)$$

Rarely, if ever, is the heat-transfer term Q nonnegligible.

Perform an isentropic expansion. That is, as the pressure decreases from P_0 to the backpressure P_b (usually ambient pressure P_a), select intermediate values of pressure P_1. At each P_1 find the temperature that keeps S constant T_{S1}. Solve for the vapor fraction x, using the entropy balance between planes 0 and 1:

$$S_0 = xS_{G1} + (1 - x - x_{sol})\,S_{L1} + x_{sol}S_{sol} \qquad (23\text{-}33)$$

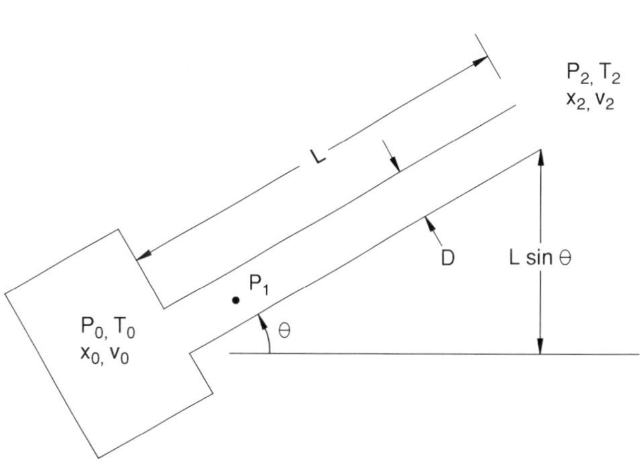

FIG. 23-31 Configuration modeled for pipe flow with elevation change.

Since the mass fraction of solids is constant, the vapor flash fraction is

$$x = \frac{S_0 - [(1 - x_{sol})S_{L1} - x_{sol}S_{sol}]_{T_{S1}}}{(S_{G1} - S_{L1})_{T_{S1}}} \quad (23\text{-}34)$$

Find the enthalpy H_2 at T_{S1} from physical property correlations. Velocity is found solving Eq. (23-32) for homogeneous flow (with $Q = 0$):

$$Gv_e = u = \sqrt{2(H_0 - H_2)} \quad (23\text{-}35)$$

Find the density ρ from an equation of state at P_1, T_{S1} for the phase densities ρ_G and ρ_L, and the reciprocal of the homogeneous specific volume v_H. Calculate the mass flux from

$$G = u\rho = \frac{u}{v} \quad (23\text{-}36)$$

Search with values of P_1 until G is maximized. The choke pressure, P_{ch} is the value of P_1 that produces a maximum value of mass flux G_{max}. The discharge rate w is given from the mass flux, a discharge coefficient C_D, and the orifice cross-sectional area A as

$$w = C_D A G_{max} \quad (23\text{-}37)$$

This approach is illustrated for two-phase flashing flow with a multicomponent mixture (mole fractions 0.477 allyl alcohol, 0.3404 allyl chloride, 0.1826 dodecane). Figure 23-32 plots the flash curve showing that as pressure decreases, flashing begins at the bubble point of 7.22 bar(a) (104.7 psia) and increases to 30.8 mass %. At the same time, temperature follows the saturation curve down. Figure 23-33 plots the two-phase density, the velocity, and the product of these, the mass flux G. In the plot G is a maximum at 8586 kg/(m^2·s) at the choke pressure just below the bubble point. This is so because the two-phase density drops quickly as flashing occurs at reduced pressures, while the velocity increases more slowly. With a discharge coefficient of 0.61 and a 2-in orifice (area of 0.002027 m^2) the discharge rate w is 10.6 kg/s.

Momentum Balance in Dimensionless Variables For pipe flow, it is necessary to solve the momentum balance. The momentum balance is simplified by using the following dimensionless variables:

$$\text{Pressure ratio: } \eta = \frac{P}{P_0}$$

$$\text{Mass flux ratio: } G_* = \frac{G}{\sqrt{P_0\rho_0}} \quad (23\text{-}38)$$

$$\text{Specific volume ratio: } \varepsilon = \frac{v}{v_0}$$

The discharge relationships are derived by solving the differential momentum balance over a tank plus pipe:

$$v\,dP + G^2 v\,dv + \left(4f\frac{dz}{D} + K_e\right)\frac{1}{2}G^2 v_L^2\,\phi_L^2 + g\sin\theta\,dz = 0 \quad (23\text{-}39)$$

The terms represent, respectively, the effect of pressure gradient, acceleration, line friction, and potential energy (static head). The effect of fittings, bends, entrance effects, etc., is included in the term K_e correlated as a number of effective "velocity heads." The inclination angle θ is the angle to the horizontal from the elevation of the pipe connection to the vessel to the discharge point. The term ϕ_L^2 is the two-phase multiplier that corrects the liquid-phase friction pressure loss to a two-phase pressure loss. Equation (23-39) is written in units of pressure/density.

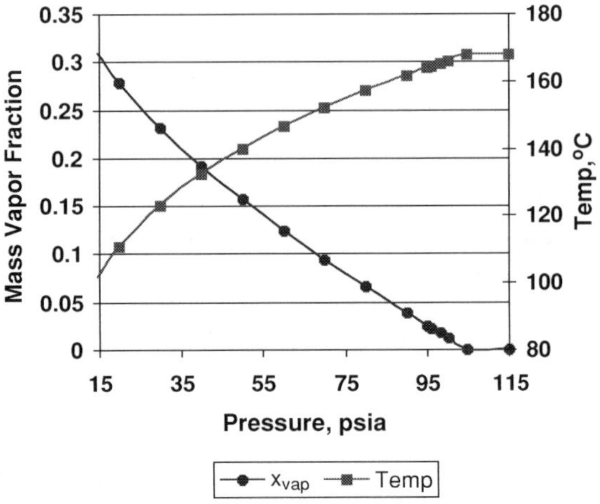

FIG. 23-32 Example flash curve for multicomponent material.

In dimensionless variables, the momentum balance is

$$\varepsilon\,d\eta + G_*^2\,\varepsilon\,d\varepsilon + N\frac{1}{2}G_*^2\,\varepsilon^2\phi_L^2 + \frac{g\sin\theta dz}{P_0 v_0} = 0 \quad (23\text{-}40)$$

where N collects the friction loss terms in terms of the number of velocity heads, or

$$N = 4f_L\frac{dz}{D} + K_e \quad (23\text{-}41)$$

The two-phase flow multiplier φ_L^2 is discussed in a later section.

The momentum balance for homogeneous flow can be factored to a form that enables integration as

$$-N = \frac{G_{*p}^2\varepsilon\,d\varepsilon + \varepsilon\,d\eta}{\frac{1}{2}G_{*p}^2\varepsilon^2 + F_I} \quad (23\text{-}42)$$

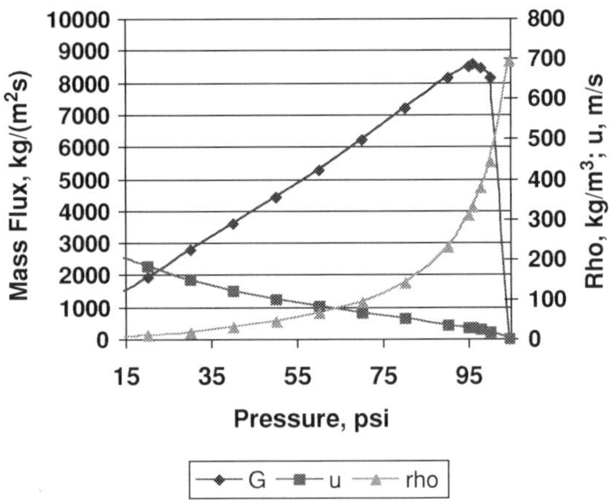

FIG. 23-33 Example of using the energy balance to find the orifice discharge flux.

by defining a pipe inclination factor F_I as

$$F_I = \frac{gD\sin\theta}{4f_{L0}P_0v_0} \tag{23-43}$$

where F_I is positive for upflow, negative for downflow, and zero for horizontal flow.

Analytical Solutions for Orifice and Pipe Flow Equation (23-42) can be solved analytically for pipe breaks and tank punctures for the following cases:

- Subcooled liquid flow
- Adiabatic expansion of ideal gases
- Flashing liquid flow without noncondensable gases ($\alpha_0 = 0$)
- Subcooled liquid mixed with noncondensable gases ($\alpha_0 > 0$) (frozen flow)
- Flashing liquid mixed with noncondensable gases (hybrid flow)

The solutions for all except hybrid flow follow.

Orifice Discharge for Gas Flow The analytic solution for discharge through an orifice of an ideal gas is derived by invoking the equation of state for adiabatic expansion of an ideal gas:

$$\frac{P}{P_0} = \left(\frac{\rho}{\rho_0}\right)^{\gamma} = \left(\frac{T}{T_0}\right)^{\gamma/(\gamma-1)} \tag{23-44}$$

where

$$\gamma = \frac{C_P}{C_V} \tag{23-45}$$

The solution is

$$G_{*g}^2 = \frac{2\gamma}{\gamma-1}\eta_2^{2/\gamma}\left(1 - \eta_2^{(\gamma-1)/\gamma}\right) \tag{23-46}$$

This solution applies for both subsonic and choked flow. If the flow is choked, the exit pressure ratio η_2 is replaced by the choked pressure ratio η_{ch}, given by

$$\eta_{ch} = \left(\frac{2}{\gamma+1}\right)^{\gamma/(\gamma-1)} \tag{23-47}$$

Equation (23-47) must be evaluated to test for choked flow in any event. When η_{ch} from Eq. (23-47) is substituted into Eq. (23-46), the general solution reduces to the choked flow solution:

$$G_{*g}^2 = \gamma\left(\frac{2}{\gamma+1}\right)^{(\gamma+1)/(\gamma-1)} \tag{23-48}$$

The discharge rate is found by using Eqs. (23-37) and (23-38).

Blowdown of Gas Discharge through Orifice An analytic solution is available for blowdown (time-dependent discharge) of an ideal gas from a tank. The time-varying mass of gas in the tank m_T is the product of the tank volume V_T and the density ρ:

$$m_T(t) = V_T\rho(t) \tag{23-49}$$

Differentiating, solving for dt, and integrating gives

$$t = -V_T\int_{P_0}^{P}\frac{d\rho}{w} \tag{23-50}$$

where w is the time-varying discharge rate. Typically, all but a small fraction of the mass in a tank is discharged at sonic flow, so a sonic flow solution is most useful. Transforming P and ρ to temperature T, using Eq. (23-44) allows Eq. (23-50) to be integrated to give a solution in terms of the initial discharge rate w_0 and the initial tank mass m_{T0}:

$$w(t) = w_0[F(t)]^{(\gamma+1)(\gamma-1)} \tag{23-51}$$

where

$$F(t) = (1 + At)^{-1} \tag{23-52a}$$

$$A = \frac{w_0(\gamma+1)}{2m_{T0}} \tag{23-52b}$$

Pipe and Orifice Flow for Subcooled Liquids Since liquids are essentially incompressible, ε is constant at ε_0, and $d\varepsilon$ is zero in Eq. (23-42). Recognizing that η_0 and ε_0 are unity, we see that integration gives

$$-N\left(\frac{1}{2}G_{*p}^2\,\varepsilon_0^2 + F_I\right) = \varepsilon_0\int_{\eta_0}^{\eta_2}d\eta = \eta_2 - \eta_0$$

or

$$G_{*p}^2 = 2\left(\frac{1-\eta_2}{N} - F_I\right) \tag{23-53}$$

The solution for orifice flow is a special case with N equal to unity (entrance losses only) and F_I equal to zero, giving

$$G_{*\text{ori}} = \sqrt{2(1-\eta_2)} \tag{23-54}$$

Numerical Solution for Orifice Flow With orifice flow, the last two terms of the momentum balance (line resistance and potential energy change) are negligible. The momentum balance, Eq. (23-40), reduces to

$$G_*^2\,\varepsilon\,d\varepsilon = -\varepsilon\,d\eta \tag{23-55}$$

This equation can be treated in differential form and in integral form. In differential form it becomes

$$G_{*\text{diff}} = -\left(\frac{d\varepsilon}{d\eta}\right)^{-1/2} \quad\text{or}\quad G_{\text{diff}} = -\left(\frac{dv}{dP}\right)^{-1/2} \tag{23-56}$$

For the integral form, express Eq. (23-55) as

$$\frac{1}{2}G_*^2\,d\varepsilon^2 = -\varepsilon\,d\eta$$

to obtain

$$G_*\,d\varepsilon = (-2\varepsilon\,d\eta)^{1/2} \tag{23-57}$$

In general, the limits of integration are from ε_0, η_0 to an arbitrary final point ε, η (recognizing that $\varepsilon_S = \varepsilon_0 = 1$). However, this method works better, using an indefinite integration of $d\varepsilon$:

$$G_{*\text{int}} = \frac{1}{\varepsilon}\left(2\int_{\eta}^{\eta_0}\varepsilon\,d\eta\right)^{1/2} \tag{23-58}$$

or in dimensional form

$$G_{\text{int}} = \frac{1}{v}\left(2\int_{p}^{p_0} v\,dP\right)^{1/2} \tag{23-59}$$

Equations (23-56) and (23-59) are readily evaluated at intermediate pressure points P_1 in the range $P_a < P_1 < P_0$, giving two curves for G. The two curves cross at the choke point, which is also the point of maximum mass flux, as illustrated in Fig. 23-34 for a mixture of isobutane (90.898 mol %) and ethylene (9.102 mol %) initially at 33.0 bara and 357.4 K. The mass flux G found by evaluating Eq. (23-59) is labeled G_{int} in the figure, and the value of G found by evaluating Eq. (23-56) is labeled G_{diff}. The G_{int} curve has a maximum value of 20,637 kg/(m²·s) at the choke pressure of 20.64 bara.

The disadvantage of this solution is that it is sensitive to inaccuracies in the physical properties correlations used to evaluate the flash fraction and specific volumes. Figure 23-34 is found by using DIPPR properties (Daubert and Danner, 1989). The mass flux found does not agree well with that found by using the energy balance method that gives 24,330 kg/(m²·s). In addition, the rightmost branch of the G_{diff} curve is anomalously low and crosses the G_{int} curve spuriously to the right of the maximum flux. By using more accurate physical properties developed by the STRAPP program of NIST (Ely and Huber, 1990), the maximum value of G_{int} agrees with that found by the energy balance method and the right-hand branch of the G_{diff} curve does not fall low and cause this confusion. For this reason the method is not recommended for general use, but provides good confirmations when used with accurate physical properties correlations.

Omega Method Model for Compressible Flows The factored momentum balance, Eq. (23-42), can be analytically integrated after first relating the dimensionless specific volume ε to the dimensionless pressure ratio η. A method to do this, designated the *omega method,* was suggested by Leung (1986):

$$\varepsilon = \begin{cases} \omega\left(\dfrac{\eta_s}{\eta} - 1\right) + 1 & \text{if } \dfrac{\eta_s}{\eta} > 1 \\[2ex] 1.0 & \text{if } \dfrac{\eta_s}{\eta} \le 1 \end{cases} \tag{23-60}$$

Equation (23-60) represents a linear relationship between the two- or three-phase specific volume and reciprocal pressure (v versus P^{-1} or ε versus η^{-1}) beginning at the bubble point P_S, where η is η_s. For single components, ω is found by using the Clapeyron equation to give

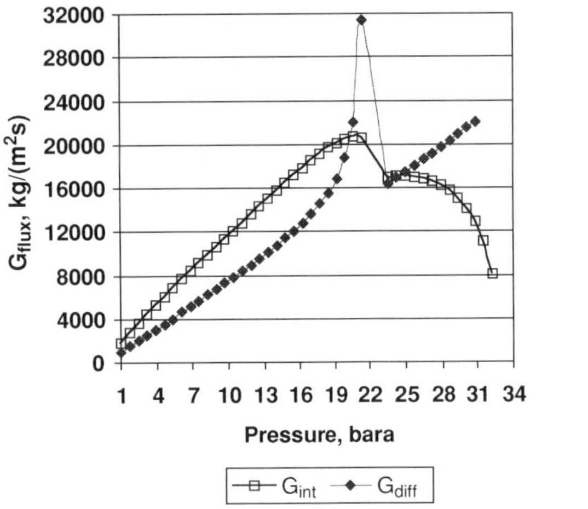

FIG. 23-34 Numerical solution of the momentum balance for orifice flow using Eqs. (23-56) and (23-59).

$$\omega = \frac{C_{pL}T_0P_S}{v_{L0}}\left[\frac{v_{GL0}(P_S)}{H_{GL0}(P_S)}\right]^2 \tag{23-61a}$$

Alternately, use the slope of the ε versus η^{-1} curve between the bubble point and a second, lower pressure at ε_2, η_2 to evaluate ω, or

$$\omega = \frac{\varepsilon_2 - 1}{\eta_s/\eta_2} - 1 \tag{23-61b}$$

This is a more convenient formula for multicomponent mixtures.

This value of ω can be called the saturation value or ω_s, since it applies only with flashing liquids (i.e., in the flashing region for pressures less than the bubble point, as seen in Fig. 23-32). A generalization defines omega to apply also with noncondensable gases by using $\alpha_0 = x_{v_0}v_{v_0}/v_0$

$$\omega = \alpha_0 + (1 - \alpha_0)\omega_s \tag{23-61c}$$

Homogeneous Equilibrium Omega Method for Orifice and Horizontal Pipe Flow The homogeneous equilibrium model (HEM) solution is obtained by substituting Eq. (23-60) for ε and integrating the momentum balance. The solution is given first for horizontal flow, with the flow inclination factor $F_I = 0$. Orifice flow is a special case in which only the orifice solution is needed. For pipe flow, two solutions are needed, the orifice flow solution giving $G_{*\text{ori}}$ and the pipe flow solution giving G_{*p}. Plane 1 in Fig. 23-31 is at the pressure ratio η_1 where the pipe pressure balance begins and the inlet orifice pressure loss ends. The final pipe solution finds η_1 so that

$$G_{*\text{ori}} = G_{*p} \tag{23-62}$$

The integration for both solutions must be conducted over the subcooled region and the flashing region. The complicating factor is that the bubble point pressure ratio η_s could fall either in the orifice flow integration span ($\eta_s > \eta_1$), giving flashing in the orifice (case 1), or in the pipe integration span ($\eta_s < \eta_1$), giving flashing in the pipe (case 2). These two options are illustrated in Fig. 23-35.

Again, over the subcooled region from η_0 to η_s, $d\varepsilon$ is zero and ε is constant at unity, ε_0. The solutions for the two cases are as follows:

Case 1: Flashing in Orifice, Two-Phase Flashing Flow in Pipe

$$G^2_{*\text{ori}} = \frac{2\{(1 - \eta_s) + [\omega\eta_s \ln(\eta_s/\eta_{\text{ch}}) - (\omega - 1)(\eta_s - \eta_{\text{ch}})]\}}{\varepsilon^2_{\text{ch}}} \tag{23-63}$$

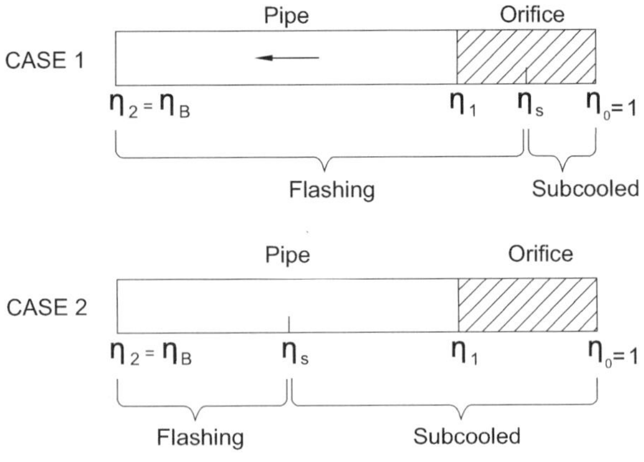

FIG. 23-35 Options for integration range of HEM omega method.

$$G_{*p}^2 = 2\left\{\frac{\dfrac{\eta_1-\eta_2}{1-\omega}+\dfrac{\omega\eta_s}{(1-\omega)^2}\ln\dfrac{\eta_2\varepsilon_2}{\eta_1\varepsilon_1}}{N+2\ln\left(\dfrac{\varepsilon_2}{\varepsilon_1}\right)}\right\} \tag{23-64}$$

Case 2: Subcooled Liquid in Orifice, Flashing in Pipe

$$G_{*\text{ori}}^2 = 2(\eta_0-\eta_1) \tag{23-65}$$

$$G_{*p}^2 = 2\left[\frac{\eta_1-\eta_s+\dfrac{\eta_s-\eta_2}{1-\omega}+\dfrac{\eta_s-\eta_2}{1-\omega}+\dfrac{\omega\eta_s}{(1-\omega)^2}\ln\left(\dfrac{\eta_2\varepsilon_2}{\eta_s\varepsilon_S}\right)}{N+2\ln\left(\dfrac{\varepsilon_2}{\varepsilon_S}\right)}\right] \tag{23-66}$$

The choke point is found by the usual maximization relationship:

$$\left(\frac{dG_*}{d\eta_2}\right)_{\eta_2=\eta_{ch}} = 0 \tag{23-67}$$

This gives an implicit equation in the choked pressure ratio η_c (shorthand for η_{ch}/η_s):

If $\omega \le 2$:

$$\eta_c^2 + \omega(\omega-2)(1-\eta_c)^2 + 2\omega^2\ln\eta_c + 2\omega^2(1-\eta_c) = 0 \tag{23-68a}$$

An explicit equation provides an adequate approximate solution for larger values of ω:

For $\omega \ge 2$:

$$\eta_c = 0.55 + 0.217\ln\omega - 0.046(\ln\omega)^2 + 0.004(\ln\omega)^3 \tag{23-68b}$$

For orifice flow, the definition of choked flow in terms of the back-pressure ratio η_b is

- Choked compressible: $\eta_b \le \eta_{ch}$ set $\eta = \eta_{ch}$
- Subsonic compressible: $\eta_b > \eta_{ch}$ set $\eta = \eta_b$
- Subcooled liquid: always subsonic

The omega method HEM solution for orifice flow is plotted in Fig. 23-36. The solution for flashing liquids without noncondensables is to the right of $\omega = 1$, and the solution for frozen flow with subcooled liquids plus noncondensables is to the left. The omega method HEM solution for horizontal pipe flow is plotted in Fig. 23-37 as the ratio of pipe mass flux to orifice mass flux.

HEM for Inclined Pipe, Discharge If a pipe leak occurs at an elevation above or below the pump or source tank the elevation change can be idealized between the starting and ending points as shown in Fig. 23-38. That is, elevation changes can be treated as an inclined pipe with a non-zero inclination factor F_I. This is an approximation to the actual piping isometrics, but is often an adequate approximation.

The HEM solution in this case is implicit in G_{*p}:

Case 1: Flashing in Orifice, Two-Phase Flashing Flow in Pipe: Use Eq. (23-63) for $G_{*\text{ori}}$:

$$N + \ln\left\{\frac{X(\eta_2)}{X(\eta_1)}\left[\frac{\eta_1}{\eta_2}\right]^2\right\} = \frac{1-\omega}{c}(\eta_1-\eta_2)$$
$$+\frac{b(1-\omega)-c\omega\eta_s}{2c^2}\ln\frac{X(\eta_2)}{X(\eta_1)} \tag{26-69}$$
$$+\frac{bc\omega\eta_s-(1-\omega)(b^2-2ac)}{2c^2}[I_0(\eta_2)-I_0(\eta_1)]$$

Case 2: Subcooled Liquid in Orifice, Flashing in Pipe: Use Eq. (23-65) for $G_{*\text{ori}}$:

$$N + \ln\left\{\frac{X(\eta_2)}{X(\eta_S)}\left[\frac{\eta_S}{\eta_2}\right]^2\right\} = \frac{\eta_1-\eta_S}{\frac{1}{2}G_*^2+F_I} + \frac{1-\omega}{c}(\eta_s-\eta_2)$$
$$+\frac{b(1-\omega)-c\omega\eta_s}{2c^2}\ln\frac{X(\eta_2)}{X(\eta_s)}$$
$$+\frac{bc\omega\eta_S-(1-\omega)(b^2-2ac)}{2c^2}[I_0(\eta_2)-I_0(\eta_S)] \tag{23-70}$$

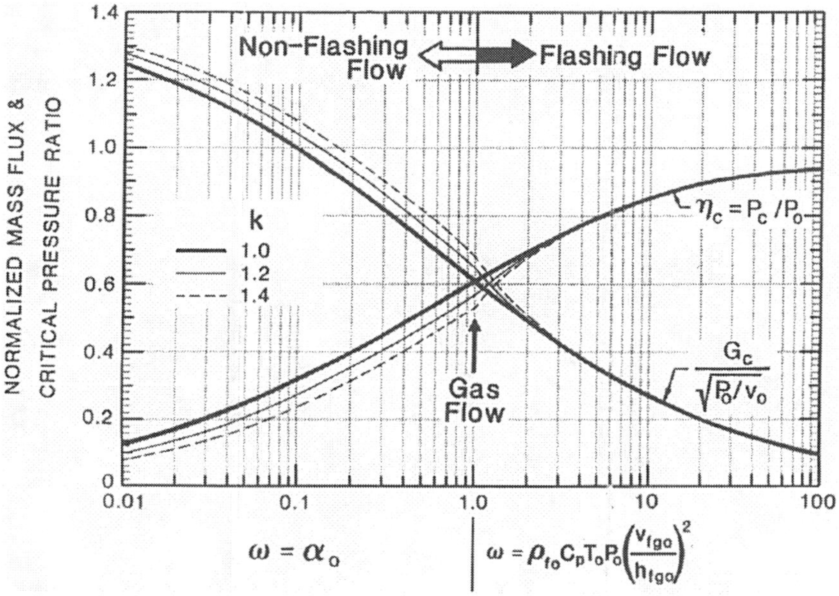

FIG. 23-36 Omega method solution for orifice flow of flashing liquids and for noncondensable gas plus subcooled liquids.

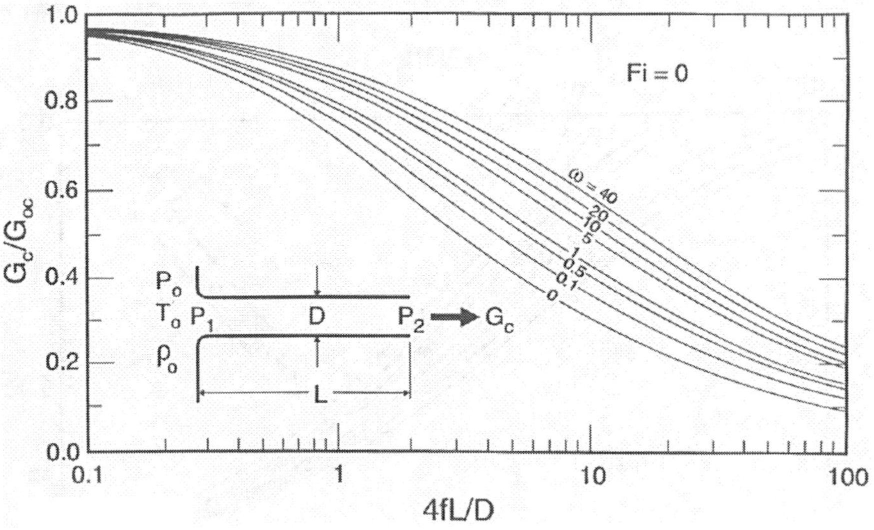

FIG. 23-37 Omega method solution for flashing liquid horizontal pipe flow.

where

$$a = \frac{1}{2} G_*^2 \omega^2 \eta_s^2 \tag{23-71a}$$

$$b = \frac{1}{2} G_*^2 2\omega (1 - \omega)\eta_s \tag{23-71b}$$

$$c = \frac{1}{2} G_*^2 (1 - \omega)^2 + F_I \tag{23-71c}$$

$$q = 4ac - b^2 \tag{23-71d}$$

$$X(\eta) = a + b\eta + c\eta^2 \tag{23-72}$$

$$I_0(\eta) = \int \frac{d\eta}{X(\eta)}$$

If $q > 0$, $F_I > 0$, upflow:

$$I_0(\eta) = \frac{2}{\sqrt{q}} \tan^{-1} \frac{2c\eta + b}{\sqrt{q}} \tag{23-73a}$$

If $q < 0$, $F_I < 0$, downflow:

$$I_0(\eta) = \ln \frac{2c\eta + b - \sqrt{-q}}{2c\eta + b + \sqrt{-q}} \tag{23-73b}$$

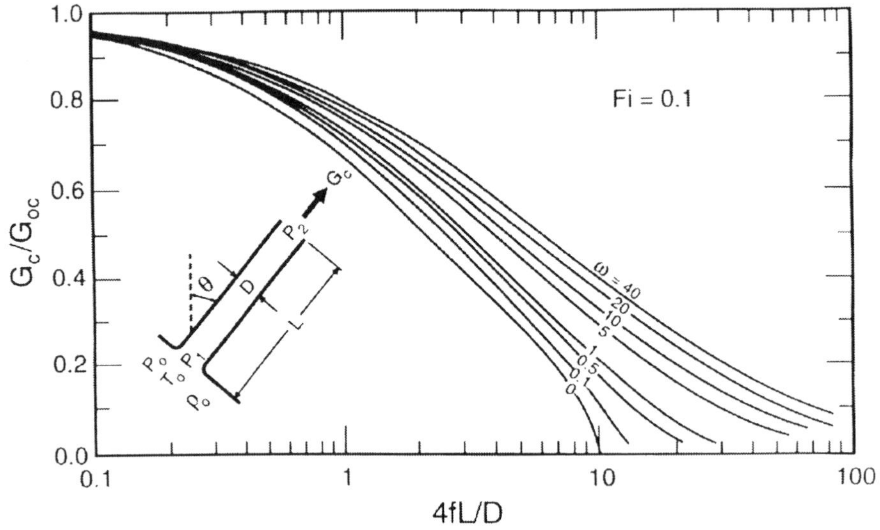

FIG. 23-38 Omega method HEM solution for inclined pipe flow at $F_I = 0.1$.

If $q = 0$, $F_l = 0$, horizontal flow:

$$I_0(\eta) = \frac{-1}{(1-\omega)\varepsilon\eta} \tag{23-73c}$$

The omega method HEM solution for inclined pipe flow is illustrated in Fig. 23-38.

Nonequilibrium Extension of Omega Method The omega method HEM tends to produce discharge rates that are low, particularly for short pipes. To correct this deficiency, Diener and Schmidt proposed a modification they term the *nonequilibrium compressibility factor N*, defined by

$$N = \left[x_0 + C_{PL0} T_0 P_0 \left(\frac{v_{GL0}}{H_{GL0}^2}\right) \ln\left[\frac{1}{\eta_C}\right]\right]^a \tag{23-74}$$

with

$$a = \begin{cases} 0.6 & \text{for orifices, control valves, short nozzles} \\ 0.4 & \text{for pressure relief valves, high – lift control valves} \\ 0 & \text{for long nozzles, office with large area ratio} \end{cases}$$

The choked pressure ratio η_c is found by using Eqs. (23-68*a*) and (23-68*b*). The nonequilibrium factor N is used to modify ω:

$$\omega = \alpha_0 + \omega_S N \tag{23-75}$$

Differences between Subcooled and Saturated Discharge for Horizontal Pipes Data by Uchaida and Narai (1966) in Figs. 23-39 and 23-40 illustrate the substantial differences between subcooled and saturated-liquid discharge rates. Discharge rates decrease with increasing pipe length in both cases, but the drop in discharge rate is much more pronounced with saturated liquids. This is so, because the flashed vapor effectively chokes the flow and decreases the two-phase density.

Accuracy of Discharge Rate Predictions Model verification is difficult because agreement with data varies substantially from one set of experimental data to another. The accuracy of the HEM and the NEM correction is illustrated in Fig. 23-41 and compared with data for saturated water by Uchaida and Narai (1966). The NEM correction using the power coefficient a of 0.6 decreases omega and increases the predicted discharge rates so that good predictions are obtained for pipe lengths greater than 0.5 m.

The accuracy of the energy balance method for discharge of flashing liquids through orifices and horizontal pipes is illustrated in Figs. 23-42 and 23-43. The ratio of predicted to observed mass flux is plotted for saturated water data by Uchaida and Narai and by Sozzi and

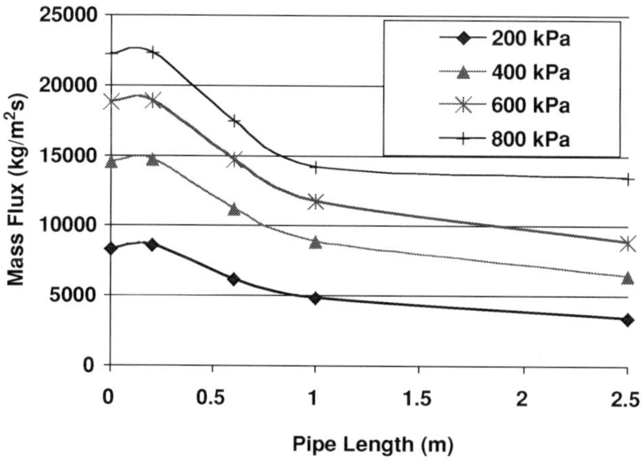

FIG. 23-39 Subcooled water data by Uchaida and Narai.

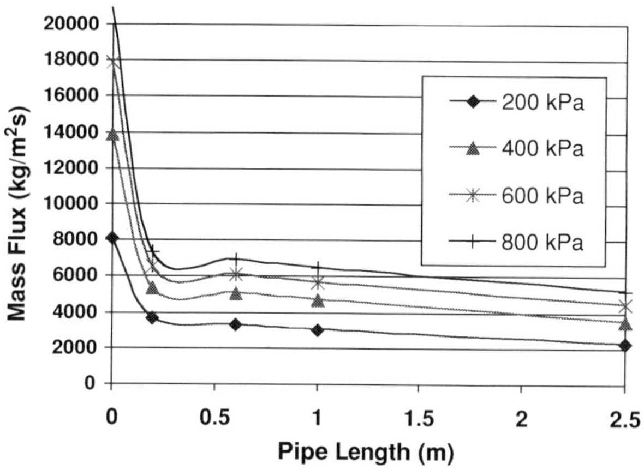

FIG. 23-40 Saturated water data by Uchaida and Narai.

Sutherland (1975). Orifice flow rates are underpredicted by about the same factor with the energy balance method and with the NEM. Discharge predictions for short (0.2-m) pipes are overpredicted by the energy balance method. In this region, the assumption of homogeneous equilibrium is not justified. A model that takes slip velocity into account may improve predictions for short pipes.

ATMOSPHERIC DISPERSION

Nomenclature

$<C>_1$	(Ensemble) time-averaged concentration for averaging time t_1, mass per volume
d_0	Displacement height, length
E	Airborne contaminant mass rate in a plume, mass per time
E_t	Total airborne contaminant mass in a puff, mass
g	Acceleration due to gravity, length squared per time
k	Von Karman's constant (typically 0.4)
L	Monin-Obukhov length, length
p	Index reflecting decrease in time-averaged concentration with averaging time
t_d	Along wind (x direction) dispersion time scale, time
t_h	Time scale for contaminant to pose a particular hazard, time
t_s	Source time scale, length of time for contaminant to become airborne, time
t_t	Time scale for a contaminant cloud to reach hazard endpoint distance x_h, time
u	(Characteristic) wind speed at elevation z, length per time
u_r	Wind speed at reference height z_r, length per time
u_*	Friction velocity, length per time
x_v	Virtual source distance upwind of real source, length
x	(Downwind) distance in wind direction, length
x_h	Hazard endpoint distance, length
z_o	Surface roughness, length
z_r	Reference height, length
	Greek symbols
α	Monin-Obukhov length coefficient
ε	Height of ground covering, length
ρ	Air-contaminant mixture density at concentration C, mixture mass per mixture volume
ρ_a	Ambient air density, mass per volume
$\sigma_x, \sigma_y, \sigma_z$	Dispersion coefficients in x, y, z directions, length
σ_θ	Wind direction standard deviation, angle
	Subscripts
x	Along wind direction
y	Lateral direction
z	Vertical direction

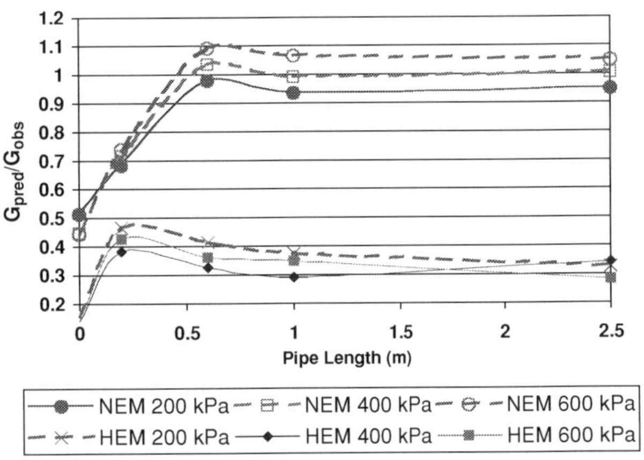

FIG. 23-41 Accuracy of Omega method HEM and NEM correction compared with saturated water data by Uchaida and Narai.

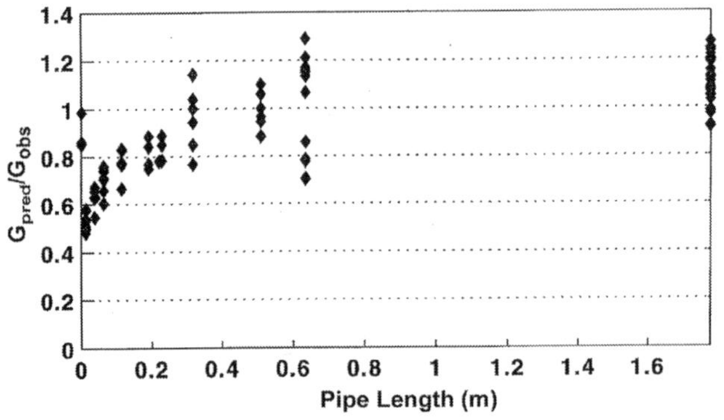

FIG. 23-43 Accuracy of energy balance method for flashing liquid discharge through orifices and horizontal pipes.

GENERAL REFERENCES: Britter and McQuaid, *Workbook on the Dispersion of Dense Gases,* Health and Safety Executive Report 17/1988, Sheffield, U.K., 1988. Mannan, *Lees' Loss Prevention in the Process Industries,* 3d ed., Chap. 15, Elsevier Butterworth-Heinemann, Oxford, U.K., 2005. Panofsky and Dutton, *Atmospheric Turbulence,* Wiley, New York, 1984. Pasquill and Smith, *Atmospheric Diffusion,* 3d ed., Ellis Horwood Limited, Chichester, U.K., 1983. Seinfeld, *Atmospheric Chemistry and Physics of Air Pollution,* Chaps. 12–15, Wiley, New York, 1986. Turner, *Workbook of Atmospheric Dispersion Estimates,* U.S. Department of Health, Education, and Welfare, 1970.

Introduction Atmospheric dispersion models predict the dilution of an airborne contaminant after its release (and depressurization) to the atmosphere. The discussion in this section focuses on episodic releases representing acute biological, toxic, or flammable hazards. Physical and mathematical dispersion models discussed here are typically used for (1) forensic purposes (in comparison with actual events such as experiments, field trials, or accidents); (2) regulatory purposes, such as estimating the consequences of releases of toxic or flammable materials for siting requirements; or (3) planning purposes. Release consequences may be used to estimate the risk to a facility and workforce or the surrounding population. All such estimates can be used to identify appropriate emergency response or mitigation measures and the priority with which such measures should be considered.

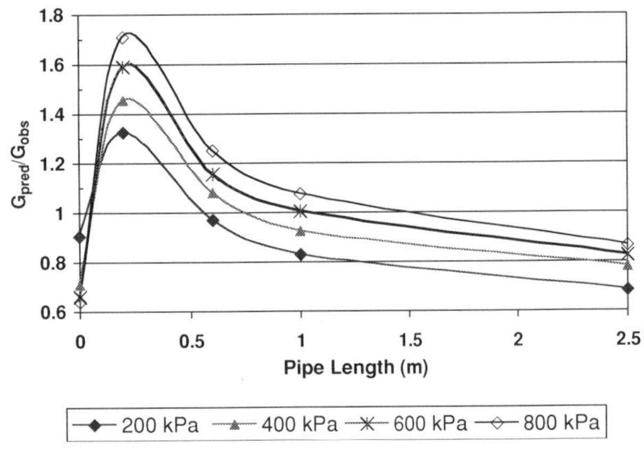

FIG. 23-42 Accuracy of energy balance method compared with saturated water data by Uchaida and Narai.

Parameters Affecting Atmospheric Dispersion The parameters important to atmospheric dispersion can be divided roughly into three categories: contaminant source; atmospheric and terrain properties; and contaminant interaction with the atmosphere.

Contaminant Source The contaminant source includes such factors as the following:

1. *Rate or total amount of contaminant that becomes airborne.* For a continuous contaminant release, the contaminant concentration at a fixed downwind distance is roughly proportional to the rate at which the contaminant becomes airborne. Likewise, the contaminant concentration at a fixed downwind distance is roughly proportional to the amount of contaminant released if the release is instantaneous. However, materials released from containment may not immediately become airborne. For example, a liquid below its boiling point stored at atmospheric pressure will form a liquid pool when released. The rate at which the contaminant becomes airborne will depend on the heat-transfer rate to the liquid pool and the mass-transfer rate from the liquid pool to air. However, the same liquid stored at a higher pressure may form an aerosol so that the liquid phase is airborne. If the aerosol droplet size formed in the release is too large to remain suspended, the liquid phase will fall to the ground (rainout), and the resulting liquid pool evaporation will again be dictated by heat- and mass-transfer considerations.

2. *Release momentum.* For jet releases, the amount of air entrained in an unobstructed jet is proportional to the jet velocity. Depending on the orientation of the jet relative to nearby obstructions, the momentum of a jet can be dissipated without significant air entrainment. The degree of initial air entrainment can be an important determinant of the hazard extent, particularly for flammable hazards. It would be (possibly overly) conservative to assume the source momentum is dissipated without air dilution. Explosive releases are high-momentum, instantaneous releases. For explosive releases, a rough first approximation is to assume that the mass of contaminant in the explosion is mixed with 10 times that mass of air.

3. *Release buoyancy.* Contaminant release buoyancy is determined by the initial contaminant temperature and molecular weight, and whether there is a suspended liquid phase as in an aerosol. Lighter-than-air contaminants will rise and be more readily dispersed. Denser-than-air contaminants will tend to stay near ground level, and the atmospheric dispersion of such materials can be importantly determined by buoyancy effects. The degree of importance is typically quantified by a Richardson number (proportional to the density difference with air and the quantity or rate of contaminant release and inversely proportional to the square of the wind speed); the higher the Richardson number, the more important the effect of buoyancy on the atmospheric dispersion (see Denser-than-Air Dispersion Models — Britter-McQuaid below). Denser-than-air contaminants can actually displace the atmospheric flow field. In contrast, contaminants which do not perturb the atmospheric flow field and can be thought of as simply following the airflow are termed *passive* contaminants.

4. *Other source conditions.* The source height and the source area are also important source characteristics. The greatest impact is typically associated with ground-level sources, so elevated sources are not considered in the model discussion below.

Atmospheric and Terrain Parameters In addition to terrain parameters, basic atmospheric parameters are listed below. There can be many other meteorological effects which can be important in some circumstances (e.g., inversion layers) and are beyond the scope of this introduction.

1. *Wind direction.* Wind direction is the most important determinant of the location of hazard zones with notable exceptions involving high-momentum releases or releases where buoyancy is important (particularly for denser-than-air contaminants involving terrain effects such as valleys and slopes especially under low-wind-speed conditions). Near ground level, lateral wind direction variability is much larger than vertical variability (typical of flat-plate boundary layers) and measured with the standard deviation σ_θ which is a function of atmospheric stability.

2. *Atmospheric stability.* Atmospheric stability is often characterized by Pasquill stability class (ranging from A through F). Atmospheric stability classes are broad classifications which are used to characterize the continuum of atmospheric turbulence available to the (mixing) dispersion process. Neutral atmospheric stability (class D) occurs most often and indicates that the vertical momentum flux is not influenced by the vertical heat flux in the lowest atmospheric layers. Unstable atmospheric conditions (instability increasing from classes C through A) indicate the atmospheric vertical heat flux is enhancing the vertical mixing of the atmosphere, such as when ground surface heating due to insolation (incident solar radiation) enhances vertical atmospheric mixing (turbulence). Stable atmospheric conditions (stability increasing from classes E through F) indicate the atmospheric vertical heat flux is suppressing the vertical mixing of the atmosphere, such as when ground surface cooling at night suppresses vertical mixing (turbulence). All atmospheric dispersion models rely in some way on measured atmospheric parameters, and these parameters are quite frequently correlated on the basis of Pasquill stability class. Consequently, atmospheric dispersion model results should be viewed as only representative of the atmospheric conditions present in a real or hypothetical release. Atmospheric stability class is an important determinant of concentration at a fixed downwind distance in dispersion models. For passive contaminants, concentration at a fixed downwind distance can be roughly an order of magnitude higher for F stability in comparison to D stability. Table 23-24 illustrates the relationship between atmospheric stability class and other meteorological conditions. Determination of Pasquill stability class should be made with detailed measurements such as discussed by Golder ("Relations among Stability Parameters in the Surface Layer," *Boundary Layer Meteorol.*, 3, pp. 47–58, 1972).

TABLE 23-24 Typical Atmospheric Stability Classes in Terms of Wind Speed, Insolation, and State of the Sky

Surface wind speed m/s	Insolation			Night	
	Strong	Moderate	Slight	Thinly overcast or ≥4/8 low cloud	≤3/8 cloud
<2	A	A–B	B	—	—
2–3	A–B	B	C	E	F
3–5	B	B–C	C	D	E
5–6	C	C–D	D	D	D
>6	C	D	D	D	D

For A–B, take the average of values for A and B, etc. Pasquill and Smith relate insolation to conditions in England. Seinfeld (*Atmospheric Chemistry and Physics of Air Pollution*, Wiley, New York, 1986) classifies insolation greater than 700 W/m² as strong, less than 350 W/m² as slight, and between these limits as moderate. Night refers to the period from 1 h before sunset to 1 h after dawn. The neutral class D should be used, regardless of wind speed, for overcast conditions during day or night and for all sky conditions during the hour preceding or following the night period.

SOURCE: Pasquill and Smith, *Atmospheric Diffusion*, 3d ed., Ellis Horwood Limited, Chichester, U.K., 1983.

3. *General terrain characteristics.* General terrain characteristics are used to describe features with length scales much smaller than the depth of the contaminant cloud or the height of the characteristic wind speed. General terrain characteristics influencing the shape of the vertical wind speed profile are parameterized with the surface roughness and displacement height. The surface roughness z_0 is roughly proportional to the height of the ground covering ε (and $z_0 \sim 0.05\varepsilon$ to 0.15ε) and can be used to infer the amount of vertical turbulence (mixing) in the atmospheric flow (higher obstacles enhance vertical turbulence above the obstacle height). However, obstacles can also displace the atmospheric flow; the displacement height is the height at which the (average) ambient wind speed is negligible. If the displacement height is negligible (flow blockage is not important), contaminant concentration at a fixed downwind distance decreases with increased surface roughness (an order-of-magnitude increase in surface roughness decreases the concentration by roughly a factor of 3).

4. *Wind speed.* Because wind speed varies with height above the ground, the reference wind speed u_r must be specified at a particular height z_r (typically 10 m provided local flow obstructions are not important at this height). For a continuous contaminant release, the contaminant concentration at a fixed downwind distance is roughly inversely proportional to the wind speed. As indicated above, denser-than-air effects are less important at higher wind speeds. The wind speed profile is affected by atmospheric stability, surface roughness z_0, and displacement height d_0 (the height below which the wind speed is negligible). The vertical profile of the wind speed is logarithmic

$$u = \frac{u_*}{k}\left[\ln\left(\frac{z + d_0}{z_0}\right) + \alpha\left(\frac{z}{L}\right)\right] \tag{23-76}$$

for neutral and stable conditions when $z >> z_0$ and d_0; k is von Karman's constant (typically 0.4), u_* is the friction velocity (u_*^2 characterizes the atmospheric vertical mixing rate), L is the Monin-Obukhov length (a measure of stability class), and α is the Monin-Obukhov length coefficient ($\alpha = 5.2$ for stable conditions). For D stability, $L = \infty$ so $z/L = 0$, and the second term in brackets is zero. Since Eq. (23-76) must also hold for u_r and z_r, u_* is proportional to u_r (all other things being equal), so reduction of the wind speed by a factor of 2 reduces the vertical atmospheric mixing by a factor of 4.

5. *Flow obstructions.* In contrast to general terrain characteristics, flow obstructions have length scales much larger than the depth or width of the contaminant cloud or the height of the characteristic wind speed. Flow obstructions can increase or decrease contaminant concentration depending on location. Flow obstructions increase concentration by delaying the dispersal of the contaminant cloud; e.g., inside a dike, concentration is higher, but downwind of the dike, concentrations can be smaller; and the downwind side of flow obstructions can temporarily trap higher concentrations. Obstructions also distort the wind speed profile given above.

Contaminant Interaction with the Atmosphere Contaminant interaction with the atmosphere is important for several reasons:

1. There are chemical reactions between the released contaminant and ambient air or surfaces. If the released contaminant reacts, any reacted material can no longer be considered airborne (although the reaction products may also be hazardous), and so chemical reactions effectively reduce the rate or amount of airborne contaminant. Some reactions can be characterized as dry or wet deposition.

2. Phase changes are typically associated with the evaporation of any suspended liquid phase in an aerosol release. As air is mixed with an aerosol, equilibrium constraints cause additional evaporation of the liquid phase which reduces the temperature of the liquid phase (and the vapor phase if thermal equilibrium is maintained).

3. Ground-to-contaminant cloud heat transfer acts to warm the contaminant cloud if the cloud temperature is lower than ambient. Ground-to-cloud heat transfer can be important for cold clouds at ground level because the buoyancy of the contaminant cloud can be significantly reduced for cold clouds with contaminant molecular weight less than that of air. At higher wind speeds, heat transfer is by forced convection, and even though such conditions produce higher heat-transfer coefficients than do low-wind-speed conditions, heat

transfer is typically more important at low-wind-speed conditions because of two effects: (1) At low wind speeds, the amount of air entrainment is reduced so that the heat-transfer driving force is larger, and (2) at low wind speeds, the contact time between the contaminant cloud and the ground is longer.

Atmospheric Dispersion Models Atmospheric dispersion models generally fall into the categories discussed below. Regardless of the modeling approach, models should be verified that the appropriate physical phenomena are being modeled and validated by comparison with relevant data (at field and laboratory scale). The choice of modeling techniques may be influenced by the expected distance to the level of concern.

1. *Physical or wind tunnel models* Wind tunnel models have long been used to study the atmospheric flow around structures such as buildings and bridges to predict pressure loading and local velocities. Wind tunnel measurement of contaminant concentrations for release scenarios can be used to estimate hazard zones. However, wind tunnel models are generally considered to be incapable of simultaneously scaling mechanical turbulence and thermally induced turbulence (verification issue). Wind tunnel experiments can be very useful when considering validation of mathematical models. Wind tunnel models typically do not account for the lateral variation in wind direction.

2. *Empirical models* Empirical models rely on the correlation of atmospheric dispersion data for characteristic release types. Two examples of empirically based models are the Pasquill-Gifford model (for passive contaminants) and the Britter-McQuaid model (for denser-than-air contaminants) both of which are described below. Empirical models can be useful for the validation of other mathematical models but are limited to the characteristic release scenarios considered in the correlation. Selected empirical models are discussed in greater detail below because they can provide a reasonable first approximation of the hazard extent for many release scenarios and can be used as screening tools to indicate which release scenarios are most important to consider.

3. *First principle mathematical models* These models solve the basic conservation equations for mass and momentum in their form as partial differential equations (PDEs) along with some method of turbulence closure and appropriate initial and boundary conditions. Such models have become more common with the steady increase in computing power and sophistication of numerical algorithms. However, there are many potential problems that must be addressed. In the verification process, the PDEs being solved must adequately represent the physics of the dispersion process especially for processes such as ground-to-cloud heat transfer, phase changes for condensed phases, and chemical reactions. Also, turbulence closure methods (and associated boundary and initial conditions) must be appropriate for the dispersion processes present, especially for denser-than-air contaminants. Regardless of the algorithm for solving the PDEs, any solution must demonstrate resolution independence (i.e., the numerical solution must be independent of grid spacing or time step). Finally, models should be validated against relevant information for the scenario considered. Despite decreased computational costs, such models still require a significant investment for investigating a release scenario.

4. *Simplified mathematical models* These models typically begin with the basic conservation equations of the first principle models but make simplifying assumptions (typically related to similarity theory) to reduce the problem to the solution of (simultaneous) ordinary differential equations. In the verification process, such models must also address the relevant physical phenomenon as well as be validated for the application being considered. Such models are typically easily solved on a computer with typically less user interaction than required for the solution of PDEs. Simplified mathematical models may also be used as screening tools to identify the most important release scenarios; however, other modeling approaches should be considered only if they address and have been validated for the important aspects of the scenario under consideration.

All mathematical models predict (ensemble) time-averaged cloud behavior for a particular set of release conditions. To illustrate in very broad terms, consider a set of trials (field experiments) with continuous contaminant releases (as plumes) that are conducted under identical atmospheric conditions. Suppose that you could measure the concentration on the plume centerline at a given downwind distance with a reasonably fast concentration sensor. Owing to the turbulent nature of the atmosphere and dispersion process, the measured concentration on the plume centerline at a given downwind location observed in each trial would not be identical. If these measurements are averaged during the period for which the contaminant is present, the average of measurements will not change after a sufficient number of trials; this is an ensemble average. This ensemble average reflects the instantaneous (ensemble average) concentration provided the averaging time of the sensor is sufficiently fast. If one considers the difference between any one measured data set and this ensemble average, the measurements will show peak concentrations higher than the average (mean). Peak-to-mean concentration values depend on many factors, but for many purposes, a peak-to-mean concentration ratio of 2 can be assumed. For this hypothetical example, the concentrations were assumed to be measured on the plume centerline. However, the plume centerline of a passive contaminant does not remain at the same ground location due to variation in the wind direction related to large-scale atmospheric turbulence, as illustrated in Fig. 23-44 for a ground-level release on flat, unobstructed terrain. This effect is termed *plume meander.* (Note that denser-than-air contaminant plumes would exhibit less of this effect because such plumes may actually displace the

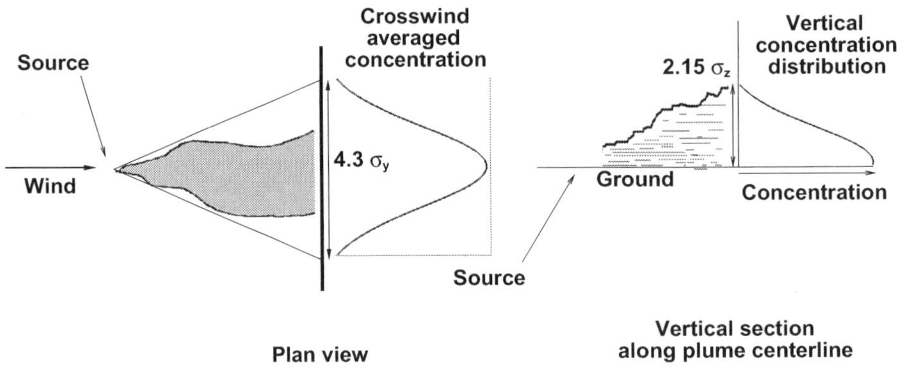

FIG. 23-44 Schematic representation of time-averaged distribution and spread for a continuous plume. σ_y and σ_z are the statistical measures of crosswind and vertical dimensions; $4.3\sigma_y$ is the width corresponding to a concentration 0.1 of the central value when the distribution is of gaussian form (a corresponding cloud height is $2.15\sigma_z$). (*Redrawn from Pasquill and Smith,* Atmospheric Diffusion, *3d ed., Ellis Horwood Limited, Chichester, U.K., 1983*).

atmospheric flow field.) For a fixed ground location, the concentration sensor will be at various locations in the plume (effectively moving in and out of the plume). For a constant average wind direction, concentrations at a fixed ground location will again approach an (ensemble) average over a sufficiently long averaging time (10-min averaging time has proved to be standard). In this case, concentrations are (ensemble-averaged) 10-min concentrations. (Other properties of the concentration distribution are considered in the literature.) For the effect of plume meander, the relationship between concentrations for various averaging times for ground-level plumes can be approximated as

$$\frac{\langle C\rangle_2}{\langle C\rangle_1} = \left(\frac{t_1}{t_2}\right)^p \tag{23-77}$$

where $\langle C\rangle_1$ and $\langle C\rangle_2$ are (ensemble) time-averaged concentrations with averaging times t_1 and t_2 and p is an index typically taken to be around 0.2 for passive plumes. In addition to other more sophisticated approaches, some dispersion models adjust the dispersion coefficients to account for the effect of plume meander. Equation (23-77) shows that concentration decreases as averaging time increases. Based on comparison between passive puff and (10-min) plume coefficients, the averaging time associated with puff coefficients is about 20 s, which is the smallest value that should properly be used in Eq. (23-77). There is no correction to puff coefficients or concentrations associated with plume meander. Note that this is only a rough illustration of the processes present which are discussed in much greater detail in the literature.

Basic Scenario Time Scales There are several competing time scales that are important:

1. The source time scale t_s describes the length of time for the contaminant to become airborne; the source time scale is also limited by the inventory of contaminant.

2. The hazard endpoint time scale t_h describes the length of time required for the contaminant to pose a hazard. There are many different time scales associated with various toxicity levels (e.g., TLV-C ceiling limit values are never to be exceeded, TLV-STEL values are not to be exceeded in a 15-min period, etc.). Time scales associated with flammability hazards reflect the maximum local concentration (and also typically including peak-to-mean concentration ratios) and for reasons discussed above are considered representative of dispersion model averaging times of around 20 s.

3. The travel time t_t is the time required for a contaminant cloud to reach the endpoint distance x_e. As a first approximation, $t_t = x_e/(u_r/2)$ for ground-level clouds; for elevated releases, $t_t = x_h/u$ where u is the characteristic wind speed.

The source time scale t_s must be greater than the travel time t_t for a steady-state plume to be possible. Other time scale restrictions are considered for the models discussed here.

Scenario Development and Simulation The typical procedure for assessing the consequences of an airborne contaminant release is as follows:

1. Identify release scenarios by which containment can be lost along with the hazards of that scenario. Hazards may differ depending on the physical state of the released contaminant and the circumstances of the scenario. Hazards arise from the properties of the released material such as biological agents, toxic materials, or flammable materials; flammable or reactive materials may pose an explosion hazard depending on their reactivity and the degree to which the air-contaminant mixture is confined. Scenarios include a description of the applicable atmospheric conditions (which may be dictated by regulatory requirements).

2. Develop an appropriate source model to define the source description (see previous subsection) for each scenario.

3. Use an appropriate atmospheric dispersion model to assess the consequences or risk of each scenario. For screening purposes, atmospheric dispersion models which are less costly may be used to identify the most important scenarios; examples are discussed in the following subsection. More expensive modeling procedures can be applied to the most important scenarios provided such procedures are more appropriate and accurate. Screening methods may also be useful in considering the validity of more complicated models.

4. Determine if the resulting consequences or risk to people and property is acceptable. For unacceptable scenarios, mitigation measures should be applied such as discussed by Prugh and Johnson (*Guidelines for Vapor Release Mitigation*, AIChE, New York, 1988), and amended scenarios should be reassessed. If mitigation measures cannot sufficiently reduce the consequences or risk, the appropriate business and ethical decision would be to discontinue such operations.

Passive Contaminant (Pasquill-Gifford) Dispersion Models The gaussian dispersion model is based on the assumption of a passive contaminant release (i.e., the released contaminant does not change the atmospheric flow field). Based on the theoretical model of a passive contaminant, the spatial distribution would have a gaussian distribution with characteristic length scales. Using extensive observations of (steady-state) plumes, the Pasquill-Gifford dispersion model correlates characteristic vertical and lateral length scales (or dispersion coefficients σ_z and σ_y, respectively) with atmospheric stability class; other correlations have also been proposed for the plume dispersion coefficients taking other effects into account such as the effect of surface roughness z_0. Less extensive observations of instantaneously released puffs have been used to characterize the length scales of puffs (with the additional length scale σ_x to characterize the along wind direction); σ_x is often assumed to be approximated by σ_y in puff models. Although the Pasquill-Gifford approach provides for the prediction of the concentration distribution, the discussion below is limited to the maximum predicted concentration since this is most important for hazard assessment purposes.

Pasquill-Gifford plume model At a given downwind distance x, the maximum (average) concentration for a (continuous) passive plume from a point source is

$$\langle C\rangle_1 = \frac{E}{\pi\sigma_y\sigma_z u} \tag{23-78}$$

where E is the mass rate at which the contaminant becomes airborne and u is the characteristic wind speed (typically taken to be u_r). Pasquill-Gifford plume dispersion coefficients as a function of downwind distance and atmospheric stability are available from many sources (Seinfeld, *Atmospheric Chemistry and Physics of Air Pollution*, Wiley, New York, 1986; Mannan, *Lees' Loss Prevention in the Process Industries*, 3d ed., Chap. 15, Elsevier Butterworth-Heinemann, Oxford, U.K., 2005; Griffiths, "Errors in the Use of the Briggs Parameterization for Atmospheric Dispersion Coefficients," *Atmospheric Environment*, vol. 28, no. 17, pp. 2861–2865, 1994). Passive dispersion coefficients are typically not provided for distances less than 100 m or greater than a few kilometers because predicted concentrations outside this range must be viewed with some caution (e.g., meteorological conditions may not persist over such large time scales, and at such long distances, large-scale meteorological and terrain features may dictate plume behavior in ways not accounted for by this simple approach).

Note that the predicted values of σ_y and σ_z are sensitive to the *specification of* atmospheric stability. Between D and F stability classes, σ_y for D stability is roughly three times greater than for F stability, and σ_z is roughly two times greater. Since $\langle C\rangle_1$ is inversely proportional to $\sigma_y\sigma_z$, the predicted $\langle C\rangle_1$ for F stability is roughly six times greater than for D stability.

Pasquill-Gifford puff model At a given downwind distance x, the maximum (average) concentration for a (instantaneous) passive puff from a point source is

$$\langle C\rangle = \frac{2E_t}{(2\pi)^{3/2}\sigma_x\sigma_y\sigma_z} \tag{23-79}$$

where E_t is the total contaminant mass that becomes airborne. Pasquill-Gifford puff dispersion coefficients as function of downwind distance and atmospheric stability are available from many sources (e.g., Mannan, *Less' Loss Prevention in the Process Industries*, 3d ed., Chap. 15, Elsevier Butterworth-Heinemann, Oxford, U.K., 2005). As indicated above, σ_x is often assumed to be approximated by σ_y in puff models. As for passive plumes, note that passive puff dispersion coefficients are not provided for distances of less than 100 m (where near source effects will be important), and predicting concentrations for distances longer than a few kilometers must be viewed with some caution for the same reasons cited above.

Note that the predicted values of σ_y and σ_z are sensitive to the *specification of* atmospheric stability for puffs as well. Between D and F stability classes, σ_y for D stability is roughly three times greater than for F stability, and σ_z is roughly 10 times greater. Since $\langle C \rangle$ is inversely proportional to $\sigma_y^2\sigma_z$, the predicted $\langle C \rangle$ for F stability is roughly 40 times greater than for *D* stability.

Virtual sources As indicated above, the gaussian model was formulated for an idealized point source, and such an approach may be unnecessarily conservative (predict an unrealistically large concentration) for a real release. There are formulations for area sources, but such models are more cumbersome than the point source models above. For point source models, methods using a virtual source have been proposed in the past which essentially use the maximum concentration of the real source to determine the location of an equivalent upwind point source that would give the same maximum concentration at the real source. Such an approach will tend to overcompensate and unrealistically reduce the predicted concentration because a real source has lateral and along-wind extent (not a maximum concentration at a point). Consequently, the modeled concentration can be assumed to be bounded above, using the point source formulas in Eq. (23-78) or (23-79), and bounded below by concentrations predicted by using a virtual source approach.

Assuming the source concentration C_s is known, the virtual distance is found by using the known source concentration to find the virtual source distance. For a plume, solve Eq. (23-78) for the product $\sigma_y\sigma_z$, then determine the virtual source distance x_v by iterative solution (or trial and error) using E and C_s. For a puff, solve Eq. (23-79) for the product $\sigma_x\sigma_y\sigma_z$ (or $\sigma_y^2\sigma_z$); then determine the virtual source distance x_v by iterative solution (or trial and error) using E_t and C_s. The dispersion coefficients at distance x_e will now represent a distance from the real source of $x_e - x_v$.

Passive puff or plume In addition to the restriction on plumes discussed above, there is an along-wind dispersion time scale given by $t_d = 2\sigma_x/u_r$ where σ_x is evaluated at the endpoint distance x_e. The release can usually be considered a plume if $t_s > 2.5\ t_d$, where t_s is the source time scale defined above, and the release can be considered a puff if $t_d > t_s$. For $t_d \le t_s \le 2.5\ t_d$, neither puff nor plume models are entirely appropriate; the predicted concentration is considered the largest of the puff and plume predictions.

Denser-than-Air Contaminant (Britter-McQuaid) Dispersion Models Britter and McQuaid (*Workbook on the Dispersion of Dense Gases,* Health and Safety Executive Report 17/1988, Sheffield, U.K., 1988) proposed a correlation for estimating the dispersion of denser-than-air contaminants from area sources for plume and puff releases. Their objective was to produce correlations which predicted the distance to a given concentration level within a factor of 2. Their analysis identified the dominant independent variables as (1) density of released contaminant after depressurization to atmospheric pressure ρ_s; (2) volumetric rate E/ρ_s (or total volume E_t/ρ_s) of contaminant released; (3) characteristic wind speed u_r (typically taken to be at 10-m elevation z_r); and (4) characteristic source dimension D_s. Based at least in part on the fact that presently available field test data for denser-than-air contaminants do not clearly indicate the importance of these parameters, Britter and McQuaid considered some independent variables to be of lesser importance, and these parameters were not considered in the correlation, including surface roughness, atmospheric stability, and exact source dimensions. (Many models for denser-than-air behavior indicate these parameters to be important.) Other effects were not included in the analysis including dilution due to source momentum and condensation of ambient humidity; such effects may be of crucial importance for contaminants that have a molecular weight less than that of air including, e.g., liquefied natural gas (LNG), ammonia, and hydrogen fluoride.

Britter and McQuaid provide correlations for denser-than-air (continuous) plumes and (instantaneous) puffs released at ambient temperature. Since many materials of practical interest are released below ambient temperature, Britter and McQuaid provide guidance as to how to predict the limiting cases for such releases.

Britter and McQuaid report that the averaging time t_1 for the plume correlation is 10 min, and Eq. (23-77) should be used with $p = 0.12$ for other averaging times (and limited to averaging times no shorter than about 20 s as for relases).

Denser-than-air puff or plume Britter and McQuaid use the ratio of the source duration to the travel time to distinguish between plumes and puffs with a slightly different definition of travel time: $t_t = x_e/(0.4u_r)$. The release can be considered a plume if $t_s > t_t$, where t_s is the source time scale defined above, and the release can be considered a puff if $t_s < t_t/4$. For $t_t/4 \le t_s \le t_t$, neither puff nor plume models are entirely appropriate; the predicted concentration is considered the largest of the puff and plume predictions.

Recommended Procedure for Screening Estimates The recommended procedure for making concentration estimates at a specified downwind distance with the simplified models discussed here is as follows:

1. For a given release scenario, estimate the state of the released contaminant after it has depressurized and become airborne (including any initial dilution). The initial mole fraction of hazardous components will be applied to the final reported concentrations and hazardous endpoint concentrations throughout the process. If source momentum is important (as in a jet release or for plume rise), other models are available that can address these considerations. Disregarding the dilution due to source momentum will likely result in higher concentrations downwind, but not always.
2. Consider the important time scales involved, and decide whether a puff or plume model is indicated. If this choice is unclear, assume a plume release.
3. Determine whether denser-than-air behavior is important.
4. When denser-than-air effects are important, use the Britter-McQuaid (plume or puff) models. Otherwise, assume the release is passive and use the Pasquill-Gifford (plume or puff) models. Adjust values for the virtual source correction(s) as appropriate.
5. Adjust values for the averaging time correction for plume predictions with Eq. (23-77). Note that the index p for use in the averaging time correction depends on the model used. If the hazard time scale t_h is different from the model averaging time scale (10 min for plumes), then the predicted concentration should be adjusted to t_h but only if $t_s \ge t_h$; if $t_s < t_h$, then adjust the predicted concentration to t_s.
6. For plume predictions, confirm that plume behavior applies by consideration of appropriate time scales. If plume behavior is not justified, revise the calculations with the puff model and recheck the dispersion time scale. Report the appropriate concentration or distance.

ESTIMATION OF DAMAGE EFFECTS

Nomenclature

A	Projected area of fragment, breach area, or fragment cross-sectional area
a_a	Sound velocity in atmosphere
a_e	Sound velocity in high-pressure gas prior to vessel failure
B	Batch energy availability
d	Fragment diameter
E	Explosion energy available to generate blast and fragment kinetic energy, etc.
E_p	Critical perforation energy ($\frac{1}{2} MV_f^2$)
E_y	Young's modulus of elasticity
F	Dimensionless initial fragment acceleration
F	$P_eAR/Ma_e = P_eR/ma_e$ for vessel completely shattered into many small fragments
g	Acceleration due to gravity
h	Vessel wall thickness
k	Ratio of vessel outside diameter to internal diameter
L	Length of cylindrical vessel
M	Fragment mass
m	Mass per unit area of vessel shell
N	Length of cylindrical vessel forming rocketing tub fragment
N/m²	Unit of pressure in SI system, N/m²; also called pascal (Pa). One psi = 6.89476 × 10³ Pa or 6.89476 kPa.
P	Liquid pressure
P_a	Atmospheric pressure

Nomenclature *(Concluded)*

P_b	Dynamic vessel burst pressure
P_c	Pressure at expanding gas contact surface
P_e	Pressure at vessel failure
P_{inc}	Incident (side-on) blast pressure
P_r	Normally reflected (face-on) blast pressure
R	Vessel radius
r	Fragment radius = $(A/\pi)^{0.5}$
R_g	Range of fragment
t	Steel target thickness
U	$U_f + U_m$
u	Ultimate tensile strength of target steel
U_f	Fluid compression energy
U_m	Elastic strain energy in vessel walls
V	Volume of gas
V_f	Fragment velocity
V_L	Liquid volume
W	Equivalent mass of TNT
w	Unsupported span of steel target
X	Distance from wall of vessel to target
	Greek letters
β_T	Fluid compressibility
γ	Ratio of specific heats of gas C_p/C_v
ϕ_0	Standard steady-state availability
ν	Poisson's ratio of vessel steel
	Subscripts
0	Reference state
1	Initial state
a	Environmental state
$1 \rightarrow a$	Denotes the path from state 1 to the environmental ambient state a

General References: Baker, Cox et al., "Explosion Hazards and Evaluation," *Fundamental Studies in Engineering 5,* Elsevier Science Publishing, New York, 1983. Kinney and Graham, *Explosive Shocks in Air,* 2d ed., Springer-Verlag, New York, 1985. Petes, *Annals, New York Academy of Sciences 1968,* vol. 152, pp. 283–316. Holden, *Assessment of Missile Hazards: Review of Incident Experience Relevant to Major Hazard Plant,* UKAEA SRD/HSE/R477, November 1988. Lees, *Loss Prevention in the Process Industries,* Butterworths, London, 1996. Leslie and Birk, "State of the Art Review of Pressurized Liquefied Gas Container Failure Modes and Associated Projectile Hazards," *Journal of Hazard Materials* **28,** 1991, pp. 329–365. *ASCE Structural Analysis and Design of Nuclear Plant Facilities Manual and Reports on Engineering Practice* no. 58, 1980. Pritchard and Roberts, "Blast Effects from Vapour Cloud Explosions: A Decade of Progress," *Safety Science,* vol. 16 (3,4) 1993, pp. 527–548. "Explosions in the Process Industries," Major Haz. Monograph Series, *I. Chem. E.* (U.K.), 1994.

The availability of energy from an explosion can be approximately calculated in most cases but the method used depends upon the nature of the explosion.

Inert, Ideal Gas-Filled Vessels The energy available for external work following the rapid disintegration of the vessel is calculated by assuming that the gas within the vessel expands adiabatically to atmospheric pressure.

$$E = \frac{P_e V}{(\gamma - 1)} \left\{ \left[1 - \left(\frac{P_a}{P_e} \right)^{(\gamma-1)/\gamma} \right] + (\gamma - 1)\left(\frac{P_a}{P_e} \right)\left[1 - \left(\frac{P_a}{P_e} \right)^{-1/\gamma} \right] \right\} \tag{23-80}$$

In the case of thick-walled HP vessels, the strain energy in the vessel shell can contribute to the available energy, but for vessels below about 20 MN/m² (200 barg) it is negligible and can be ignored. If a Mollier chart for the gas is available, the adiabatic energy can be measured directly. This is the preferred method, but in many cases the relevant chart is not available.

The available energy is dissipated in several ways, e.g., the strain energy to failure, plastic strain energy in the fragments, kinetic energy of the fragments, blast wave generation, kinetic energy of vessel contents, heat energy in vessel contents, etc. For damage estimation purposes, the energy distribution can be simplified to:

$E(\int p\, dv)$

30% blast 40% fragment kinetic energy 30% other dissipative mechanisms

Blast Characteristics Accurate calculation of the magnitude of the blast wave from an exploding pressure vessel is not possible, but it may be estimated from several approximate methods that are available.

One method of estimating the blast wave parameters is to use the TNT equivalent method, which assumes that the damage potential of the blast wave from a fragmenting pressure vessel can be approximated by the blast from an equivalent mass of trinitrotoluene (TNT). The method is not valid for the region within a few vessel diameters from the vessel. However, a rough approximation can be made outside this region by calculating an equivalent mass of TNT and utilizing its well-known blast properties. The term *equivalent mass* means the mass of TNT which would produce a similar damage pattern to that of the blast from the ruptured vessel. The energy of detonation of TNT is 4.5 MJ/kg (1.5×10^{-6} ft·lb/lb), so the TNT equivalent mass W is given by $W = 0.3E/4.5$ kg. Standard TNT data (Dept. of the Army, Navy, and Air Force, "Structures to resist the effects of accidental explosions," TM5-1300, NAVFAC P-397, AFM 88-22, U.S. Gov. Printing Office, vol. 2, November 1990, Figs. 2-7 and 2-15; or Kingery and Pannill, Memorandum Report No. 1518, Ballistic Research Laboratories, Aberdeen Proving Ground, U.S., April 1964) can then be used to determine the blast parameters of interest (Fig. 23-45). This method has limitations in the far field where the peak incident overpressure is less than 4 kN/m² (0.5 psi). In this region, local terrain and weather effects become significant.

The blast parameters also depend upon the physical location of the vessel. If the vessel is located close to or on the ground, then *surface-burst* data should be used. In other circumstances where the vessel is high in the air, either *free-air* or *air-burst* blast data may be used. These data are best presented in the form of *height-of-burst* curves (Petes, "Blast and Fragmentation Characteristics," *Ann. of New York Acad. of Sciences,* vol. 152, art. 1, fig. 3, 1968, p. 287). For incident blast pressures of 3×10^5 N/m² (3 bar) or less, using surface-burst data may overestimate the blast pressure by about 33 percent. Generally, pressure vessel ruptures rarely cause ground craters, so no allowance for cratering should be made.

Fragment Formation The way in which a vessel breaks up into several fragments as a consequence of an explosion or metal failure is

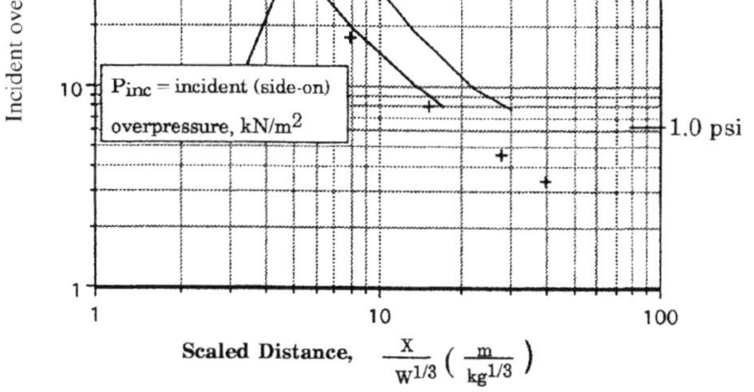

FIG. 23-45 Incident overpressure vs. scaled distance, surface burst. (*The "+" points are from Kingery and Pannill, Memo Report 1518 BRL. Adapted from Department of Army, Navy, and Air Force TM5-1300, NAVFAC P-397, AFM 88-22.*)

impossible to predict. Consequently, in most cases it is necessary to assume several failure geometries and to assess the effect of each. The number of fragments formed is strongly dependent upon the nature of the explosion and the vessel design. For high-speed explosions—e.g., detonations or condensed phase explosions—the vessel frequently shatters into many fragments, but for slower-speed explosions—e.g., deflagrations and BLEVEs—generally fewer than ten fragments are formed, and frequently less than five.

Initial Fragment Velocity (V_f) The process of energy transfer from the expanding gas to the vessel fragments is not efficient and seldom exceeds 40 percent of the available energy. According to Baum ("The Velocity of Missiles Generated by the Disintegration of Gas Pressurized Vessels and Pipes," *Journal of Press. Vessel Technology,* Trans. ASME, vol. 106, November 1984, pp. 362–368), there is an upper limit to the fragment velocity, which is taken to be the velocity of the contact surface between the expanding high-pressure gas and the surrounding atmospheric air. This is referred to as the *zero-mass* fragment velocity and, for most industrial low- to medium-pressure vessels, is less than about 1.3 Mach. It is calculated using ideal gas, one-dimensional shock tube theory and is given by the equation for the shock tube contact surface velocity (Wright, *Shock Tubes,* Methuen & Co., London, 1961).

$$\frac{V_f}{a_0} = -\frac{2}{(\gamma-1)}\left[\left(\frac{P_c}{P_e}\right)^{(\gamma-1)/2\gamma} - 1\right] \qquad (23\text{-}81)$$

where P_c is determined from the relationship:

$$\frac{a_a(1-\mu_a)\left(\dfrac{P_c}{P_a}-1\right)}{a_e\left[(1+\mu_a)\left(\dfrac{P_c}{P_a}+\mu_a\right)\right]^{1/2}} = \frac{2}{(\gamma-1)}\left[1-\left(\frac{P_a}{P_e}\cdot\frac{P_c}{P_a}\right)^{(\gamma-1)/2\gamma}\right] \qquad (23\text{-}82)$$

and

$$\mu_a = \frac{\gamma_a - 1}{\gamma_a + 1} \qquad (23\text{-}83)$$

where P_c = pressure at expanding gas contact surface
a_a = sound velocity at ambient conditions
a_e = sound velocity in gas prior to vessel failure

The value of a_e may be approximated using physical property data for the specific gas at the temperature and pressure at the start of the expansion. Equation (23-82) must be solved using a trial-and-error method. Most fragments never achieve the zero-mass velocity and their velocity can be assessed using the correlations of Baum ("Disruptive Failure of Pressure Vessels: Preliminary Design Guidelines for Fragments Velocity and the Extent of the Hazard Zone," *J. Pressure Vessel Technology,* Trans. ASME, vol. 110, May 1988, pp. 168–176; Baum, "Rupture of a Gas-Pressurized Cylindrical Pressure Vessel. The Velocity of Rocketing Fragments," *J. Loss Prev. Process Ind.,* vol. 4, January 1991, pp. 73–86; Baum, "Velocity of a Single Small Missile Ejected from a Vessel Containing High Pressure Gas," *J. Loss Prev. Process Ind.,* vol. 6, no. 4, 1993, pp. 251–264).

Baum provides correlations for several vessel failure modes.

Vessel Filled with Reactive Gas Mixtures Most cases of damage arise not from the vessel failing at its normal operating pressure but because of an unexpected exothermic reaction occurring within the vessel. This usually is a decomposition, polymerization, deflagration, runaway reaction, or oxidation reaction. In assessing the damage potential of such incidents, the peak explosion or reaction pressure can often be calculated, and if this peak pressure P_e is then inserted into Eq. (23-80), the available energy can be assessed and the blast and fragment hazard determined. Where the expected peak explosion pressure P_e is greatly in excess of the vessel dynamic burst pressure, it is sufficient to increase the burst pressure to allow for the increase in vessel pressure during the period necessary for both the vessel to rupture and the fragments to be removed from the path of the expanding vessel contents. Where the gas pressure in the vessel is rising rapidly, the gas may reach a much higher pressure than the estimated dynamic burst pressure of the vessel. This effect is similar to the accumulation on a relief valve. It is, therefore, conservative to assume that the gas reaches the pressure calculated on the assumption of complete reaction. The reaction is assumed to go to completion before the containing vessel fails. However, there are reactions where it is simpler to calculate the energy availability using thermodynamic methods. The maximum energy released in an explosion can be assessed from the change in the Helmholz free energy ($-\Delta H = -\Delta E + T\Delta S$), but if the required data is not available, it may be necessary to use the Gibbs free energy ($\Delta F = \Delta H - T\Delta S$), which—especially in the case of reactions with little or no molal change, e.g., hydrocarbon/air oxidation—is similar to the Helmholz energy. It may sometimes be more convenient to calculate the batch energy availability [$B = \phi_0 + \Delta\phi_{1\to a} + \Delta(PV)_{1\to a} - P_a\Delta V_{1\to a}$] (Crowl, "Calculating the Energy of Explosion Using Thermodynamic Availability," *J. Loss Prev. Process Ind.,* **5,** no. 2, 1992, p. 109), which for an ideal gas becomes

$$B_1 = f_0 + f_{1\to a} + nRT_1\left[\left(\frac{P_a}{P_1}\right) - 1\right] \qquad (23\text{-}84)$$

The energy partition between blast wave energy and fragment kinetic energy is as described in paragraph 1.

Vessels Completely Filled with an Inert High-Pressure Liquid* A typical example is the pressure testing of vessels with water. The energy available to cause damage is the sum of the liquid compression energy and the strain energy in the vessel shell. The sudden release of this energy on vessel failure generally creates flying fragments but rarely any significant blast effects.

The fluid compression energy up to about 150 MN/m² (22,000 psig) can be estimated from $U_f = \frac{1}{2}\beta_T P^2 V_L$, where β_T is the liquid bulk compressibility, P is the liquid pressure, and V_L is the liquid volume. At higher pressure, this simple equation becomes too conservative and more complex methods of calculating the fluid compression energy are required. The elastic strain energy for cylindrical vessels, ignoring end closures, can be estimated from:

$$U_m = \frac{P^2 V_L}{2E(k^2-1)}\left[3(1-2\nu) + 2k^2(1+\nu)\right] \qquad (23\text{-}85)$$

where P = pressure of liquid
V_L = volume of liquid
E_y = Young's modulus of elasticity
ν = Poisson's ratio

Energy available $U = U_f + U_m$.

Only a small fraction of U is available to provide kinetic energy to the fragments. There are few data available, but in five incidents analyzed by High (unpublished data), no fraction was greater than 0.15. The fragment initial velocity can be assessed from $0.15\ U = \frac{1}{2}MV_f^2$.

Distance Traveled by Fragments There is no method available to estimate the distance traveled by an irregularly shaped, possibly tumbling, subsonic fragment projected at an unknown angle. A conservative approach is to assume that all the fragments are projected at an angle of 45° to the horizontal and to ignore the aerodynamic effects of drag and/or lift. The range R_g is then given by $R_g = V_f^2/g$, where g = gravitational acceleration.

This is too conservative to provide anything more than an upper bound. Some limited guidance is given by Scilly and Crowther ("Methodology for Predicting Domino Effects from Pressure Vessel Fragmentation," *Proc. Hazards Ident. and Risk Anal., Human Factors and Human Reliability in Process Safety,* Orlando, Fla., 15 Jan 1992, p. 5, sponsored by AIChE and HSE), where the range, for vessels with walls less than 20 mm (0.79 in), is 2.8 P_b with the range in meters and P_b as the vessel burst pressure in bars. Other sources are Baker (*Explosion Hazards and Evaluation,* Elsevier, 1983, p. 492) and Chemical

*An excellent review of the necessary precautions to be taken is given by Saville, "Pressure Test Safety," HSE Contract Research Report 168/1998, HMSO, U.K., 1998.

Propulsion Information Agency (*Hazards of Chemical Rockets and Propellants Handbook*, vol. 1, NTIS, Virginia, May 1972, pp. 2-56, 2-60).

Fragment Striking Velocity It is generally impossible to assess the fragment velocity, trajectory, angle of incidence, and fragment attitude at the moment of striking a target; consequently, the conservative view is taken that the fragment strikes the target at right angles, in the attitude to give the greatest penetration, with a velocity equal to the initial velocity.

Damage Potential of Fragments In designing protection for fragment impact, there are two failure modes to be considered: local response and overall response. Local response includes penetration/perforation in the region of the impact. Overall response includes the bending and shear stresses in the total target element; i.e., will the whole target element fail regardless of whether the element is penetrated or perforated?

Local Failure The penetration or perforation of most industrial targets cannot be assessed using theoretical analysis methods, and recourse is made to using one of the many empirical equations. In using the equations, it is essential that the parameters of the empirical equation embrace the conditions of the actual fragment.

The penetrability of a fragment depends on its *kinetic energy density* (KED), given by

$$\text{KED} = \frac{1}{2}\frac{MV_f^2}{A} \tag{23-86}$$

where A is the fragment cross-sectional area. The KED is a useful comparative measure of a fragment's penetrability when comparing like with like. Several equations are given in the following sections.

Ballistics Research Laboratory (BRL) Equation for Steel Targets

$$E_p = 1.4 \times 10^9\,(dt)^{1.5} \tag{23-87}$$

where d is the fragment diameter, t is the steel target plate thickness, and E_p is the critical perforation energy in SI units (kg, m, m/s, J), when applied to fragments between 1 kg and 19.8 kg, impacting targets 1 mm to 25 mm (1 in) thick plate with velocities from 10 m/s to 100 m/s. Neilson (*Procedures for the Design of Impact Protection of Off-shore Risers and ESV's*, U.K. AEA [ed.], 1990) found a large scatter in the results, but most were within ±30 percent.

Stanford Research Institute (SRI) Equation for Steel Targets

$$E = \frac{dut^2}{10.3}\left(42.7 + \frac{w}{t}\right) \tag{23-88}$$

where w is the unsupported span of the target plate (m) and u the ultimate tensile strength of the target steel (N/m²). The parameters for this equation are given by Brown ("Energy Release Protection for Pressurized Systems," part II, "Review of Studies into Impact/Terminal Ballistics," *Applied Mechanics Review*, vol. 39, no. 2, 1986, pp. 177–201) as $0.05 \le d \le 0.25m$, $414 \le u \le 482$ MN/m² for a fragment mass between 4.5 and 50 kg.

Overall Response The transition from local to overall response is difficult to define. High-velocity impact implies that the boundary conditions of the target have little influence on the local response (excluding reflected shock waves). If the fragment is small relative to the target, local response will dominate, but fragments that are of the same order of size as the target will produce an overall response. It is often necessary to consider both overall and local response. Low values of KED are associated with overall response. Design methods for dynamically applied loads are given by Newark ("An Engineering Approach to Blast Resistant Design," ASCE New York, 1953), Baker (see General References), or ASCE (*Manual and Reports on Engineering Practice*, no. 58, 1980).

Response to Blast Waves The effect of blast waves upon equipment and people is difficult to assess because there is no single blast wave parameter which can fully describe the damage potential of the blast. Some targets respond more strongly to the peak incident overpressure and others to the impulse ($\int p\,dt$) of the blast. The blast parameters are usually based on the conservative assumption that the blast strikes the target normal to its surface, so that normal reflection parameters are used.

TABLE 23-25 Damage Effects

Incident, psi	Pressure, kPa	Damage effects
10	70	Damage to most refineries would be severe, although some pumps, compressors, and heat exchangers could be salvaged. All conventional brick buildings would be totally destroyed. Rail wagons (rail cars) overturned. Storage tanks ruptured. Fatalities certain.
5.0	34	Brick buildings severely damaged, 75% external wall collapse. Fired heaters badly damaged. Storage tanks leak from base. Threshold for eardrum damage to people. Domino or knock-on radius. Pipe bridges may move.
2.0	14	Doors and windows removed. Some frame distortion to steel frame buildings and cladding removed. Some electrical/instrument cables broken.*
1.0	7	Lethal glass fragments. Limit for public housing, schools, etc.
0.3	2	About 50% domestic glass broken.

*1% probability electrical cables broken at 2.0 psi inc. 99% probability electrical cables broken at 3.6 psi inc.

The pressure exerted by the blast wave on the target depends upon the orientation of the target. If the target surface faces the blast, then the target will experience the reflected or face-on blast pressure P_r, but if the target surface is side-on to the blast, then the target will experience the incident or side-on blast pressure P_{inc}. The reflected blast pressure is never less than double the incident pressure and can, for ideal gases, be as high as eight times the incident pressure. For most industrial targets where the incident pressure is less than about 17 kN/m² (25 psi), the reflected pressure is not more than 2.5 times the incident pressure.

Response of Equipment The response of equipment to blast is usually a combination of two effects: one is the displacement of the equipment as a single entity and the other is the failure of the equipment itself. The displacement of the equipment is an important consideration for small, unsecured items—e.g., empty drums, gas cylinders, empty containers. Most damage results from the failure in part or totally of the equipment or containing structure itself.

The blast parameters are usually based on the conservative assumption that the blast strikes the target normal to its surface, so that normal reflection parameters may be used.

The response of a target is a function of the ratio of the blast wave duration and the natural period of vibration of the target (T/T_n). Neither of these parameters can be closely defined.

Calculating the specific response of a specific target can generally be done only approximately. Accuracy is not justified when the blast properties are not well defined. A guide to the damage potential of condensed phase explosive blast is given in Table 23-25 (Scilly and High, "The Blast Effect of Explosions," *Loss Prevention and Safety Promotion in the Process Industries*, European Fed. of Chem. Eng., 337 Event, France, September 1986, table 2). Nuclear data is available (Table 23-26) (Walker, "Estimating Production and Repair Effort in Blast-damaged Petroleum Refineries," *Stanford Research Inst.*, July 1969, fig. 5, p. 45), which is based upon long positive-duration blast (±6 s). This suggests that the Walker data will be conservative for the much shorter duration blast from accidental industrial explosions.

A blast incident overpressure of 35 kN/m² (5 psi) is often used to define the region beyond which the damage caused will be minor and not lead to significant involvement of plant and equipment beyond the 35 kN/m² boundary.

Response of People The greatest hazard to people from blast is generally from the deceleration mechanism after people have been blown off their feet and they become missiles. This occurs at an incident overpressure of about 27 kN/m² (4.0 psi) for long positive-duration

TABLE 23-26 Blast Overpressure Effects on Vulnerable Refinery Parts

Equipment	Overpressure (psi)																								
	0.5	1.0	1.5	2.0	2.5	3.0	3.5	4.0	4.5	5.0	5.5	6.0	6.5	7.0	7.5	8.0	8.5	9.0	9.5	10.0	12.0	14.0	16.0	18.0	20.0	>20.0
Control house: steel roof	*a*	*c*	*d*				*h*																			
Control house: concrete roof	*a*	*e p*	*d*				*n*													*e*						
Cooling tower	*b*		*f*				*o*																			
Tank: cone roof		*d*				*k*							*u*													
Instrument cubicle			*a*			*i m*						*t*														
Fired heater				*g*	*i*					*t*																
Reactor: chemical				*a*				*i*					*p*					*t*								
Filter				*h*					*i*										*v*		*t*					
Regenerator						*i*				*i p*					*t*											
Tank: floating roof						*k*							*u*												*d*	
Reactor: cracking							*i*							*i*							*t*					
Pipe supports							*p*					*s o*														
Utilities: gas meter									*q*																	
Utilities: electric transformer									*h*						*l*					*t*						
Electric motor										*h*								*l*								*v*
Blower										*q*										*t*						
Fractionation column											*r*			*t*												
Pressure vessel: horizontal												*p i*						*t*								
Utilities: gas regulator												*i*								*m q*						
Extraction column													*i*							*v*	*t*					
Steam turbine															*l*						*m*	*s*				*v*
Heat exchanger															*i*			*t*								
Tank: sphere																*i*						*i*	*t*			
Pressure vessel: vertical																					*i*	*t*				
Pump																					*i*		*v*			

CODE:

a	Windows and gauges break.	*h*	Debris-missile damage occurs.	*p*	Frame deforms.
b	Louvers fall at 0.3–0.5 psi.	*i*	Unit moves and pipes break.	*q*	Case is damaged.
c	Switchgear is damaged from roof collapse.	*j*	Bracing fails.	*r*	Frame cracks.
d	Roof collapses.	*k*	Unit uplifts (half-filled).	*s*	Piping breaks.
e	Instruments are damaged.	*l*	Power lines are severed.	*t*	Unit overturns or is destroyed.
f	Inner parts are damaged.	*m*	Controls are damaged.	*u*	Unit uplifts (0.9 filled).
g	Brick cracks.	*n*	Block walls fail.	*v*	Unit moves on foundation.
		o	Frame collapses.		

SOURCE: F. E. Walker, "Estimating Production and Repair Effort in Blast Damaged Petroleum Refineries," *SRI*, July 1969.

nuclear weapon blasts. People have more blast resistance than most equipment and can survive incident overpressures of 180 kN/m² (27 psi) (Bowen, Fletcher, and Richmond, DASA-2113, Washington, D.C., October 1968), even for long-duration blasts.

PROJECT REVIEW AND AUDIT PROCESSES

GENERAL REFERENCES: Center for Chemical Process Safety (CCPS), *Guidelines for Hazard Evaluation Procedures, Second Edition with Worked Examples,* AIChE, September 1992. CCPS, *Guidelines for Technical Management of Chemical Process Safety,* AIChE, 1989. CCPS, *Guidelines for Auditing Process Safety Management Systems,* AIChE, 1993.

Introduction Review and audit processes are used in the chemical process industry to evaluate, examine, and verify the design of process equipment, operating procedures, and management systems. These processes assure compliance with company standards and guidelines as well as government regulations. Reviews and audits can encompass the areas of process and personnel safety, environmental and industrial hygiene protection, quality assurance, maintenance procedures, and so on.

A review is a critical examination or evaluation of any operation, procedure, condition, event, or equipment item. Reviews can take many forms and be identified as project reviews, design reviews, safety reviews, pre-start-up reviews, and so on. The following discussion of the review process will deal with project reviews associated with capital projects and focus on the area of process safety.

An audit is a formal, methodical examination and verification of an operation, procedure, condition, event, or series of transactions. The verification element of an audit makes it distinctive from a review. A project review will *recommend* design, procedural, maintenance, and management practices to minimize hazards and reduce risk while meeting company standards and government regulations. An audit will *verify* that the design, the procedures, and the management systems are actually in place, and are being maintained and used as intended. In fact, it is not uncommon for an audit to be done on a review process, to verify that the elements of the review process are being followed.

Project Review Process The scope of capital projects can be large, involving the construction of new plants with new technologies and products, or small, involving minor changes to existing facilities. In either case, project safety reviews can be used to evaluate and examine the process design, operating procedures, and process control scheme for process hazards, conformance to company standards and guidelines, and compliance with government regulations. Some objectives of the review process (CCPS, 1992, p. 53) are: (1) identify equipment or process changes that could introduce hazards, (2) evaluate the design basis of control and safety systems, (3) evaluate operating procedures for necessary revisions, (4) evaluate the application of new technology and any subsequent hazards, (5) review the adequacy of maintenance and safety inspections, and (6) evaluate the consequences of process deviations and determine if they are acceptable (CCPS, 1989, p. 46).

The project review process should be integrated with the development of the project from the conceptual stage to the start-up stage (CCPS, 1989, p. 46). Figure 23-46 depicts the various stages of a capital project. The size and complexity of a project will determine if the project progresses through all these stages and, in the same manner, determine the number and type of reviews that are needed. The earlier in a project that a review can be used to identify required changes, the less costly the change will be to implement.

As the project progresses, more information is available; therefore, the review technique used can be different at each stage of the project. The use of various hazard evaluation techniques, such as checklist analyses, relative rankings, what-if analyses, and hazard and operability studies, is documented in *Guidelines for Hazard Evaluation Procedures: Second Edition with Worked Examples* (CCPS, 1992). The need to use more quantitative techniques for hazard evaluation may be identified during these reviews, and become an action item for the project team.

The project review process involves multiple steps that should be defined in management guidelines (CCPS, 1993, pp. 57–61). The steps include: (1) review policy, (2) review scheduling, (3) review technique, (4) review team representation, (5) review documentation, (6) review follow-up, (7) review follow-up verification, and (8) review procedures change management. These steps define how a review, whether it be a safety review, environmental review, pre-start-up review, or whatever, is conducted and how closure of review action items is achieved.

Review Policy The review policy should establish when project safety reviews should be done. All capital projects, large or small, should have one or more safety reviews during the course of the project. The number and types of review should be stated in a management policy. Any reasons for exceptions to the policy should be documented as well. The policy should address not only projects internal to a company, but also any joint ventures or turnkey projects by outside firms.

Review Scheduling A review scheduling procedure should be established that documents who is responsible for initiating the review and when the review(s) should occur during the project. The scheduling needs to balance availability of process information, review technique used, and the impact of potential review action items on project costs (i.e., early enough to minimize the cost of any potential changes

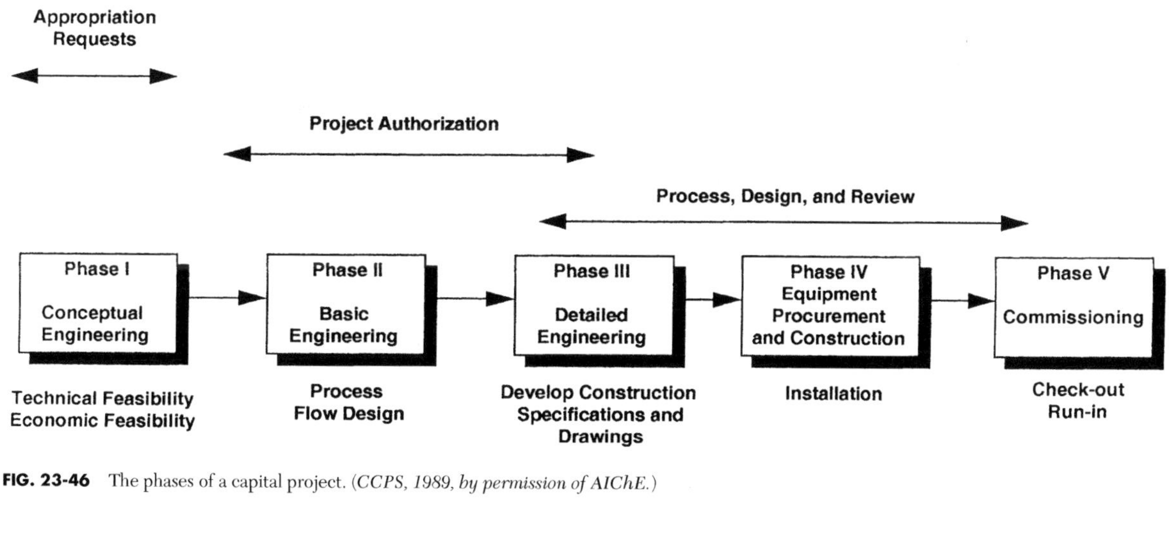

FIG. 23-46 The phases of a capital project. (*CCPS, 1989, by permission of AIChE.*)

to the process). The actual amount of time needed for the review should also be stated in the procedure. On the basis of the number of project reviews required and the estimated time needed for each review, the project cost estimate should include the cost for project reviews as part of the total cost for the project.

Review Team The project review should be conducted by a functionally diverse team. The team should consist of a team leader to organize and lead the team review, a scribe or secretary to record and issue a review summary with action items, and functional experts in fields relevant to the project such as safety, environmental, and industrial hygiene (CCPS, 1993, p. 58). The team leader should be experienced in the use of the selected review technique with leadership skills and no direct involvement with the project under review. The review procedure should address the minimum requirements for team leaders and team members. Some typical requirements could be years of experience, educational background, and training in the review technique. Responsibilities should be clearly defined for initiating the review, assigning the review team, recording the team findings, and monitoring follow-up of team recommendations.

Review Techniques The review techniques used at the various stages of a project should be selected based on the amount of process information and detail available. Figure 23-47 depicts some typical review techniques at the various stages of a capital project. A detailed description, including the type and amount of process information required, for each review technique can be found in *Guidelines for Hazard Evaluation Procedures: Second Edition with Worked Examples* (CCPS, 1992). The process information required for the review should be defined and documented in the review guidelines. Up-to-date and accurate process information is essential to conducting a successful review.

Review Documentation The project review team leader has the ultimate responsibility for documenting the results of the project review. This responsibility may be delegated to a team scribe or secretary to record the review minutes and issue a summary report with listed action items. The action items could address exceptions to company or industry standards and government regulations, review team recommendations based on experience and knowledge, and further issues for study that could not be resolved during the review session.

The summary report should have a standard format and could contain a short project scope summary, a listing of review team members by function, a listing of project team members present, a meeting agenda or checklist of topics reviewed by the team, and a list of concerns and action items for project team follow-up. The distribution list for the summary report should be established and include the review team, project team, and any personnel outside the project team who have follow-up responsibilities for any of the action items. Also, include on the distribution list any appropriate management personnel, whether they be project team supervisors, manufacturing managers, or engineering managers. The documentation for the review should be archived in a process plant file with the appropriate records retention time (e.g., the life of the plant).

Review Follow-up An important element (maybe the most important) of the review process is the follow-up to action items. The project review will result in a list of potential concerns and action items, but, without follow-up, the issues will never be resolved and implemented. A person(s) should be assigned to each action item, preferably at the time of the project review. The person(s) assigned should have a combination of knowledge, resources, and authority to do a proper job in following up on the action items (CCPS, 1993, p. 59). The total action item list should not be assigned to one person, since it may overwhelm one individual. Depending on the number of action items generated, prioritizing the action item list may be helpful and a responsibility the review team can assume.

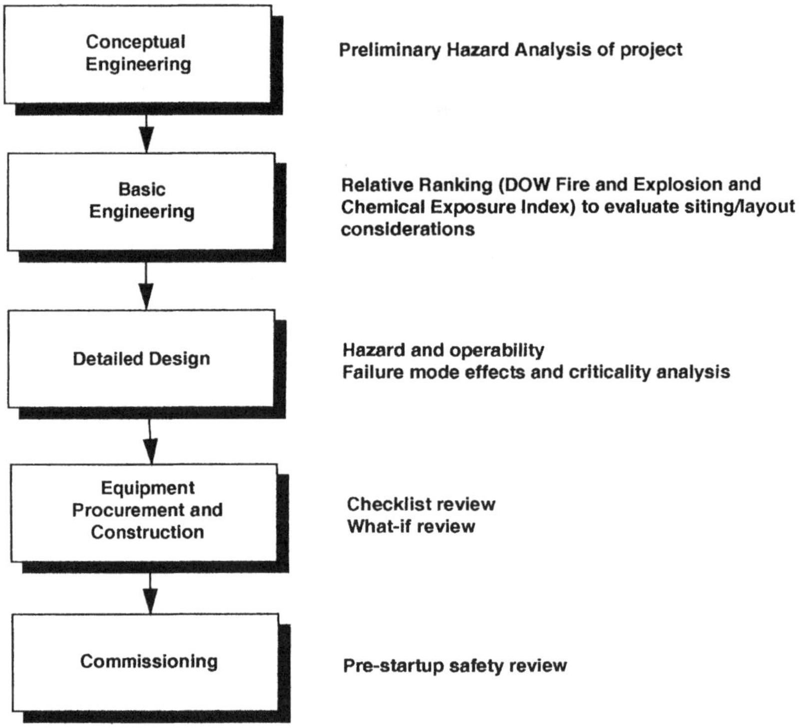

FIG. 23-47 Hazard evaluation at various project stages. (*CCPS, 1993, by permission of AIChE.*)

Progress on the action items should be documented in periodic progress reports to the review team leader or others assigned that functional responsibility. If no one is assigned the responsibility of tracking this progress, completion of the action list will probably be relegated to a lower priority and not be done.

Changes made to the process as a result of the project review may require a similar review before implementation, especially if the change is significant.

Review Follow-up Verification Responsibility should be assigned to verify that any process changes were actually made in the field. This verification can be done by a review team as part of a process pre-start-up review. It could also be part of the project team management responsibility or assigned to a particular functional (i.e., safety and loss prevention) representative. The closure of the review process is complete once implementation is verified.

On rare occasions, the resolution of project review concerns or action items is a point of contention between review team and project team members. In such a case, a management structure must be in place to arbitrate such disputes.

Review Procedure Change Management The project review process can require changes in policy and procedures at certain times. Therefore, the procedures should provide a management-of-change mechanism for suggesting changes and assign a person responsible for initiating and implementing any necessary changes.

Audit Process Audits in the chemical process industry can be focused on process safety, process safety management, environmental, and health areas. The discussion in this section will focus on the process safety and process safety management area, but it should be recognized that the process can be applied to the other areas as well. "Process safety audits are intended to provide management with increased assurance that operating facilities and process units have been designed, constructed, operated, and maintained such that the safety and health of employees, customers, communities, and the environment are being properly protected" (CCPS, 1989, p. 133). Process safety management system audits "provide increased assurance that operating units have appropriate systems in place to manage process risk" (CCPS, 1989, p. 130).

The key steps in the audit process are outlined according to preaudit activities, audit activities, and postaudit activities in Fig. 23-48. These activities are described in detail in *Guidelines for Auditing Process Safety Management Systems* (CCSP, 1993) and will be only briefly discussed in this section.

Preaudit Process Prior to the actual on-site audit, some preliminary activities should take place. These activities include selecting the facilities to be audited, scheduling the audit, selecting the audit team, and planning the audit. The selection criteria may be random, based on potential hazards of the facilities or the value of the facilities from a business standpoint. Audit scheduling must account for the availability of key facility personnel and audit team members, operational mode of the facility (i.e., it should be in normal operation), and the lead time required to obtain background information that may require advance visits to the facility and preaudit interviews. The audit team members should possess the technical training and experience to understand the facilities being audited. They should be knowledgeable in the auditing process and in the appropriate regulations and standards that will apply to the facilities. They should also be impartial and objective about audit findings. The audit plan should define the audit scope (what parts of the facility will be covered, what topics, who will do it, etc.), develop an audit protocol that is a step-by-step guide to how the audit is performed, identify any priority topics for coverage, and develop an employee interview schedule.

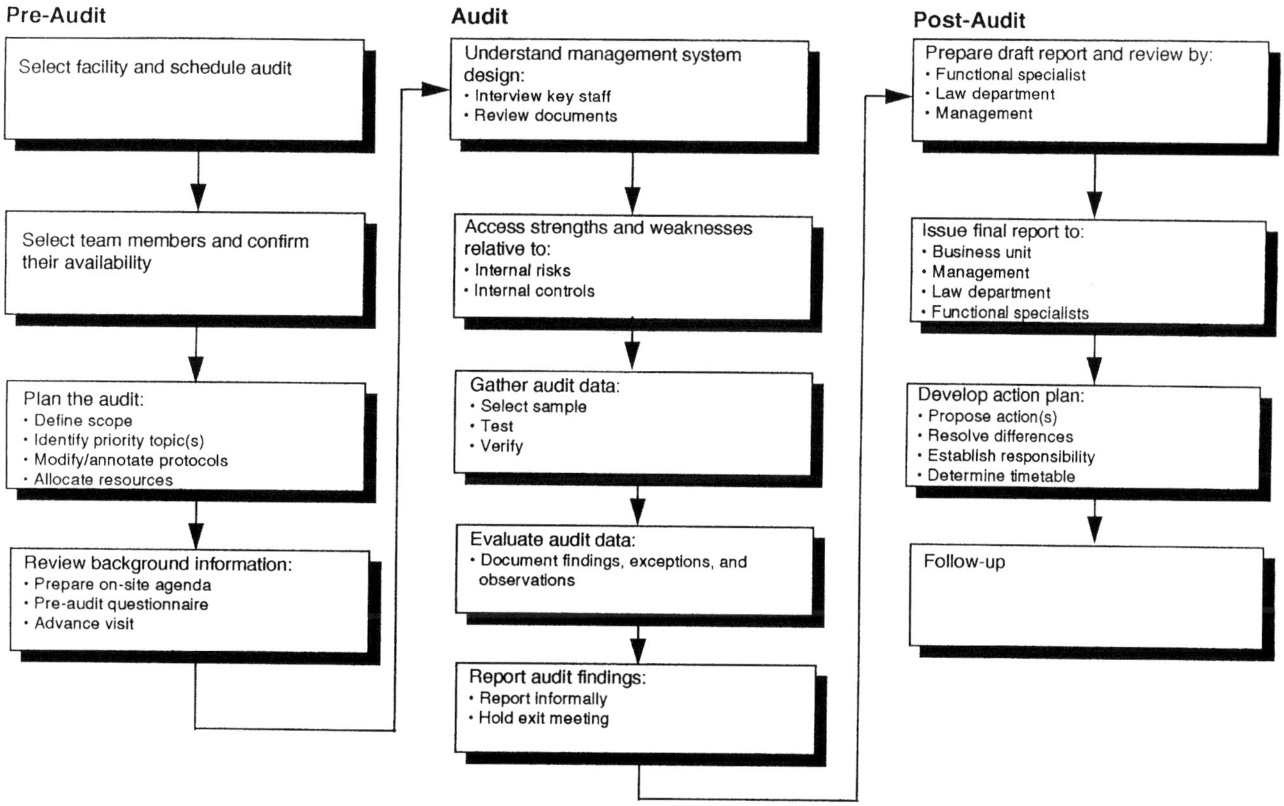

FIG. 23-48 Typical steps in the process safety management audit process. (*CCSP, 1993, by permission of AIChE.*)

On-Site Audit Process An opening meeting with key facility personnel is held at which the audit team covers the objectives and approach for the audit, and the facility personnel provide an overview of the site operations including site safety rules and a site tour. The on-site audit process should then follow five basic steps that include: (1) understanding management systems, (2) evaluating management systems, (3) gathering audit information, (4) evaluating audit information, and (5) reporting audit findings (CCSP, 1993, p. 17).

An understanding of the management systems in place to control and direct the process safety of the facility can be obtained from reading engineering and administrative standards, guidelines, and procedures that should be available in the background information supplied prior to the on-site audit. Informal procedures and guidelines used by the facility may only be discovered in interviews with staff management and operations management. This understanding of the formal and informal management systems is a critical step in the audit process.

The next process step evaluates the process safety management systems to determine if they are adequate to achieve the desired results, and if they are used as intended. This evaluation is highly subjective on the auditor's part. This step sets the stage for the rest of the audit, guiding the auditor's information gathering and focusing attention on critical areas.

Gathering audit data can be accomplished through observations, documents, and interviews. The data obtained is used to verify and validate that the process safety management systems are implemented and functioning as designed. Data gathering can be aided by the use of audit samples, where a representative number of items are audited to draw a conclusion, and by using self-evaluation questionnaires.

The audit data can now be evaluated, resulting in audit findings (i.e., conclusions both positive and negative). The audit team should confirm that sufficient data has been collected to support each finding. Additional data may need to be gathered if the team decides a preliminary finding needs to be strengthened. The conclusions drawn from the data evaluation should be a team consensus.

The reporting step of the on-site audit should be planned to avoid any surprises to facility personnel. Reporting sessions should be held at the end of each audit day to inform facility personnel of the findings, clear up any misunderstandings of the data, and help redirect the audit team, if necessary. The on-site audit should end with a well-planned exit or closeout meeting between the audit team and facility personnel. All the findings of the audit team should be presented at this meeting. This verbal report is the opportunity for clarification of any ambiguities and determination of the final disposition of the findings (written audit report, for local attention only, etc.).

Postaudit Process The postaudit process consists of preparation of a draft report, preparation of a final report, development of action plans, and follow-up. A draft report of the audit findings should be prepared shortly after the completion of the on-site audit. The draft report usually undergoes review and comment by facility personnel involved with the audit, experienced auditors not involved with the subject audit, functional specialists, and attorneys. The review of the draft report is done to assure that a clear, concise, and accurate report is issued, and not to modify or change the findings. Once this review procedure is completed, a final report can be issued and distributed based on a distribution list provided by the facility personnel. The final audit report should be issued in a timely manner and meet the time requirement specified in the audit plan.

An action plan should be developed by the appropriate personnel of the audited facility to address any deficiencies stated in the audit report. Action plans should state what is to be done, who is responsible for getting it done, and when it is to be completed. Rationale for not taking any action for any of the stated deficiencies should also be documented. The action plan is an important step in closing the audit process.

It would not be unusual for some action plans to take a long time to complete. When extended implementation time is necessary, a follow-up mechanism should be used to document progress and show that an effort is being made to resolve the issues. Periodic (i.e., quarterly, semiannually) progress reports should be used as a follow-up method to ensure implementation. Future audits of the facility should include confirmation of the implementation of previous audit action plans.

The final audit report, action plans, progress reports, and any closure report should be retained by the facility based on the facility record retention policy. Typically, these items will be retained until future audit documentation replaces them. In some cases, audit records are retained for the life of the plant.

SAFETY EQUIPMENT, PROCESS DESIGN, AND OPERATION

PRESSURE RELIEF SYSTEMS

General References: Center for Chemical Process Safety (CCPS), *Guidelines for Pressure Relief and Effluent Handling Systems,* American Institute of Chemical Engineers, New York, 1998. DIERS Project Manual, *Emergency Relief System Design Using DIERS Technology,* American Institute of Chemical Engineers, New York, 1992. Leung, "Simplified Vent Sizing Equations for Emergency Relief Requirements in Reactors and Storage Vessels," *AIChE J.* 32(10), pp. 1622–1634 (1986a). Leung, "A Generalized Correlation for One-Component Homogeneous Equilibrium Flashing Choked Flow," *AIChE J.* 32(10), pp. 1743–1746 (1986b). Leung, "Easily Size Relief Devices and Piping for Two-Phase Flow," *Chem. Eng. Prog.* 92(12), pp. 28–50 (1996). See more references under Code, Standards, and Guidelines subsection.

Introduction All process designs should attempt to arrive at an inherently safe facility. Incorporating safety features that are intrinsic (built-in) rather than extrinsic (added-on) to the basic design, together with the use of high-integrity equipment and piping, provides the first lines of defense against the dramatic, often catastrophic effects of an overpressure and subsequent rupture. In recent years, many companies have incorporated the principles of depressurizing or instrumental shutdown of key equipment as a means to control a release and avoid the actuation of pressure relief devices. This minimizes the probability of failure of the device, because, once used, the device may no longer be dependable. Since maintenance of relief devices can be sporadic, this redundancy provides yet another layer of safety. However, regardless of the number of lines of defense and depressurizing systems in place, overpressure protection must still be provided. Emergency pressure relief systems are intended to provide the last line of protection and thus must be designed for high reliability, even though they will have to function infrequently.

Self-actuated pressure relief systems must be designed to limit the pressure rise that can occur as a result of overcompressing, overfilling, or overheating either an inert or a chemically reactive medium in a closed system. Pressure generation is usually the result of either expansion of a single-phase medium (by material addition and/or heating) or a shift of the phase equilibrium in a multiphase medium (as a result of composition and/or temperature changes, particularly in the case of a reactive system). These mechanisms of pressure generation differ from what is commonly referred to as explosion venting. Events such as dust explosions and flammable vapor deflagrations propagate nonuniformly from a point of initiation, generating pressure or shock waves. Such venting problems are not included in these discussions.

Relief System Terminology Refer to API-RP520 Part I for complete terminology.

Accumulation The rise of pressure above the MAWP of the protected system, usually expressed as a percentage of the gauge MAWP. Maximum allowable accumulations are established by applicable codes for emergency operating and fire contingencies.

Backpressure The pressure existing at the outlet of a relief device. The value under no-flow conditions is superimposed backpressure. The value under flowing conditions consists of both superimposed backpressure and built-up pressure due to piping pressure drop.

Blowdown The reduction in flowing pressure below the set point required for a PRV to close.

Design pressure The design pressure used to determine the minimum thickness of a vessel component and possibly used in place of MAWP where the latter has not been established. It is equal to or less than the MAWP. It is the pressure specified on the equipment purchase order.

Maximum allowable working pressure (MAWP) The maximum allowed pressure at the top of the vessel in its normal operating position at the operating temperature specified for that pressure.

Overpressure A pressure increase above the set point during relief flow, usually expressed as a percentage of the differential set pressure.

Pressure relief valve (PRV) A pressure relief device designed to open and relieve excess pressure and to reclose after normal conditions have been restored. *PRV* is a generic term applied to *relief valve* (set up for liquid flow), *safety valve* (set up for gas or vapor flow), and *safety relief valve* (set up for either liquid or compressible flow).

Relieving pressure Set pressure plus the overpressure.

Rupture disk A non-reclosing pressure relief device actuated by static differential pressure and designed to function by the bursting of a pressure-containing disk.

Set pressure The inlet gauge pressure at which a PRV will start to open (or a rupture disk will burst) under service conditions of temperature and backpressure.

Codes, Standards, and Guidelines Industry practice is to conform to the applicable regulations, codes, and recommended practices. In many cases, these will provide different guidelines. A suggested approach would be to review all applicable codes, standards, and recommended practices prior to choosing a design basis. The Design Institute for Emergency Relief Systems (DIERS) was established by AIChE to address sizing aspects of relief system for two-phase, vapor-liquid flashing flow regimes. The DIERS Project Manual (*Emergency Relief System Design Using DIERS Technology*, 1982) and the CCPS Guidelines (*Guidelines for Pressure Relief and Effluent Handling Systems*, 1998) are the generally accepted industry standard for two-phase relief venting.

NFPA 30 and API Standard 2000 provide guidance for design of overpressure protection involving storage tanks that operate at or near atmospheric pressure. In particular, NFPA 30 focuses on flammability issues, while API 2000 addresses both pressure and vacuum requirements. The ASME code (Sections I and VIII) and API RP 520 are the primary references for pressure relief device sizing requirements.

Designers of emergency pressure relief systems should be familiar with the following list of regulations, codes of practice, and industry standards and guidelines in the United States.

API RP 520. *Sizing, Selection, and Installation of Pressure-Relieving Devices in Refineries*. Part I, *Sizing and Selection,* 7th ed., January 2000, and Part II, *Installation*, 4th ed., December 1994. American Petroleum Institute, Washington.

API RP 521, 1997. *Guide for Pressure-Relieving and Depressuring Systems*, 4th ed. American Petroleum Institute, Washington.

API STD 526, 1995. *Flanged Steel Pressure Relief Valves*, 4th ed. American Petroleum Institute, Washington.

API STD 2000, 1998. *Venting Atmospheric and Low-Pressure Storage Tanks, Nonrefrigerated and Refrigerated*, 5th ed. American Petroleum Institute, Washington.

API RP 2001, 1984. *Fire Protection in Refineries*. American Petroleum Institute, Washington.

ASME, 2001. *Boiler and Pressure Vessel Code*, Section I, *Power Boilers*, and Section VIII, *Pressure Vessels*. American Society of Mechanical Engineers, New York.

ASME, 1988. *Performance Test Code PTC-25, Safety and Relief Valves*. American Society of Mechanical Engineers, New York.

CCPS, 1993. *Engineering Design for Process Safety*. American Institute of Chemical Engineers, New York.

CCPS, 1998. *Guidelines for Pressure Relief and Effluent Handling Systems*, American Institute of Chemical Engineers, New York.

DIERS, 1992. *Emergency Relief System Design Using DIERS Technology*, DIERS Project Manual. American Institute of Chemical Engineers, New York.

National Board of Boiler and Pressure Vessel Inspectors, 2004. *Pressure Relieving Device Certifications (Red Book NB-18)*. National Board of Boiler and Pressure Vessel Inspectors, Columbus, Ohio.

NFPA 30, 2000. *Flammable and Combustible Liquids Code*. National Fire Protection Association, Quincy, Mass.

OSHA 1910.106, 2005, *Flammable and Combustible Liquids, Regulations (Standards—29 CFR)*. U.S. Dept. of Labor, Occupational Safety and Health Administration, Washington.

Relief Design Scenarios The most difficult part of designing an adequate emergency pressure relief system lies in determining the emergency events (credible design scenarios) for which to design. The difficulty arises primarily because the identification of credible design scenarios usually involves highly subjective judgments, which are often influenced by economic situations. Unfortunately, there exists no universally accepted list of credible design scenarios. Relief systems must be designed for the credible chain of events that results in the most severe venting requirements (worst credible scenario). Credibility is judged primarily by the number and the time frame of causative failures required to generate the postulated emergency. Only totally independent equipment or human failures should be considered when judging credibility. A failure resulting from another failure is an effect, rather than an independent causative factor. A suggested guideline for assessing credibility as a function of the number and time frames of independent causative events is as follows:

- Any single failure is *credible.*
- Two or more simultaneous failures are *not credible*.
- Two events in sequence are *credible*.
- Three or more events in sequence are *not credible*.

The first step in scenario selection is to identify all the credible emergencies by using the preceding guidelines (or a similar set). This is perhaps best accomplished by identifying all the possible sources of pressure and vacuum. Table 23-27 lists a number of commonly existing pressure and vacuum sources.

Fire The main consequence of fire exposure is heat input causing thermal expansion, vaporization, or thermally induced runaway reaction and decomposition resulting in a pressure rise. An additional result of fire exposure is the possibility of overheating the wall of the equipment in the vapor space where the wall is not cooled by the liquid. In this case, the vessel wall may fail due to the high temperature, even though the relief system is operating. Hence API RP-521 recommends vapor depressurizing facilities for high-pressure services (greater than 17 bar or 250 psig). Guidelines for estimating the heat input from a fire are found in API Recommended Practices, NFPA 30 (for bulk storage tanks), OSHA 1910.106, and corporate engineering standards. In determining the heat input from fire exposure, NFPA allows credit for application of water spray to a vessel; API allows no such credit.

Pressure vessels (including heat exchangers and air coolers) in a plant handling flammable fluids are subject to potential exposure to external fire. A vessel or group of vessels which could be exposed to a

TABLE 23-27 Common Sources of Pressure and Vacuum

Heat Related
- Fire
- Out-of-control heaters and coolers
- Ambient temperature changes
- Runaway chemical reactions

Equipment and Systems
- Pumps and compressors
- Heaters and coolers
- Vaporizers and condensers
- Vent manifold interconnections
- Utility headers (steam, air, water, etc.)

Physical Changes
- Gas absorption (e.g., HCl in water)
- Thermal expansion
- Vapor condensation

pool fire must be protected by pressure relief device(s). Additional protection to reduce the device relief load can be provided by insulation, water spray, drainage, or remote-controlled depressurizing devices. Plant layout should consider spacing requirements, such as those set forth by NFPA, API, Industrial Risk Insurers, or Factory Mutual, and must include accessibility for firefighting personnel and equipment. Several pieces of equipment located adjacent to each other that cannot be isolated by shutoff valves can be protected by a common relief device, providing the interconnecting piping is large enough to handle the required relief load and the relief set pressure is no higher than the minimum MAWP of these pieces of equipment.

Operational Failures A number of scenarios of various operational failures may result in the generation of overpressure conditions:

- *Blocked outlet* Operation or maintenance errors (especially following a plant turnaround) can block the outlet of a liquid or vapor stream from a piece of process equipment, resulting in an overpressure condition.
- *Opening a manual valve* Manual valves which are normally closed to isolate two or more pieces of equipment or process streams can be inadvertently opened, causing the release of a high-pressure stream or resulting in vacuum conditions.
- *Cooling water failure* The loss of cooling water is one of the more commonly encountered causes of overpressurization. Different scenarios should be considered for this event, depending on whether the failure affects a single piece of equipment (or process unit) or is plantwide.
- *Power failure* The loss of power will shut down all motor-driven rotating equipment, including pumps, compressors, air coolers, and vessel agitators.
- *Instrument air failure* The consequences of the loss of instrument air should be evaluated in conjunction with the failure mode of the control valve actuators. It should not be assumed that the correct air failure response will occur on these control valves, as some valves may stick in their last operating position.
- *Thermal expansion* Equipment and pipelines that are liquid-full under normal operating conditions are subject to hydraulic expansion if the temperature increases. Common sources of heat that can result in high pressures due to thermal expansion include solar radiation, steam or other heated tracing, heating coils, and heat transfer from other pieces of equipment.
- *Vacuum* Vacuum conditions in process equipment can develop due to a wide variety of situations, including instrument malfunction, draining or removing liquid with venting, shutting off purge steam without pressurizing with noncondensable vapors, extreme cold ambient temperatures resulting in subatmospheric vapor pressures, and water addition to vessels that have been steam-purged. If vacuum conditions can develop, then either the equipment must be designed for vacuum conditions or a vacuum relief system must be installed.

Equipment Failure Most equipment failures that can lead to overpressure situations involve the rupture or break of internal tubes inside heat exchangers and other vessels and the failure of valves and regulators. Heat exchangers and other vessels should be protected with a relief system of sufficient capacity to avoid overpressure in case of internal failure. API RP 521 presents guidance in determining these requirements, including criteria for deciding when a full tube rupture is likely. In cases involving the failure of control valves and regulators, it is important to evaluate both the fail-open and fail-closed positions.

Runaway Reactions Runaway temperature and pressure in process vessels can occur as a result of many factors, including loss of cooling, feed or quench failure, excessive feed rates or temperatures, contaminants, catalyst problems, and agitation failure. Of major concern is the high rate of energy release in runaway reactions and/or formation of gaseous products, which generally cause a rapid pressure rise in the equipment. To properly assess these effects, the reaction kinetics must be either known or obtained experimentally (see "Reactivity Testing" in the earlier "Chemical Reactivity" subsection for a description of adiabatic calorimeters for runaway reactions). In general, a lower relief set pressure (much below the equipment MAWP) is desirable for these runaway reaction systems in order to relieve the system at a lower reaction rate, since most reactions are Arrhenius in behavior.

TABLE 23-28 Summary of Device Characteristics

	Reclosing devices		Nonreclosing devices
	Relief valves	Disk-valve combinations	Rupture disks
Fluid above normal boiling point	+	+	–
Toxic fluids	+	+	–
Corrosive fluids	–	+	+
Cost	–	–	+
Minimum pipe size	–	–	+
Testing and maintenance	–	–	+
Won't fatigue and fail low	+	+	–
Opens quickly and fully	–	–	+

NOTE: + indicates advantageous
– indicates disadvantageous

Pressure Relief Devices The most common method of overpressure protection is through the use of safety relief valves and/or rupture disks which discharge into a containment vessel, a disposal system, or directly to the atmosphere. Table 23-28 summarizes some of the device characteristics and the advantages.

Safety Relief Valves (SRVs) Conventional safety relief valves are used in systems where built-up backpressures typically do not exceed 10 percent of the set pressure. The spring setting of the valve is reduced by the amount of superimposed backpressure expected. Higher built-up backpressures can result in a complete loss of continuous valve relief capacity. The designer must examine the effects of other relieving devices connected to a common header on the performance of each valve. Some mechanical considerations of conventional relief valves are presented in the ASME code; however, the manufacturer should be consulted for specific details.

Balanced safety relief valves may be used in systems where built-up and/or superimposed backpressure is high or variable. In general, the capacity of a balanced valve is not significantly affected by backpressures below 30 percent of the set pressure. Most manufacturers recommend keeping the backpressure on balanced valves below 45 to 50 percent of the set pressure. Consult API-526 and valve manufacturers for the maximum outlet pressure limit for bellows-type SRVs.

For both conventional and balanced SRVs, the inlet pressure loss, including the mounting nozzle entrance loss, rupture disk flow resistance, and inlet pipe friction, is recommended to stay below 3 percent of the differential set pressure, or else valve instability may occur, resulting in degraded relief capacity.

Pilot-Operated Relief Valves In a pilot-operated relief valve, the main valve is combined with and controlled by a smaller, self-actuating pressure relief valve. The pilot is a spring-loaded valve that senses the process pressure and opens the main valve by lowering the pressure on the top of an unbalanced piston, diaphragm, or bellows of the main valve. Once the process pressure is lowered to the blowdown pressure, the pilot closes the main valve by permitting the pressure in the top of the main valve to increase. Pilot-operated relief valves are commonly used in clean, low-pressure services and in services where a large relieving area at high set pressures is required. The set pressure of this type of valve can be close to the operating pressure. Pilot-operated valves are frequently chosen when operating pressures are within 5 percent of set pressures and a close tolerance valve is required.

Rupture Disks A rupture disk is a non-reclosing device designed to function by the bursting of a pressure-retaining disk. This assembly consists of a thin, circular membrane usually made of metal, plastic, or graphite that is firmly clamped in a disk holder. When the process reaches the bursting pressure of the disk, the disk ruptures and releases the pressure. Rupture disks can be installed alone or in combination with other types of devices. Once blown, rupture disks do not reseat; thus, the entire contents of the upstream process equipment will be vented. Rupture disks of nonfragmented type are commonly used in series (upstream) with a safety relief valve to prevent corrosive fluids from contacting the metal parts of the valve. In addition, this combination is a reclosing system.

The burst tolerances of rupture disks are typically about ± 5 percent for set pressures above 2.76 barg (40 psig). Consult API-520-I (Section 2.3.6.1) on the proper selection and burst setting of the rupture disks.

Pressure Vacuum Relief Valves For applications involving atmospheric and low-pressure storage tanks, pressure-vacuum relief valves (PVRVs) are used to provide pressure relief. Such devices are not included in the scope of ASME Section VIII and thus are not Code-certified; hence they are used mostly in non-Coded vessels (<1.03-barg or 15-psig design). These units combine both a pressure and a vacuum relief valve into a single assembly that mounts on a nozzle on top of the tank and are usually sized to handle the normal in-breathing and out-breathing requirements.

Sizing of Pressure Relief Systems A critical point in design is to determine whether the relief system must be sized for single-phase or two-phase relief flow. Two-phase flow frequently occurs during a runaway, but it can also occur in nonreactive systems such as vessels with gas spargers, vessels experiencing high heat input rates, or systems containing known foaming agents such as latex. The *drift flux* methodology (Zuber and Findlay, "Average Volumetric Concentration in Two-Phase Flow Systems," *Trans. ASME J. Heat Transfer,* pp. 453–468, November 1965; Wallis, *One-Dimensional Two-Phase Flow,* McGraw-Hill, New York, 1969) has been extended and applied to both the volumetric heating case (uniform vapor generation throughout the liquid) and wall heating case (vaporization occurring only at vessel wall) in the DIERS (Design Institute for Emergency Relief Systems) study (DIERS, 1992). The DIERS methodology is important as a means of addressing situations, such as two-phase flow, not covered adequately by current ASME and API methods. The recent CCPS Guidelines (CCPS, 1998) is the best source of updated information on these methods. For a top-vented vessel, the important mechanism for the liquid carryover resulting in two-phase relief is *boilover.* The vessel hydrodynamic model based on drift flux formulation is used to estimate the quality (i.e., vapor mass fraction) entering the vent system. The churn-turbulent regime vessel model is generally reserved for nonfoaming and nonviscous liquids. This regime would yield the highest degree of vapor-liquid disengagement. The bubbly regime vessel model is generally applied to foamy liquids and viscous systems. The bubbly regime would yield only limited disengagement in the vessel. Finally, a conservative but often realistic case for runaway reactive systems is the homogeneous (or uniform froth) vessel model. This model assumes no vapor-liquid disengagement inside the vessel, an idealized assumption, but it leads to much simpler sizing equations.

Required Relief Rate The required relief rate is the vent rate W (kg) required to remove the volume being generated within the protected equipment when the equipment is at its highest allowed pressure:

$$W_{\text{req}} = \frac{\text{net volume generation rate}}{\text{specific volume of vent stream}} \qquad (23\text{-}89)$$

For steady-state design scenarios, the required vent rate, once determined, provides the capacity information needed to properly size the relief device and associated piping. For situations that are transient (e.g., two-phase venting of a runaway reactor), the required vent rate would require the simultaneous solution of the applicable material and energy balances on the equipment together with the in-vessel hydrodynamic model. Special cases yielding simplified solutions are given below. For clarity, nonreactive systems and reactive systems are presented separately.

Nonreactive Systems

Consider constant flow into protected equipment (blocked outlet). For the steady-state design scenario with a constant flow of fluid W_{in} (kg/s) from a pressure source that is above the maximum allowed pressure in the protected equipment, volume is being generated within the equipment at a rate of $W_{\text{in}}/\rho_{\text{in}}$, where ρ_{in} (kg/m^3) is the incoming fluid density evaluated at the maximum allowed pressure. Denoting ρ_{out} (kg/m^3) as the vent stream fluid density, Eq. (23-89) then yields the required vent rate:

$$W_{\text{req}} = \frac{W_{\text{in}}}{\rho_{\text{in}}}\,\rho_{\text{out}} \qquad (23\text{-}90)$$

Consider constant heat input into protected equipment. If the addition of heat to the equipment does not cause the fluid to boil, then the volume generation rate is the thermal expansion rate of the fluid:

$$W_{\text{req}} = \frac{q\beta}{C_p} \qquad (23\text{-}91)$$

where q is the heat input rate (J/s), β is the coefficient of volumetric expansion at constant pressure (°C^{-1}), and C_p is the constant-pressure specific heat [J/(kg·°C)].

The properties are evaluated at maximum allowed pressure conditions. For liquids, β can typically be evaluated from the specific volume change over a 5°C temperature increment. For ideal gases, Eq. (23-91) becomes

$$W_{\text{req}} = \frac{q}{C_p T} \qquad (23\text{-}92)$$

where T is the absolute temperature (K).

If the fluid is at its boiling point, then volume is generated through the phase change that occurs upon vaporization. For *nonfoamy fluids,* vents sized for all-vapor relief are adequate, even if some initial two-phase venting is predicted. The required vent rate based on the volumetric rate balance criterion of Eq. (23-89) is given for a single-component fluid as (Leung, 1986a)

$$W_{\text{req}} = \frac{q}{h_{fg}}\left(\frac{v_{fg}}{v_g}\right) \qquad (23\text{-}93)$$

where h_{fg} is the latent heat of vaporization (J/kg), v_g is the vapor specific volume (m^3/kg), v_{fg} is the specific volume increase upon vaporization $= v_g - v_f$ (m^3/kg), and v_f is the liquid specific volume (m^3/kg).

Near the critical region, the property ratio h_{fg}/v_{fg} can be replaced by $T(dP/dT)_{\text{sat}}$ via the Clapeyron relation since both h_{fg} and v_{fg} are approaching zero. Here $(dP/dT)_{\text{sat}}$ is the slope of the vapor-pressure curve and has units of Pa/K. For multicomponent fluids, Eq. (23–93) is evaluated for each major component (> 10 % wt), and the largest single-component venting requirement is used. Refer to CCPS Guidelines (1998) for more complex schemes.

For *foamy fluids,* the homogeneous vessel model with volumetric heating assumption is used (Leung, 1986a):

$$W(T_p - T_s) = \frac{q}{C_p}\left(\ln\frac{m_0 q v_{fg}}{WVh_{fg}} - 1\right) + \frac{WVh_{fg}}{m_0 C_p v_{fg}} \qquad (23\text{-}94)$$

where W is the vent rate (kg/s), T_p is the temperature at peak pressure (°C), T_s is the temperature at set pressure (°C), m_0 is the initial mass in system (kg), V is the equipment (vessel) volume (m^3), q is the heat input rate (J/s), and C_p is the constant-pressure liquid specific heat [J/(kg·°C).

Note that Eq. (23-94) is implicit in W; a trial-and-error solution method is required. This equation is applicable for single-component or pseudo-one-component systems. For the latter, h_{fg} is defined to be the enthalpy (heat) required to vaporize a unit mass of liquid at equilibrium vapor composition.

Analytical equations have been presented for both the bubbly regime and churn-turbulent regime; see Leung (Leung, "Overpressure During Emergency Relief Venting in Bubbly and Churn-Turbulent Flow," *AIChE J.* 33(6), pp. 952–958, 1987) and Dalessandro (Dalessandro, "Emergency Venting Requirements for Tempered Systems Considering Partial Vapor-Liquid Separation with Disengagement Parameters Greater than Unity, Part I: Model Development, Part II: Application," *Process Safety Prog.* 23, pp. 1–15 and 86–98, 2004).

Reactive Systems Exothermic runaway (uncontrolled chemical) reactions add considerable complexity. Note that external heating (such as a fire) on a vessel containing unstable compounds would fall into this category. The added complications are due to the following:

1. Reaction rate varies exponentially with temperature.
2. Exothermic heat release rates are time-varying.
3. Volatile or noncondensable gas can be generated (in decomposition).
4. Composition change causes shift in boiling point curve.
5. Viscosity increases due to polymerization.

In dealing with such complexity, DIERS has developed a special bench-scale apparatus commercialized by Fauske & Associates, LLC, as the Vent Sizing Package 2 (VSP2™). The design and operation are described in several references [Fauske and Leung, "New Experimental Technique for Characterizing Runaway Chemical Reactions," *Chem. Eng. Prog.,* **80**(8), pp. 39–46, 1985; Leung, Fauske, and Fisher, "Thermal Runaway Reactions in a Low Thermal Inertia Apparatus," *Thermochimica Acta,* **104**, pp. 13–29, 1986; DIERS, 1992; CCPS, 1998]. This adiabatic calorimeter has been extensively used to obtain the above key information for pressure relief design. Upset scenarios such as external heating, overcharge of reactants, or loss of cooling can be directly simulated in this apparatus. Two special-case solutions are given here.

For boiling liquid reactive systems, an analytical method yielding closed-form integral equations for homogeneous vessel two-phase relief has been used for some time (Leung, 1986a). Simplifying assumptions are that (1) the vent rate is taken as constant with time, (2) average values of heat release rate per unit mass and physical properties can be used, (3) there is constant or decreasing system volatility, and (4) a pseudo-one-component approach is taken. The required vent rate is

$$W_{req} = \frac{m_0 C_p (dT/dt)_{av}}{\{[(V/m_0)/(h_{fg}/v_{fg})]^{1/2} + (C_p\,\Delta T)^{1/2}\}^2} \tag{23-95}$$

where $\Delta T = T_p - T_s$ is the "overtemperature" (°C) corresponding to the overpressure ΔP selected, and $(dT/dt)_{av}$ is the average temperature rise rate (°C/s).

The average dT/dt is typically an arithmetic average between the value at set pressure and the value at peak allowed pressure. The properties C_p, h_{fg}, v_{fg} either can be evaluated at the set conditions or can be taken as the average values between the set condition and the peak allowed pressure condition. Alternatively, the term h_{fg}/v_{fg} in Eq. (23-95) can be replaced by $T(dP/dT)_{sat}$ via the Clapeyron relation. This holds reasonably well for a multicomponent system of near constant volatility. Such an application permits direct use of the experimental pressure-temperature data obtained from a closed-system runaway VSP2 test. This form of Eq. (23-95) has been used to demonstrate the advantageous reduction in both vent rate and vent area with allowable overpressure (Leung, 1986a).

For partial disengagement venting models (in bubbly and churn-turbulent regimes), refer to publications by Leung (1987) and Dalessandro (2004).

The special case of **gas generating reactive systems** is for a decomposition reaction with negligible gas solubility in the liquid and no significant latent heat effects. Such a reaction will proceed to the maximum rate whether vented or not. The main purpose of venting is to limit the pressure rise in these "gassy" systems. The maximum gas generation rate is usually obtained from the pressure rise rate data in the DIERS-type apparatus. At the peak allowed pressure P_m, the volumetric gas generation rate for the protected equipment is given by

$$Q_{g,\max} = \frac{m_0}{m_t}\frac{V_t}{P_m}\left(\frac{dP}{dt}\right)_{\max} \tag{23-96}$$

where m_0 is the mass in protected equipment (kg), m_t is the sample mass in test apparatus (kg), V_t is the volume occupied by gas in test apparatus (m^3), P_m is the peak allowed pressure in equipment (Pa, psia), and $(dP/dt)_{\max}$ is the maximum pressure rise rate measured in test apparatus (Pa/s, psi/s). Note that consistent pressure units can be used for P_m and dP/dt here. The required vent rate based on the volumetric rate balance criterion of Eq. (23-89) is given by

$$W^0_{\text{req}} = \frac{Q_{g,\max}}{V/m_0} \tag{23-97}$$

However, this vent rate assumes homogeneous vessel venting at the peak allowed pressure with no prior loss of mass. A less conservative design would consider the mass loss during the overpressure venting interval [Leung, "Venting of Runaway Reactions with Gas Generation," *AIchE J.,* **38**(5), pp. 723–732, 1992].

Relief System Flow Capacity This subsection establishes the relation between relief system configuration and flow capacity in order to determine the size required for a given vent rate. The mass flow rate W through a given relief system geometry, in general, requires a trial-and-error approach when the system configuration contains more than a single diameter. The generalized approach is to assume a flow rate W—this is usually governed by the limiting area such as the orifice or nozzle in a pressure relief valve (PRV) or the rupture disk (RD)—and calculate the resulting pressure profiles down the system until the final discharge pressure matches the specified value. If choked flow is encountered at any area enlargement location in the vent system, then the pressure just after the choking location is determined by calculating back up the larger pipe from a downstream point of specified pressure.

The treatment of vent flow calculations in most typical relief system configurations involves two classes of computational models: flow through nozzles and frictional flow in pipes.

The flow path in well-formed nozzles [ideal (***frictionless***) flow] follows smoothly along the nozzle contour without flow separation. The effects of small imperfections and small frictional losses are accounted for by correcting the ideal nozzle flow by an empirically determined coefficient of discharge K_d. This applies to PRV geometry. The acceleration of a fluid initially at rest to flowing conditions in an ideal nozzle is given by

$$-\frac{G^2 v}{2} = \int_{P_0}^{P_1} v\,dP \tag{23-98}$$

where P_0 is the stagnation pressure of the fluid, P_1 is the static pressure at the minimum flow area called the *throat*, G is the mass flux [kg/(m^2·s)], and v is the fluid specific volume (m^3/kg). If the fluid is compressible, the flow will increase to a maximum value as the downstream pressure is reduced, and any further decrease in the downstream pressure will not affect the flow. This maximum flow condition is referred to as the *critical* (or *choked*) condition. The flow and pressure at this condition are related by

$$G_c = \sqrt{\frac{-1}{(dv/dP)_s}} \tag{23-99}$$

The subscript s denotes an isentropic path for ideal nozzle flow. For *ideal gas* with Pv^k = constant, substitution of this isentropic expansion law into Eq. (23-98) yields the following critical pressure ratio P_c/P_0 and critical flow rate G_c:

$$\frac{P_c}{P_0} = \left(\frac{2}{k+1}\right)^{k/(k-1)} \tag{23-100}$$

$$G_c = \left[k\left(\frac{2}{k+1}\right)^{(k+1)/(k-1)}\right]^{1/2}\sqrt{\frac{P_0}{v_0}} \tag{23-101}$$

where k is the isentropic expansion exponent for ideal gas ($k = C_p/C_v$), P_c is the choking pressure at throat (Pa), P_0 is the stagnation inlet pressure (Pa), and v_0 is the ideal gas specific volume at P_0 (m^3/kg).

Thus for a PRV system, the following relief rate formula (for choked flow) is given in SI units (kg/s, m^2, N/m^2, K) in API-520-I (2000):

$$W = K_d K_b A P_0\left(\frac{M_w}{RT_0 Z}\right)^{1/2}\left[k\left(\frac{2}{k+1}\right)^{(k+1)/(k-1)}\right]^{1/2} \tag{23-102}$$

where K_d is the discharge coefficient of PRV nozzle (orifice), K_b is the capacity correction factor due to backpressure (consult API-520-I), A is the PRV orifice area (m^2), M_w is the molecular weight, R is the gas constant, T_0 is the absolute temperature at inlet (K), and Z is the compressibility. For *nonideal gas* at near thermodynamic critical or supercritical region, special treatments other than simply applying the Z factor as in Eq. (23-102) are necessary [see Leung and Eptein, "A Generalized Critical Flow Model for Non-ideal Gases," *AIChE J.* *34*(9), pp. 1568–1572, 1988; and CCPS, 1998].

The treatment for two-phase relief is more involved due to the presence of slip and thermodynamic nonequilibrium between the two phases. However, for flow capacity evaluation, the traditional homogeneous (no slip) equilibrium model (HEM) is recommended (DIERS, 1992; CCPS, 1998). For a two-phase flashing system (both single-component and multicomponent), the P-v relation can be obtained from adiabatic flash calculations—strictly speaking under constant-entropy, but constant-enthalpy flash computations are adequate for use in numerical integration of Eq. (23-98). A particularly useful simplification is the *omega* method by Leung (1986b; see the earlier subsection "Discharge Rates from Punctured Lines and Vessels" for more discussion), where a P-v relation for a two-phase fluid expansion is written as

$$\frac{v}{v_0} = \omega\left(\frac{P_0}{P} - 1\right) + 1 \qquad (23\text{-}103)$$

For a single-component or a pseudo-one-component fluid, the dimensionless parameter ω is evaluated at the inlet conditions (subscript 0) from (Leung, 1996)

$$\omega = \alpha_0\left(1 - 2\frac{P_0 v_{fg0}}{h_{fg0}}\right) + \frac{C_{p0}T_0P_0}{v_0}\left(\frac{v_{fg0}}{h_{fg0}}\right)^2 \qquad (23\text{-}104)$$

For multicomponent systems with boiling range greater than 80°C, a single adiabatic flash calculation to 80 to 90 percent of the inlet pressure P_0 yields the two-phase specific volume v_1 at pressure P_1, and ω is calculated from (Nazario and Leung, "Sizing Pressure Relief Valves in Flashing and Two-Phase Service: An Alternative Procedure," *J. Loss Prev. Process Ind.* 5(5), pp. 263–269, 1992)

$$\omega = \frac{v_1/v_0 - 1}{P_0/P - 1} \qquad (23\text{-}105)$$

For the case of a noncondensable gas and a nonflashing liquid (a nonflashing two-phase fluid), ω is simply defined by (Leung, 1996)

$$\omega = \frac{\alpha_0}{\Gamma} \qquad (23\text{-}106)$$

where the two-phase isentropic expansion exponent for a thermal equilibrium process is given by

$$\Gamma = \frac{xC_{pg} + (1 - x)C_{pf}}{xC_{vg} + (1 - x)C_{pf}} \qquad (23\text{-}107)$$

Note that for most cases of interest, Γ is close to unity since the flowing gas mass fraction $x << 1$. For the case of frozen flow (i.e., no heat transfer between the two phases), Γ would be replaced by k (or C_{pg}/C_{vg}) in Eq. (23-106). However, the difference between the two limiting cases is small (< 10 percent) in terms of the flow capacity [Leung and Epstein, "A Generalized Correlation for Two-Phase Nonflashing Homogeneous Choked Flow," *Trans. ASME J. Heat Transfer* 112 (May), pp. 528–530, 1990].

With the P-v relation given by Eq. (23-103), Eq. (23-98) can be integrated to give a generalized equation for flow through an ideal nozzle

$$\frac{G}{\sqrt{P_0/v_0}} = \frac{\{-2[\omega \ln (P/P_0)] + (\omega - 1)(1 - P/P_0)\}^{0.5}}{\omega(P_0/P - 1) + 1} \qquad (23\text{-}108)$$

The maximum (critical) or choked flow condition occurs at the critical pressure ratio P_c/P_0 from solving (Leung, 1986a)

$$\left(\frac{P_c}{P_0}\right)^2 + (\omega^2 - 2\omega)\left(1 - \frac{P_c}{P_0}\right)^2 + 2\omega^2 \ln\frac{P_c}{P_0} + 2\omega^2\left(1 - \frac{P_c}{P_0}\right) = 0 \qquad (23\text{-}109)$$

The critical mass flux G_c is given by the dimensionless form

$$\frac{G_c}{\sqrt{P_0/v_0}} = \frac{P_c/P_0}{\sqrt{\omega}} \qquad (23\text{-}110)$$

If $P_c > P_b$, the backpressure, the flow is choked, and Eq. (23-110) yields the critical mass flux. Otherwise, $P_c < P_b$, flow is not choked, and P is equated to P_b in Eq. (23-108) to obtain the unchoked mass flux.

A relief rate formula for two-phase flow similar to Eq. (23-102) for gas can be written as

$$W = K_d K_b A G \qquad (23\text{-}111)$$

For the boiling-liquid system, ω is calculated from Eq. (23-104) or Eq. (23-105), while for a gas-generating system, ω is evaluated from Eq. (23-106).

For **pipe flow,** HEM requires solution of the equations of conservation of mass, energy, and momentum. The momentum equation is in differential form, which requires partitioning the pipe into segments and carrying out numerical integration. For constant-diameter pipe, these conservation equations are as follows:

Mass:

$$G = \text{constant} \qquad (23\text{-}112)$$

Energy:

$$h_0 = h + \frac{G^2v^2}{2} \qquad (23\text{-}113)$$

Momentum:

$$v\, dP + G^2\left(v\, dv + \frac{4fv^2\, dL}{2D}\right) + g \cos\theta\, dL = 0 \qquad (23\text{-}114)$$

where h_0 is enthalpy at stagnation (J/kg), h is enthalpy (J/kg), v is fluid specific volume, f is Fanning friction factor, L is flow length (m), g is acceleration due to gravity, and θ is angle of inclination from vertical.

Equation (23-114), the momentum equation, can be solved more conveniently by rewriting it and solving it numerically

$$\Delta L = -\frac{\bar{v}\,\Delta P + G^2\bar{v}\,\Delta v}{(2f/D)G^2\bar{v}^2 + g\cos\theta} \qquad (23\text{-}115)$$

where $\bar{v}$ is the average specific volume in the pressure increment ΔP and Δv is the incremental specific volume over ΔP.

In the numerical integration of Eq. (23-115) for a given pipe length L, the following steps are suggested (note that detailed thermodynamic properties are required also):

1. Either G is known or guessed.
2. Increments of pressure are taken from the initial to the final pressure.
3. Both $\bar{v}$ and Δv are obtained for each increment for a constant stagnation enthalpy process satisfying Eq. (23-113).
4. For each ΔP taken, ΔL is computed from Eq. (23-115).
5. Total length of pipe L is $\Sigma\Delta L$.
6. If ΔL is negative, then ΔP is too large.
7. A critical flow condition corresponds to $\Delta L = 0$, and the final pressure corresponds to choked pressure.
8. If $\Sigma\Delta L > L$, then G was guessed too small, and steps 1 through 7 are repeated with a larger G. If $\Sigma\Delta L < L$, then G was guessed too large; steps 1 through 7 are repeated with a smaller G.
9. A converged solution is obtained when $\Sigma\Delta L = L$ to within a given tolerance.

The above calculations can be simplified somewhat by ignoring the kinetic energy term in the energy equation, Eq. (23-113), and simply basing calculations on an isenthalpic flash. Thus a P-v relation can be obtained from such flash calculations prior to performing the above iteration steps. In the special case of turbulent pipe flow horizontal discharge ($\cos \theta = 0$), Eq. (23-114) can be integrated based on the omega P-v relation to yield [Leung and Grolmes, "The Discharge of Two-Phase Flashing Flow in a Horizontal Duct," *AIChE J.* 33(3), pp. 524–527, 1987; also errata, 34(6), p. 1030, 1988]

$$4f\frac{L}{D} = \frac{2}{G^{*2}}\left[\frac{\eta_1 - \eta_2}{1-\omega} + \frac{\omega}{(1-\omega)^2}\ln\frac{(1-\omega)\eta_2 + \omega}{(1-\omega)\eta_1 + \omega}\right] - 2\ln\left[\frac{(1-\omega)\eta_2 + \omega}{(1-\omega)\eta_1 + \omega}\left(\frac{\eta_1}{\eta_2}\right)\right] \quad (23\text{-}116)$$

where $\eta_1 \equiv P_1/P_0$, $\eta_2 \equiv P_2/P_0$, $G^* \equiv G/\sqrt{P_0/v_0}$, P_1 is the absolute static pressure at the inlet of the pipe (N/m^2), P_2 is the absolute static pressure at the outlet of the pipe (N/m^2), and P_0 is the reference pressure typically associated with the stagnation pressure in the upstream equipment. The choking criterion at pipe exit takes the form

$$G_c^* = \frac{\eta_{2c}}{\sqrt{\omega}} \quad (23\text{-}117)$$

where subscript c denotes choking. The integrated momentum equation, Eq. (23-116), yields the pressure drop $P_1 - P_2$ in the pipe implicitly as a function of ω, $4fL/D$, and G. For the case of a pipe discharge from a vessel, the inlet acceleration pressure drop $P_0 - P_1$ is given by Eq. (23-108) above with P/P_0 being P_1/P_0 or η_1. Essentially the inlet pipe section is treated as an ideal nozzle, and any irreversible loss due to sharp entrance effect is included in the overall $4fL/D$ term. Likewise, the velocity head loss associated with bends (elbows) is to be included in the $4fL/D$ term as well. For high-velocity, two-phase discharge, an average f value of 0.005 has been used in most applications (Wallis, 1969).

Choked pipe exit: $\eta_{2c} > \eta_b$ or $P_{2c} > P_b$. Equations (23-109), (23-116), and (23-117) are used to solve for G_c^*, η_1, and η_2 (likewise, for G_c, P_1, and P_2) for a given $4fL/D$ and ω value.

Unchoked pipe exit: $\eta_{2c} < \eta_b$ or $P_{2c} < P_b$. Equation (23-116) does not apply. With $\eta_2 = \eta_b$ (or $P_2 = P_b$) for an unchoked exit, only Eqs. (23-108) and (23-116) are needed to solve for G^* and η_1 (likewise, for G and P_1). Under most situations, it suffices to set $\eta_1 = 1$ and to solve for G^* readily by rearranging Eq. (23-116).

This method has been extended to inclined pipe discharge [Leung and Epstein, "The Discharge of Two-Phase Flashing Flow from an Inclined Duct," *Trans. ASME J. Heat Transfer 112* (May), pp. 524–528, 1990], which together with some useful design charts is presented in the earlier subsection "Discharge Rates from Punctured Lines and Vessels."

EMERGENCY RELIEF DEVICE EFFLUENT COLLECTION AND HANDLING

GENERAL REFERENCES: API RP 521, *Guide for Pressure Relieving and Depressurizing Systems,* 5th ed., American Petroleum Institute, Washington, 2007. API STD 537, *Flare Details for General Refinery and Petrochemical Service,* American Petroleum Institute, Washington, September 2003. AIChE-CCPS, *Guidelines for Pressure Relief and Effluent Handling Systems,* AIChE, New York, 1998. DIERS, *Emergency Relief System Design Using DIERS Technology,* AIChE, New York, 1992. Fthenakis, *Prevention and Control of Accidental Releases of Hazardous Gases,* Van Nostrand-Reinhold, New York, 1993. Grossel and Crowl, *Handbook of Highly Toxic Materials Handling and Management,* Marcel Dekker, New York, 1995. Grossel, *Journal of Loss Prevention in the Process Industries* 3(1): 112–124, 1990. Grossel, *Plant/Operations Progress* 5(3): 129–135, 1986. Keiter, *Plant/Operations Progress* 11(3): 157–163, 1992. McIntosh and Nolan, *Journal of Loss Prevention in the Process Industries* 14(1): 17–26. 2001. McIntosh and Nolan, *Journal of Loss Prevention in the Process Industries* 14(1): 27–42, 2001. Schmidt and Giesbrecht, *Process Safety Progress* 20(1): 6-16, 2001.

Introduction In determining the disposal of an effluent vent stream from an emergency relief device (safety valve or rupture disk), a number of factors must be considered, such as

1. Is the stream single-phase (gas or vapor) or multiphase (vapor-liquid or vapor-liquid-solid)?
2. Is the stream flammable or prone to deflagration?
3. Is the stream toxic?
4. Is the stream corrosive to equipment or personnel?

Some vent streams, such as light hydrocarbons, can be discharged directly to the atmosphere even though they are flammable and explosive. This can be done because the high-velocity discharge entrains sufficient air to lower the hydrocarbon concentration below the lower explosive limit (API RP 521, 2007). Toxic vapors must be sent to a flare or scrubber to render them harmless. Multiphase streams, such as those discharged as a result of a runaway reaction, e.g., must first be routed to separation or containment equipment before final discharge to a flare, a scrubber, or a quench pool tank.

This discussion is organized into three major areas: types of equipment, criteria employed in the selection of equipment, and sizing and design of the equipment.

Types of Equipment Equipment for handling emergency relief discharge streams can be divided into two classes:

1. Vapor/gas/solid-liquid separators
2. Final vapor/gas control or destruction equipment

The two most commonly used types of vapor/gas-liquid separators are

- Gravity separators—horizontal and vertical (also called blowdown drums, knockout drums, or catch tanks)
- Emergency cyclone separators

The most commonly used types of final vapor/gas control or destruction equipment are

- Quench pools/catch tanks and quench towers
- Emergency flare systems
- Emergency scrubbers (absorbers)

Gravity Separators Three types of gravity separators are commonly used in the chemical process industries (CPI): horizontal blowdown drum/catch tank, vertical blowdown drum/catch tank, and multireactor knockout drum/catch tank.

Horizontal blowdown drum/catch tank This type of drum, shown in Fig. 23-49, combines both the vapor-liquid separation and holdup functions in one vessel. Horizontal drums are commonly used where space is plentiful. The two-phase mixture usually enters at one end, and the vapor exits at the other end. To overcome reentrainment of liquid droplets, due to high inlet velocities, various devices and piping arrangements are used to provide a more uniform distribution of vapor-liquid mixtures into the separator, as shown in Fig. 23-50. For two-phase streams with very high vapor flow rates, inlets may be provided at each end, with the vapor outlet at the center of the drum, thus minimizing vapor velocities at the inlet and aiding vapor-liquid separation.

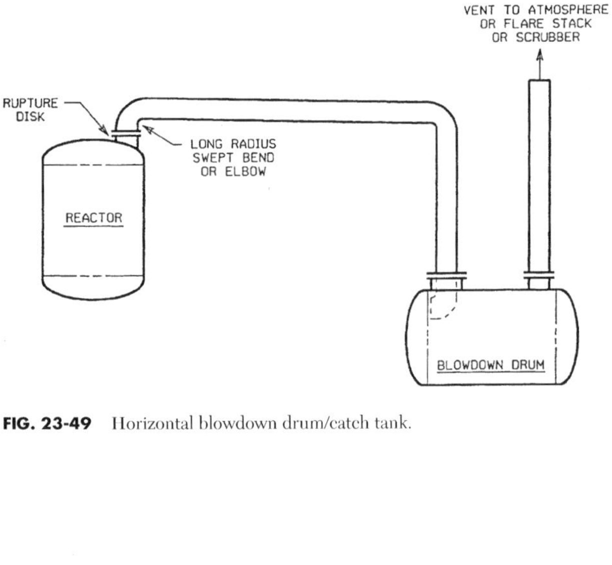

FIG. 23-49 Horizontal blowdown drum/catch tank.

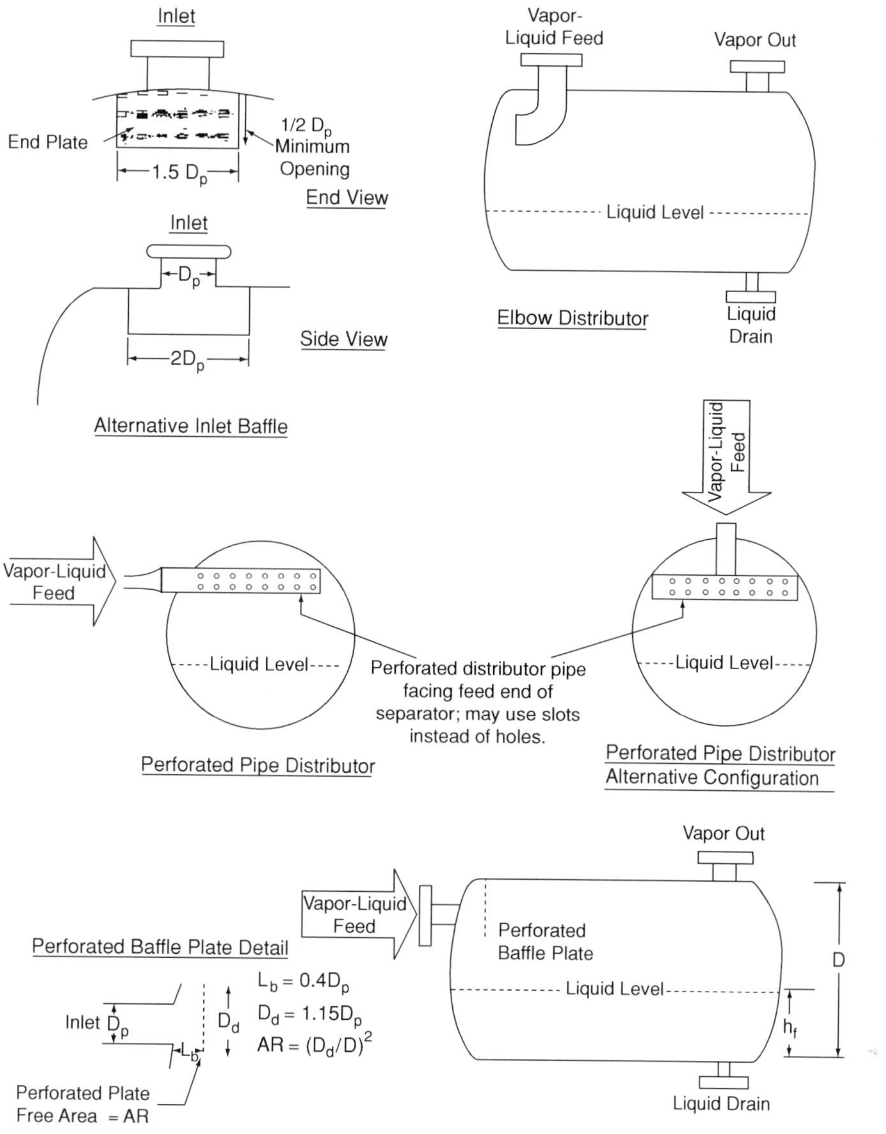

FIG. 23-50 Feed distributor devices for horizontal blowdown drums/catch tanks. (Guidelines for Pressure Relief and Effluent Handling Systems, *Center for Chemical Process Safety (CCPS) of the American Institute of Chemical Engineers (AIChE); copyright 1988 AIChE and reproduced with permission.*).

Vertical blowdown drum/catch tank This type of drum, shown in Fig. 23-51, performs the same function and operates on similar principles as horizontal separators. These separators are usually used where horizontal space is limited. The two-phase mixture enters the vessel via a nozzle on the vertical shell and is distributed by an inlet baffle chamber.

Multireactor knockout drum/catch tank This interesting system is sometimes used as the containment vessel for a series of closely spaced reactors (Speechly et al., "Principles of Total Containment System Design," presented at I.Chem.E. North West Branch Meeting, 1979).

Emergency Cyclone Separators Two types of emergency cyclone separators are commonly used, those with a separate catch tank and those with a catch tank integral with the cyclone section.

Cyclone separator with separate catch tank This type of blowdown system, shown in Figs. 23-52 and 23-53, is frequently used in chemical plants where plot plan space is limited. The cyclone performs the vapor-liquid separation, while the catch tank accumulates the liquid from the cyclone. This arrangement allows location of the cyclone knockout drum close to the reactor so that the length of the relief device discharge line can be minimized. The cyclone has internals, vital to its proper operation, which are discussed in the following subsections.

Cyclone separator with integral catch tank This type of containment system, depicted in Fig. 23-54, is similar to the aforementioned type, except that the knockout drum and catch tank are combined in one vessel shell. This design is used when the vapor rate is quite high so that the knockout drum diameter is large.

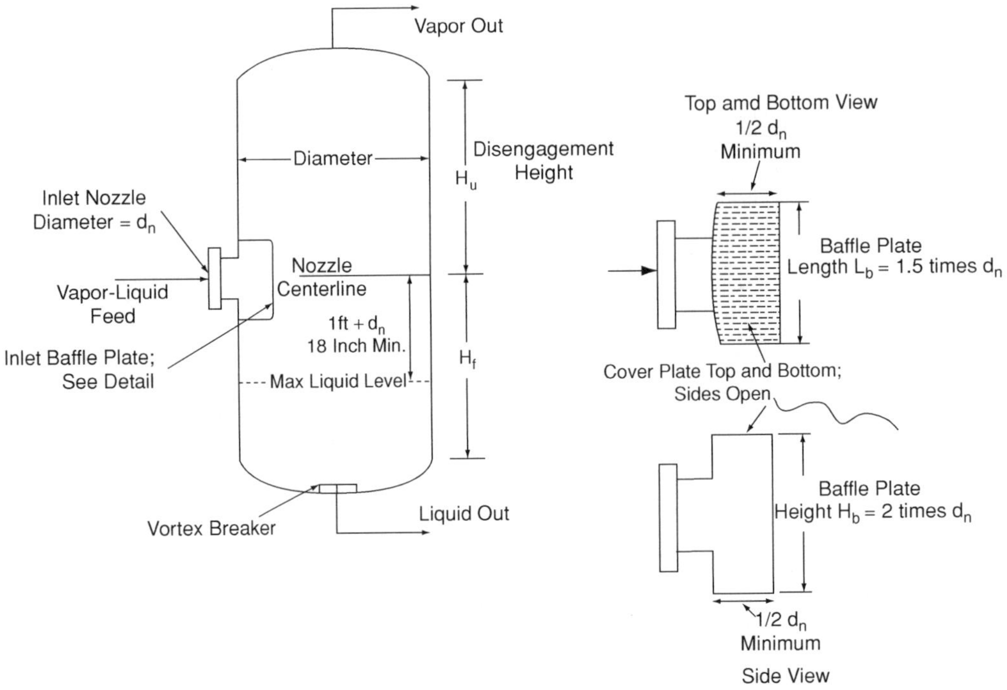

FIG. 23-51 Vertical blowdown drum/catch tank. (Guidelines for Pressure Relief and Effluent Handling Systems, *Center for Chemical Process Safety (CCPS) of the American Institute of Chemical Engineers (AIChE); copyright 1988 AIChE and reproduced with permission.*).

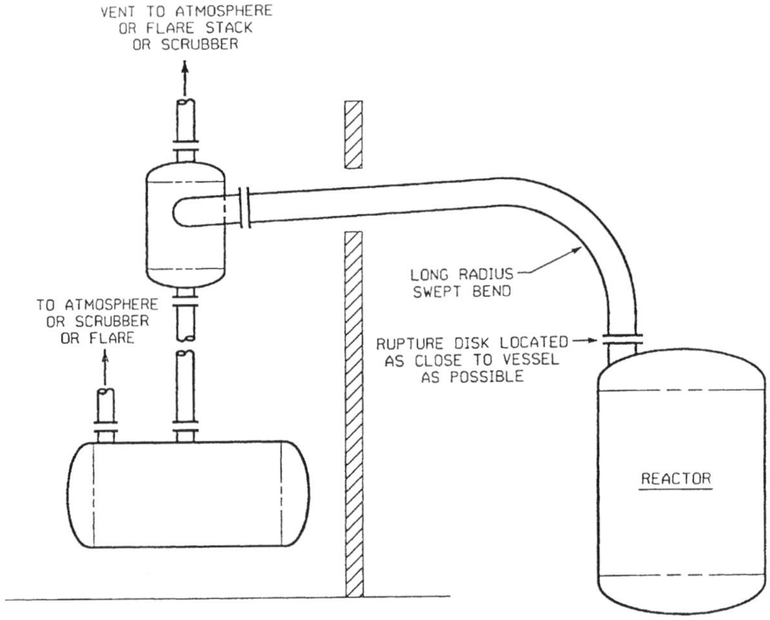

FIG. 23-52 Cyclone separator with separate liquid catch tank.

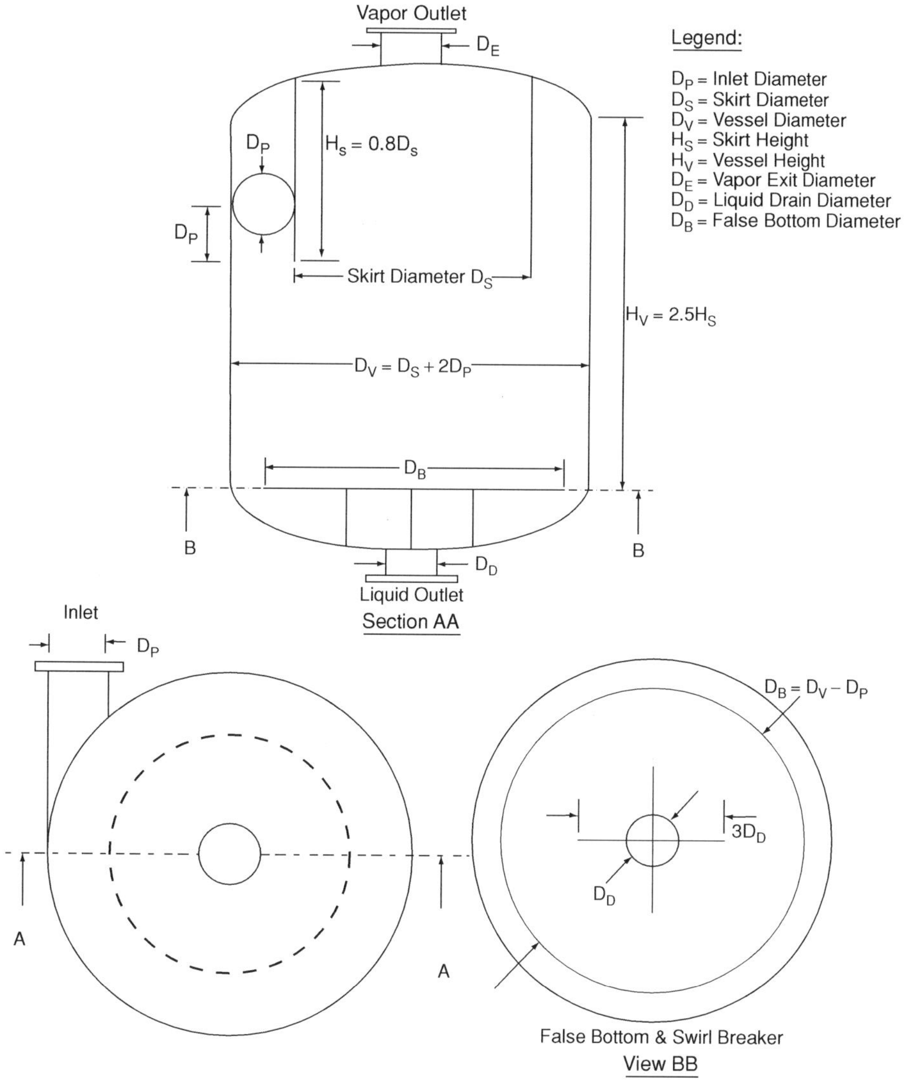

FIG. 23-53 Cyclone separator design details. (Guidelines for Pressure Relief and Effluent Handling Systems, *Center for Chemical Process Safety (CCPS) of the American Institute of Chemical Engineers (AIChE); copyright 1988 AIChE and reproduced with permission.*).

Quench Pools/Catch Tanks and Quench Towers Two types of quench systems are commonly used: the quench pool/catch tank (also called passive scrubber) and quench tower.

Quench pool/catch tank This type of system, as shown in Fig. 23-55, is used to condense, cool, react with, and/or collect a mixture of liquid and vapors discharging from a relief device by passing them through a pool of liquid in a vessel. Feed vapor and liquid (if present) are sparged into the pool of cool liquid, where the vapors are condensed and the liquid is cooled. If the feed materials are miscible with the pool liquid, they mix with and are diluted by the pool liquid; if not, the condensate, feed liquid, and pool liquid separate into layers after the emergency relief event is over. The condensed vapors, feed liquid, and quench liquid are contained in the vessel until they are sent to final disposal.

Quench pools can be used for the neutralization or killing of reactive materials. A chemical reaction can be quenched by the cooling effect in the pool, by the dilution effect if the reaction mass is miscible with the pool liquid, or by a killing or neutralizing agent added to the quench liquid. For example, discharge of chlorine or acid mixtures can be neutralized by adding sodium hydroxide to the quench liquid. However, systems involving slow to moderate reaction rates usually require more than one contacting stage, and a quench pool (equal to one stage only) may not be adequate to completely neutralize or kill the reactant in the entering relief device effluent stream. In this case, a multiple-stage emergency scrubber will be needed.

Depending on the relief stream components (with or without noncondensable gases) and quenching efficiency, this arrangement often obviates the need for a subsequent scrubber and/or flare stack. The design of the quencher arm is critical to efficient condensation and avoidance of water hammer. Figure 23-55 is the more conventional passive type quench pool used in the chemical and nuclear industry.

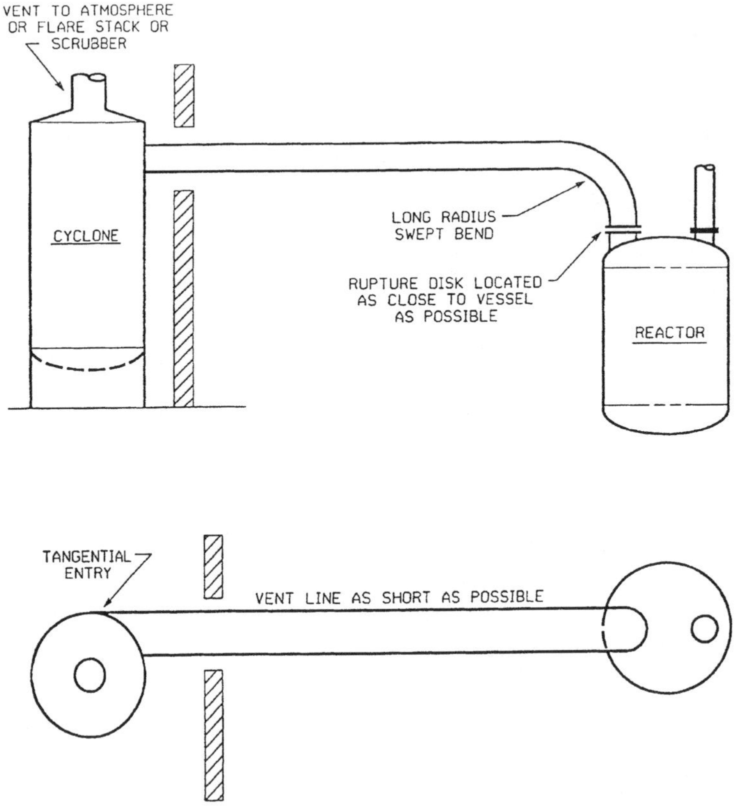

FIG. 23-54 Cyclone separator with integral catch tank.

Quench tower The type shown in Fig. 23-56, with a superimposed baffle plate section, is often used when complete condensation of the incoming vapors is not required. The exiting vapors are usually cooled to 150 to 200°F in the baffle plate section. This type is often used in petroleum refineries.

Emergency Flare Systems Single-phase effluent streams (gases or vapors) from pressure relief devices are often sent directly to emergency flares. The vapor/gas streams from gas-liquid separators are also often sent to emergency flares. Emergency flares are similar in design to normal process flares (continuous flaring), but they are only used to handle emergency releases (intermittent operation).

Flares can be grouped into two major categories, i.e., vertical and horizontal.

Vertical flares These are single-burner combusters generally oriented to fire upward. The discharge point is in an elevated position relative to the surrounding grade and/or nearby equipment. Three types of support methods are used for vertical flares:

- *Self-supported.* A mechanically and structurally designed riser supports the flame burner.
- *Guyed.* An elevated flare with the riser is supported by cables. Cables are attached to the flare riser at one or more elevations to limit the deflection of the structure. The cables (guy-wires) are typically positioned in a triangular plan to provide strong support.
- *Derrick supported.* Support is provided by a steel trussed structure that supports one or more flare risers.

The main components of an elevated flare system are the flare burner with or without smoke suppression capability, pilot(s), pilot igniter(s), support structure, and piping. A number of optional features are available, such as pilot flame detectors; air seals (buoyancy or velocity type); knockout drum; flame/detonation arrester; liquid seal; smoke suppression control system; blower(s); flow, composition, heat content, or video monitoring; ladders (caged or with safety-climbing system) and platforms; davit for tip removal; aircraft warning lights and painting; radiation heat shields; and rain shields. Detailed information about these components can be found in API STD 537.

Smokeless flaring is required by law in the United States (40 CFR 60.18, Chap. 1) for normal process flares (continuous flaring). However, smokeless flaring is not required by the EPA for emergency flaring, but local conditions and regulations may require smokeless flaring.

Air seals (also called purge reduction or gas seals) are often recommended to prevent air from entering the stack, which could possibly cause flashbacks and explosions. More information about air seals can be found in API RP 521 (2007) and the AIChE-CCPS book *Guidelines for Pressure Relief and Effluent Handling Systems.*

Horizontal flares A horizontal flare can be either an enclosed ground flare or an open field ground flare (also called a matrix flare).

Enclosed ground flares are most commonly used as a supplement to an elevated flare on the same relief system. The primary reason for an enclosed ground flare is to reduce the visual impact of flared gas combustion on a nearby community. They are often used when it is desirable that all or part of a flare load be disposed of in a way that causes the minimum of disturbance to the immediate locality. They offer many advantages in comparison to elevated flares; there is no smoke, no visible flame, no odor, no objectionable noise, and no thermal radiation (heat shield) problems. Enclosed ground flares are typically used for normal process flow (continuous) flaring, but with recent technical advances they are now also used for emergency flaring (AIChE-CCPS, 1998).

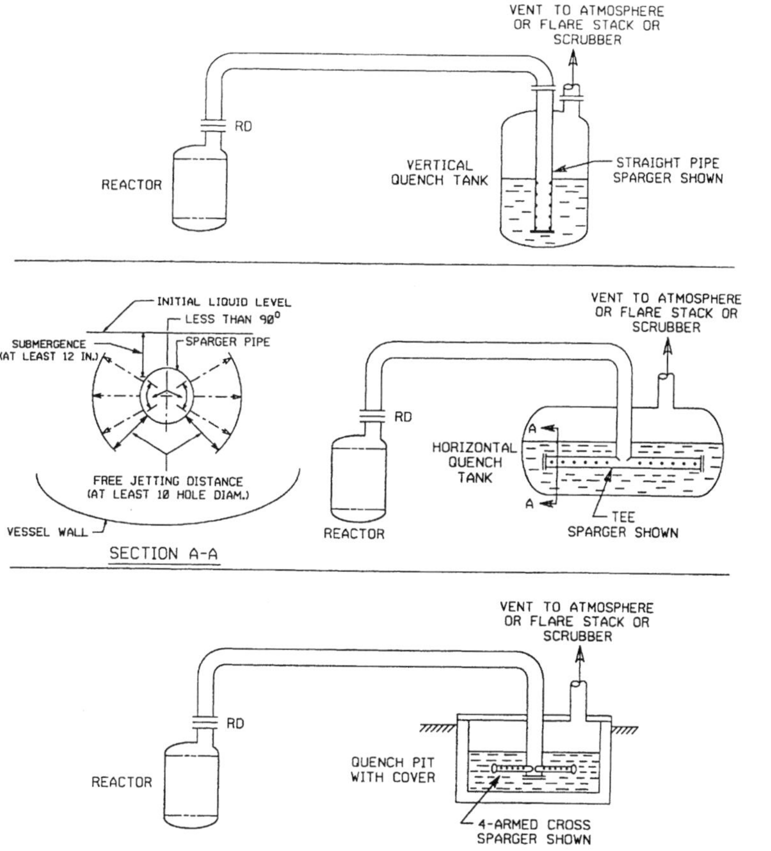

FIG. 23-55 Quench tank/catch tank.

Enclosed ground flares burn the effluent gas in multiple specially designed multitip burners placed as near to the ground as practicable to ensure good operation. The resulting flames are hidden from sight by a surrounding wall or chamber. The top of the chamber is open to the atmosphere, and allowance is made in the bottom of the chamber to permit ingress of air for combustion. Overall heights can vary according to the needs of the specific location. In some locations, a low profile of perhaps 10 m may be mandated, while in others an elevated discharge may be needed to ensure a satisfactory downwind trajectory for the flue gases. It is common for the chamber to be surrounded by a wind fence to modify the effect of crosswinds on the combustion process and to prevent unauthorized access.

An enclosed ground flare system has a number of key components: combustion chamber, burners, piping system, wind fence, and operational and safety controls. More details about enclosed ground flare system components are given in API STD 537.

Enclosed ground flares can be used to burn toxic gases, but since any noncombusted release would be near the ground, special safety dispersion systems and more stringent monitoring and control instrumentation are usually provided in this application.

There is a practical economical limit to the size of enclosed ground flares. If the maximum effluent relief rate exceeds 100 tons/h, the feasibility of an enclosed ground flare becomes questionable, and at 500 tons/h, it is impractical (AIChE-CCPS, 1998).

Open field ground flares (also called **matrix flares**) are sometimes used in lieu of elevated flares, most commonly, although not exclusively, in remote locations with relatively low population density. A primary reason for use of an open field ground (matrix) flare is to reduce the visual impact of flared gas combustion in the manner of an enclosed ground flare. The open field ground (matrix) flare, however, has a significantly larger capacity than could be practically handled in a furnacelike structure, and the visual shielding is provided by tall fencing located some distance away from the burners themselves so that the fence encloses a small "field."

An open field (matrix) ground flare consists of a number of staged multiburners mounted on a series of horizontal pipes, arranged in manifolds, in a large field or pit. Smokeless or nonsmokeless flaring can be provided. The main structural and physical design issues are concerned with protecting the piping and manifolds inside the flare field from the effects of high radiant heat loads, while still permitting normal thermal expansions. More information about ground flare components is given in API STD 537.

Emergency Scrubbers (Absorbers) Emergency relief effluent streams are often sent to scrubbers (also called absorbers, columns, or towers) for final disposal by absorption of the gas into a solvent. Some gases or vapors can be removed by physical absorption. Other gases or vapors can be removed by chemical absorption (reaction of the vent gas/vapor into a liquid that reacts with it).

A typical emergency scrubber system consists of a scrubbing column (often filled with packing), recirculating liquid pumps, solvent cooler, and in some cases (where the entering effluent gas/vapor is at a low pressure) exhaust blowers (see Fig. 23-57). Redundant equipment and instrumentation is usually provided to ensure reliable scrubber operation at all times.

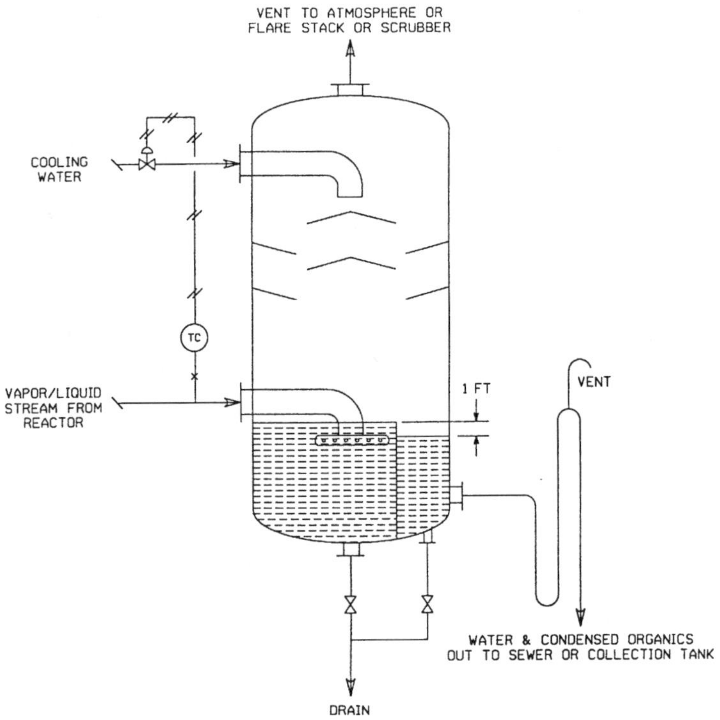

FIG. 23-56 Quench tank with direct-contact baffle tray section.

Equipment Selection Criteria and Guidelines A number of factors should be considered to determine when to select which type of vapor/gas/solid-liquid separator (blowdown drum or cyclone separator) to handle a multiphase stream from a relief device, and which final control or destruction equipment (emergency flare, emergency scrubber, or quench pool/catch tank) should be selected. These factors include the plot plan space available, the operating limitations of each type, and the physicochemical properties of the stream.

The selection of vapor/gas/solid-liquid separators and final control and destruction equipment can be helped by considering the advantages and disadvantages of each, as given below.

Gravity Separators (Horizontal and Vertical Drums/Catch Tanks)

Advantages:
- History of good performance.
- Passive.
- Excellent turndown up to the maximum design flow; lower vapor flow improves separation effectiveness.
- Low pressure drop.
- Can handle high liquid loading and provide large holdup of liquid.
- Can be combined with secondary vapor-liquid separator such as vane impingement type.
- Can handle high-viscosity liquids.
- Moderate cost.

Disadvantages:
- May be large.
- Cannot economically remove low-density (e.g., light hydrocarbon) droplets smaller than about 150 μm.
- Cannot handle stable foams.
- Reacting liquids may continue to react in separator.

Cyclones

Advantages:
- Can remove over 99 percent above 20-μm droplet diameter in high liquid-loading situations.
- Some designs able to handle foamy liquids under some conditions.
- Passive.
- May be able to handle streams containing solids or with fouling tendencies.
- Relatively low capital cost.

Disadvantages:
- Performance not as well defined for emergency services as normal process cyclones.
- Little data on handling foamy or high-viscosity mixtures.
- May require higher pressure drop than gravity separators.

Quench Pool Tanks/Catch Tanks

Advantages:
- Passive, but must have system to verify proper liquid level, and composition if pool liquid is used as a neutralizing or kill liquid.
- Can handle two-phase, foamy, and high-viscosity liquids.
- Can handle high-flow-rate discharges.
- Can handle high liquid loading with special sparger design.
- Can contain reaction-stopping or -neutralizing chemicals and antifoam agents.
- Complete containment possible at low levels of noncondensables.
- Can use variety of quench liquids.

Disadvantages:
- Pilot-plant demonstration or scale-up test possibly needed.
- Requires sufficient discharge pressure to provide adequate pressure drop (at least 10 psig) across sparger and to overcome static head of liquid in pool.
- Sparger piping to be supported to control vibration and the suddenly applied load from the relief discharge.
- Some proven sparger designs are proprietary.
- May not be suitable for systems with slow to moderate reaction rates when used as an absorber.
- If vented, entrainment and vapor may be carried out of the vessel.

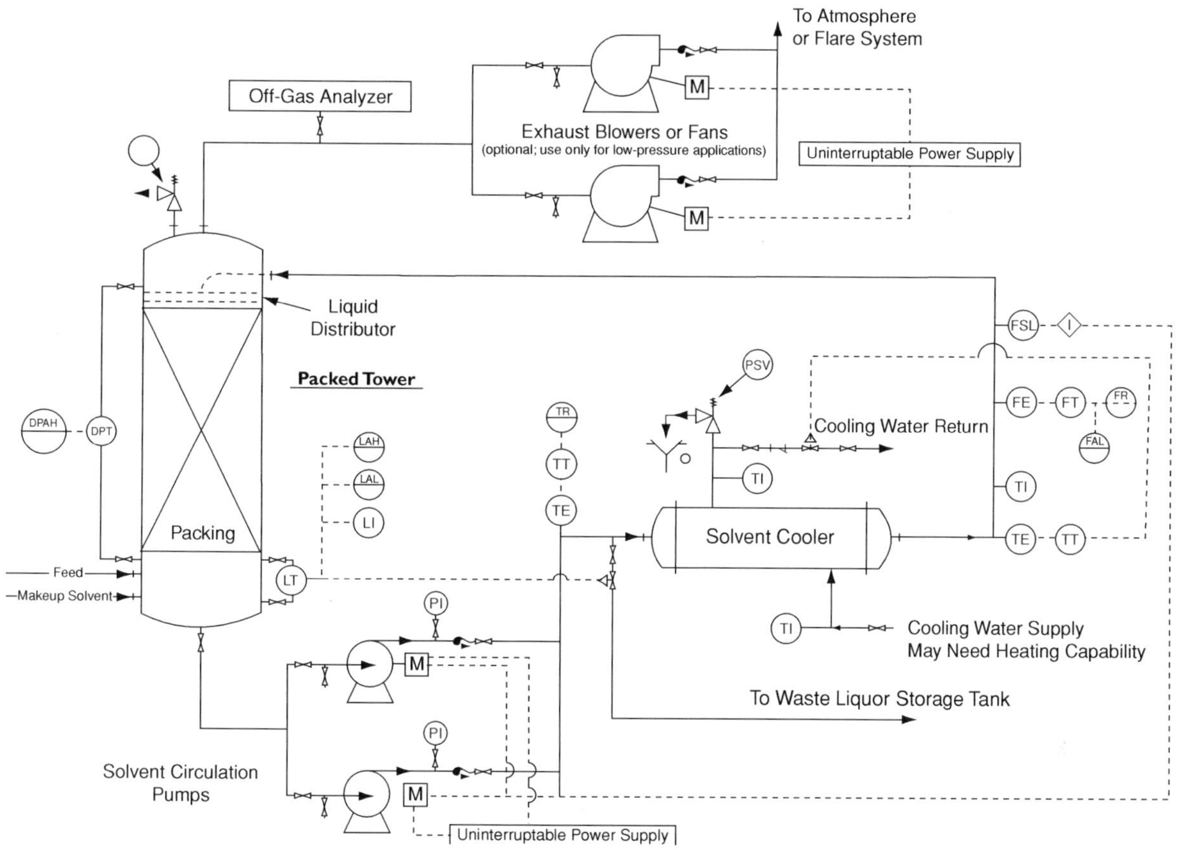

FIG. 23-57 Schematic of an emergency scrubber (absorber) system. (Guidelines for Pressure Relief and Effluent Handling Systems, *Center for Chemical Process Safety (CCPS) of the American Institute of Chemical Engineers (AIChE); copyright 1988 AIChE and reproduced with permission.).*

- Noncondensables in the feed will lower recovery efficiency, and need to be vented.
- May require large pool to handle lower-boiling materials.

Emergency Flares

Advantages:

- Can destroy over 98 percent of combustibles.
- Experimental tests not required for most chemicals.
- Commercially proven technology.
- Very reliable—onstream over 99 percent.
- Good turndown ratio.
- Smokeless operation not needed for emergency flares (only required for continuous flares).
- No generation of appreciable liquid waste as would be the case for a quench pool or an emergency scrubber, but knockout drum required upstream of the flare to remove liquids and any condensate.
- May be used as final step after other control equipment, e.g., emergency scrubber or quench pool.

Disadvantages:

- May not be suitable for some corrosive chemicals (e.g., most chlorine-containing compounds).
- May be subject to flameout.
- Required minimum heating value of 300 Btu/SCF for nonassisted flares, or 200 Btu/SCF for assisted flares.
- Not passive; require continuous pilot fuel, electricity, oxygen analyzers for collection headers, and instrumentation.
- Require backup systems to monitor flame and provide pilot burning.
- Care needed not to feed foams or liquids into the flare stack.
- Because of maintenance of flare tip on elevated flares (and even some open ground and enclosed ground flares), crane or helicopter required possibly for access. Connection often needed for backup flare system, i.e., portable trailer flare or other temporary flare.
- Elevated flares with excessive light, noise, or odors a possible public nuisance in some situations and locales.

Emergency Scrubbers

Advantages:

- Sound basis for technology.
- High turndown ratio at constant scrubbing liquid flow rate.
- Countercurrent operation that can attain high absorption (removal) efficiency and low outlet concentration of solute.
- Can handle two-phase mixtures.
- May be able to handle unstable foams (to be confirmed by testing).
- Some types of scrubbers that can handle fouling streams, and baffle trays that can handle fouling and solids-containing streams.
- Can be used for neutralization of reactive materials.
- Generally used only for gas/vapor feed, but can handle liquid by use of a separator section in the bottom of the tower.

Disadvantages:

- Possible need for spent solvent to be treated or disposed of after the emergency release.

- Not passive. Utilities and process control instrumentation must be available at all times, and must have a sufficient and reliable supply of fresh solvent for the expected duration of the emergency.
- May need large diameter to avoid flooding.
- Possible restriction of tray towers to nonfouling materials that are free of suspended solids.
- May need continuous flow of scrubbing solvent for the scrubber to be effective at the start of the effluent flow.
- *Guidelines for Pressure Relief and Effluent Handling Systems* (AIChE-CCPS, 1998) contains more information about selection criteria and factors that affect them.

Sizing and Design of Equipment This subsection presents sizing methods and design criteria for the design of vapor/gas/solid-liquid separators for normal vapor-liquid systems (low-viscosity and nonfoamy or unstable foams). They are not applicable usually to high-viscosity (newtonian and nonnewtonian) liquids and/or systems that exhibit surface-active foaming behavior, as insufficient information is available at present relating to the separation efficiency for these types of equipment. Design information is also presented for quench pool tanks which usually can handle high-viscosity liquids as well as stable foams, and for final vapor/gas control or destruction equipment.

Vapor-Liquid Gravity Separator Design Fundamentals The critical factors in the performance of a horizontal separator are the vapor residence time and the settling rate of the liquid droplets. However, two other factors enter into the design—the vapor velocity must be limited to avoid liquid entrainment, and there must be sufficient freeboard within the vessel to allow for a feed distributor. For vertical separators, the design is based on a vapor velocity that must be less than the settling velocity of the smallest droplet that is to be collected, with due allowance for turbulence and maldistribution of the feed. The vapor residence time is a function of the vapor flow rate (mass), vapor density, and volume of vapor space in the separator, based on the following:

- The vapor flow rate (mass) is the sum of the vapor feed to the separator and any vapor generated by flashing of the liquid as the feed enters the separator.
- The vapor density depends on the compositions of the vapor fed and the vapor generated, and the pressure and temperature within the separator. The pressure in the separator is essentially the sum of the pressure drop from the outlet of the separator to the ultimate disposal point of the effluent and the pressure at the downstream discharge location (e.g., the atmosphere).
- The vapor volume is the total volume of the separator minus the volume of the liquid collected; and the volume of liquid is the volume collected (separated) from the feed minus the volume of liquid flashed when the static pressure decreases as the flow rate declines. The maximum volume of liquid collected is also a function of pressure—a greater volume is collected at higher pressure, and time.

Horizontal Blowdown Drum/Catch Tank (See Fig. 23-49.) The two main criteria used in sizing horizontal blowdown drums or catch tanks are as follows:

1. The diameter must be sufficient to provide good vapor-liquid separation.

2. The total volume must be sufficient to hold the estimated amount of liquid carryover from the reactor or process vessel. For a foamy discharge, the holding volume should be greater than the reactor or process vessel volume (to be discussed below).

The recommended method is from *Guidelines for Pressure Relief and Effluent Handling Systems* (AIChE-CCPS, 1998). It is an improvement over the method presented in the 7th edition of this Handbook. The procedure involves calculating a terminal velocity for a selected droplet size, then providing enough residence time in the vapor space to allow the droplets to fall from the top of the vessel to the level of liquid collected. Also, the vapor velocity in the separator must not exceed the value above which liquid may be entrained from the liquid surface in the separator. The tank is treated as a simple horizontal cylinder, neglecting the volume of liquid in the heads.

Other approaches for sizing a horizontal separator (blowdown drum) are given in *Guidelines for Pressure Relief and Effluent Handling Systems* (AIChE-CCPS, 1998) and API RP 521 (2007).

Vertical Blowdown Drum/Catch Tank (See Fig. 23-51.) Some of the basic equations used for sizing horizontal separators can be applied to vertical separators; however, the superficial vapor velocity for vertical separators is based on the total vessel cross-sectional area, and the vapor velocity must be less than the terminal settling velocity for the droplet size selected as the basis of design.

The recommended design procedure is also from *Guidelines for Pressure Relief and Effluent Handling Systems* (AIChE-CCPS, 1998). As with a horizontal separator, the assumed value of K is expected to be suitable for most vertical separators in emergency relief services. However, if it is necessary to remove droplets smaller than 300 to 600 μm, then lower values of K should be used.

Multireactor Knockout Drum/Catch Tank The area needed for vapor disengaging is calculated by the equations given earlier in the section on horizontal blowdown drums. The diameter and length (or height) are determined by considering a number of factors:

1. The length should be sufficient to extend beyond the locations of the reactors discharging into the vessel so as to simplify discharge pipe runs (for a horizontal vessel).

2. The height should not greatly exceed the height of the building (for vertical vessels).

3. The diameter should be sufficient to allow attenuation of the shock wave leaving the deflector plate.

4. The diameter should be sufficient to allow installation of the pipes and deflector plates in such a way as not to interfere directly with one another (particularly important for vertical vessels).

5. The cost of pressure vessels increases as the diameter increases.

6. An upper limit to the diameter is set by the need to transport complete cylindrical sections from the manufacturer to the site.

7. The volume of liquid in the reactor or reactors (assuming more than one vents at the same time) must be determined.

Mechanical Design Considerations The paper by Speechly et al. ("Principles of Total Containment System Design," presented at I. Chem. E. Northwestern Branch Meeting, 1979) discusses a number of pertinent design features:

1. Each vent device discharge pipe is extended into the vessel, and its end is fitted with a deflector device. This disperses the jet stream of solids (catalyst) and liquids discharged and dissipates this force, which should otherwise be exerted on the vessel wall immediately opposite.

2. The deflector device (baffle plate) must be carefully designed as described by Woods (*Proc. Inst. Mech. Engrs.* 180, Part 3J: 245–259, 1965–1966).

3. Isolate the catch tank from both reaction loads and forces generated by thermal expansion of the pipes; the pipes can be designed to enter the vessel through a sliding gland. Depending on layout, vessels that tend to have shorter, stiffer pipes between the building and the vessel may also require flexible bellows to be incorporated in the pipes.

4. There are usually several reactors linked to a single catch tank. To ensure that rupture of a disk on one reactor does not affect the others, each reactor is fitted with a double-rupture disk assembly. The use of double-rupture disks in this application requires installation of a leak detection device in the space between the two disks.

Gravity Separator Safety Considerations and Features A pressure relief device may be required on vapor-liquid gravity separators, based on the following assumptions:

- External fire exposure, if the vapor outlet is inadequately sized for fire relief.
- The vessel outlets could be closed.
- A chemical reaction could continue in the knockout drum/catch tank.

The pressure relief device should be evaluated and sized in accordance with the latest industry standards (i.e., DIERS, API, NFPA). An open passage (nozzle or line) from the separator to the atmosphere may be considered as the pressure relief device provided that it meets the relieving capacity requirements of the ASME Boiler and Pressure Vessel Code (BPVC).

Freeze protection is not normally provided in separators intended solely for emergency release service; however, occasionally, coils or a jacket are used to keep high-viscosity liquids flowable, or to boil off volatile components. If internal coils or a heating jacket is used, it should be recognized that the source of heat might initiate a chemical reaction or increase its rate. Internal heating coils should be

- Designed for complete drainage.
- Provided with block valves to shut off the source of heat.
- Designed with a generous corrosion allowance.
- Braced to withstand impingement by the relief effluent stream.

Also the calculated liquid level in the separator should be corrected for the volume of the internal coils, and a heel of solvent might be retained in the separator to prevent freezing or to dissolve any components that can solidify or freeze.

Gravity Separator Vessel Design and Instrumentation Some of or all the following features might be appropriate in selected situations [also see API RP 521(2007)]:

- Liquid level indicator.
- High and low level alarms.
- Temperature indicator (with high-temperature alarm if deemed necessary).
- Pressure gauge.
- Pressure relief device.
- Manhole(s) for tank entry and cleaning.
- Pump for transferring the accumulated liquid to subsequent rework, recovery, or disposal.
- Antivortex baffle above the liquid outlet to the pump.
- Bracing of wall stiffeners to allow for possible "jet force" impingement on the vessel wall, and for intermittent buffeting and vibration if two-phase slug flow can occur.

For separators that normally operate at about atmospheric pressure, consider designing the vessel for 50 or 75 psig minimum MAWP to avoid vessel damage by an internal deflagration. For higher normal operating pressures, the design basis should be adjusted accordingly, consistent with the ASME BPVC, NFPA 68, and NFPA 69.

It should be realized that the separator may not be used for long periods, and the process control instrumentation and auxiliaries must receive regular inspection, testing, and maintenance.

Emergency Cyclone Separators Sizing of cyclone separators to handle unsteady-state discharge from emergency pressure relief devices is somewhat different from normal steady-state service (process cyclones). Emergency cyclones are designed usually for higher inlet velocity and pressure drop. Inlet velocity and cyclone pressure drop are limited only by the allowable backpressure on the pressure relief device. Emergency cyclone separator designs may allow some entrained mist or small droplets to leave the cyclone. This can be acceptable if the consequences of the release of the droplets are not significant.

Emergency cyclones often handle flashing vapor-liquid mixtures, and more vapor may be generated from flashing liquid because of the pressure drop in the cyclone; thus the mass flow rate of vapor leaving the cyclone is often greater than that entering. The cyclone size is determined by the outlet vapor volumetric flow rate. Since the discharge flow rate and ratio of vapor to liquid often vary, the cyclone design calculations should be made for conditions occurring at various times during the overpressure event.

Cyclone separator with separate catch tank (See Figs. 23-52 and 23-53.) The sizing of a cyclone knockout drum for emergency relief systems is somewhat different from that of a cyclone separator for normal process service for the following reasons:

1. In normal process service, the superficial vapor velocity at the inlet of tangential-entry vapor-liquid separators is limited to about 120 to 150 ft/s. Higher velocities may lead to

a. Excessive pressure drop in the separator and in the inlet piping.

b. Generation of fine mist in the inlet piping, which escapes collection in the separator.

2. Inlet velocity restrictions do not apply in the design of separators for emergency relief systems because

a. Pressure drop is usually not a penalty.

b. Escape of fine mist can usually be tolerated.

Sizing procedure The cyclone is sized by choosing a superficial F factor for the skirt in the range of 8.0 to 5.0 in USCS units (10 to 6 in SI units). The higher value may be used for waterlike liquids; the lower value for liquids such as molasses. If design F factors exceed the range of 5 to 8, the liquid draining down the skirt is entrained and escapes with the vapor. These F factors were determined in small-scale lab experiments using water and a high-polymer solution as the test liquids. The high-polymer solution had a viscosity that was molasseslike, probably in the range of 1500 cP. There were no liquids of intermediate viscosity used in the tests.

The F factors of 5.0 and 8.0 are conservative in the opinion of the researcher who performed the experiments (private communication from E. I. duPont de Nemours Co., Inc., to the DIERS Project).

The sizing procedure is presented in *Guidelines for Pressure Relief and Effluent Handling Systems* (AIChE-CCPS, 1998). Figure 23-53 shows the dimensions of a cyclone separator designed in accordance with this procedure. If liquid is continuously drained from the cyclone to a separate accumulator, a vortex breaker and false bottom should be used (Fig. 23-53, view BB).

If the liquid contents of the vented vessel are to be retained in the separator for subsequent disposition, the holdup capacity may be increased by increasing the height of the vessel to increase the total volume by an amount equal to the vented liquid volume.

If the pressure drop is too high, it may be necessary to increase the inlet line size for a distance of about 3 to 5 diameters upstream of the cyclone.

A sizing method similar to the one recommended above is proposed by Schmidt and Giesbrecht (2001), based on experimental work done at BASF in Germany.

Cyclone separator with integral catch tank (See Fig. 23-54.) The diameter of the knockout drum is calculated by the criteria given in the preceding subsection and Fig. 23-53. Since the liquid is also to be retained in the vessel, extend the shell height below the normal bottom tangent line to increase the total volume by an amount equal to the volume of the liquid carried over.

Quench Tank (See Figs. 23-55 and 23-56.) There is very little information in the open technical literature on the design of quench tanks in the chemical industry. Most of the information available deals with the design of quench tanks (also called suppression pools) for condensation of steam or steam-water mixtures from nuclear reactor safety valves. Information and criteria from quench tanks in the nuclear industry often can be used for the design of quench tanks in the chemical industry. *Guidelines for Pressure Relief and Effluent Handling Systems* (AIChE-CCPS, 1998) provides more information for chemical industry quench tank design. The following subsections summarize some of this information. Pertinent criteria for quench tank sizing and design are presented below:

Operating pressure There are three modes of operation of a quench tank: atmospheric pressure operation, nonvented operation, and controlled venting operation. Atmospheric operation is usually feasible when the effluent being emitted has a bubble point well above the maximum ambient temperature. A very small quantity of vapor escapes with the air that is displaced as the tank fills with the emergency discharge (typically about 0.2 percent of the reactor contents). Depending on the toxic or flammable properties of the vapor, the vent from the quench tank can be routed to the atmosphere or must be sent to a scrubber or flare.

In nonvented operation, no material is vented to the atmosphere, and this design is used when complete containment of the discharge is required. It is also used when the discharge mixture bubble point is close to or below the maximum ambient temperature and the concentration of noncondensable gas in the feed stream is very low. The tank design pressure is relatively high since the initial air in the tank is compressed by the rising liquid level, adding to the vapor pressure. The designer must take into consideration that the quench tank backpressure must be limited so as not to adversely affect the reactor relief system.

In controlled venting operation, the quench tank pressure is maintained at a desired level by a pressure controller/control valve system or pressure relief valve. This mode of operation is used when the discharge mixture bubble point is close to or below the maximum ambient temperature and it is desired to limit the maximum quench tank pressure.

Quench liquid selection The choice of the appropriate quench liquid depends on a number of factors. Water is usually the first quench liquid to consider. If water is selected as the quench liquid, the tank should be located indoors, if possible, to avoid freezing problems if the facility is in a cold climate. If the tank has to be located outdoors in a cold climate, the addition of antifreeze is preferable to heat-tracing the tank, since overheating of the tank can result from tracing, thus reducing its effectiveness.

If other quench liquids are required, the liquid should have as many of the following properties as possible: compatibility with the discharge effluent, low vapor pressure, high specific heat, low viscosity, low flammability, low freezing point, high thermal conductivity, immiscibility with the discharge effluent, low cost, and ready availability.

Quench liquid quantity A good discussion of the factors determining the quantity of quench liquid required is presented by CCPS (AIChE-CCPS, 1998).

When water is used as the quench medium and the effluent stream is a hydrocarbon or organic, separate liquid phases are often formed. In this case, heat transfer is the predominating mechanism during the quench. To achieve effective heat transfer, there must be a sufficient difference between the quench liquid temperature and the bubble point of the incoming effluent stream. The minimum temperature difference occurs at the end of the discharge, when the quench pool temperature is highest. A rule of thumb, from industry practice, is to allow a 10 to 20°C (18 to 36°F) ΔT. For atmospheric tank operation, the final quench liquid temperature is then set 10 to 20°C (18 to 36°F) below the normal boiling point of the final quench pool mixture. For nonvented or controlled venting operation, the final boiling point is elevated, permitting a greater design temperature rise and the use of less quench liquid. Therefore, the quench pool final temperature must be set 10 to 20°C (18 to 36°F) lower than the saturated temperature of the discharge effluent at the design maximum quench tank pressure.

The minimum capacity of quench liquid can be estimated by a heat balance, given the final quench pool temperature, using an equation given by AIChE-CCPS (1998). An equation is also presented there that includes the heat of reaction. In some cases, an experiment is necessary to confirm that the reaction indeed stops in the quench pool. An equation is also presented for the heat balance needed to calculate the minimum amount of quench liquid when the feed contains noncondensable gases.

It is good practice to provide 10 to 20 percent more quench liquid than the minimum amount calculated.

Quench tank volume The total volume of the quench tank should be equal to the sum of the following volumes:

- Quench liquid required
- Liquid entering in the multiphase effluent stream
- Liquid condensed from vapors in entering effluent stream
- Freeboard for noncondensables (a minimum of 10 percent is recommended)

Quench tank geometry Quench tanks can have any of the following three types of geometry:

- Horizontal cylindrical vessel
- Vertical cylindrical vessel
- Concrete pit (usually rectangular)

Usually, the geometry is determined by space limitations. Both horizontal and vertical cylindrical vessels are designed as pressure vessels and for pressures up to 50 to 75 psig. An L/D ratio of 2 to 3 results in an economic design.

Sparger design The effluent stream should be discharged into the quench liquid by means of a sparger, which breaks it up into small jets to provide good heat and mass transfer. The sparger design must also incorporate the following capabilities:

- Maximize momentum-induced recirculation in the quench pool.
- Provide adequate flow area (cross section for pressure relief without imposing high backpressure).
- Minimize shock due to vapor bubble collapse.
- Minimize unbalanced momentum forces.

Figure 23-55 shows conventional quench tank sparger arrangements. As can be seen in this figure, the sparger can be of the following types:

- Vertical straight-pipe sparger
- Tee sparger
- Four-armed cross-sparger

Alternative sparger designs are shown in AIChE-CCPS (1998).

The following design criteria are recommended:

1. The flow area of the manifold and/or distributor piping should be at least 2 times the total area of the sparger holes. This generally ensures that flow through the sparger holes will be uniform.
2. The number of holes should provide at least 0.2 hole per square foot (2.2 holes per square meter) of pool cross-sectional area, distributed evenly over the cross section of the pool.
3. Holes should be uniformly spaced in the distributor pipe—adjacent holes should be spaced at least 2 to 3 hole diameters apart, center to center.
4. Holes should discharge from the bottom or bottom half of the distributor pipe so that they can serve also as drains for liquid. Separate drain holes may be needed for the manifold pipe and in each distributor pipe.
5. Holes should be as small as possible consistent with consideration of possible plugging or fouling when handling liquids that contain polymers or suspended solids. For effluent streams consisting of only liquid and vapor, hole diameters ranging from ⅛ to ½ in (3 to 13 mm) are typical. Larger hole diameters (up to 2 in) may be required if the blowdown stream contains solids (polymers and/or catalyst). Smaller holes tend to minimize vibration and water hammer.
6. Sonic hole velocity is desirable in smaller holes and is essential in ½- to 2-in holes. A minimum sparger pressure drop of 10 psi should be used.
7. The quencher arm should be anchored to prevent pipe whip. It should also extend to the length (for horizontal vessels) or the height (for vertical vessels) of the vessel to evenly distribute the vapors in the pool.

When quenching effluents discharged by safety valves, it is preferable to use a straight, vertical sparger with holes in the end cap as well as in the pipe sidewalls. This is recommended to minimize the possibility of liquid hammer, which can occur more readily in horizontal spargers. The liquid hammer usually occurs for the following reasons: As the relief valve opens for the first time, the pressure spike is cushioned by the air trapped in the vent line. This air is blown out. If the valve recloses, the line may cool, causing slugs of condensate to accumulate. When the valve reopens, the slugs will accelerate to very high velocities and impact any elbows and end caps of the sparger. In severe cases, the sparger-arm end caps can be knocked off. The preceding recommendation avoids turns, and the holes in the end cap provide some relief from the pressure spike.

Many aspects of quench pool design (vessel, spargers, mechanical design loads, etc.) and operation are covered in *Guidelines for Pressure Relief and Effluent Handling Systems* (AIChE-CCPS, 1998), and it should be consulted when a quench pool has to be designed.

Mass-Transfer Contact Section Where there is a strong possibility that not all the incoming vapors will be condensed in the pool (the feed stream contains a large amount of noncondensable gases), a direct-contact mass-transfer section is superimposed on the quench tank. This can be a baffle tray section (as shown in Fig. 23-56) or a packed column section.

The design of direct-contact mass-transfer columns is discussed in detail by Scheiman [*Petro Chemical Engr.* 37(3): 29–33, 1965; ibid. 37(4): 75, 78–79] and Fair (*Chem. Eng.*, June 12, 1972).

Emergency Flare Systems A discussion of design criteria, principles, and practices for a number of flare system components is presented below.

Flare tip diameter The flare tip diameter is sized on a velocity basis, but the pressure drop must also be checked. Flare tip velocities can be chosen on the basis of EPA requirements or API recommendations. EPA recommendations are given in the AIChE-CCPS book (1998). API RP 521 states that tips may be based on velocities of 0.5 Mach number or higher, if pressure drop, noise, and other factors permit.

Flare height and thermal radiation The height of an elevated flare is based on the minimum distance from the flare flame to an object whose exposure to thermal radiation must be limited. Industrial flares are normally designed so that personnel in the vicinity are not exposed to a heat intensity greater than 1500 to 2000 Btu/(h·ft^2) when flaring at the maximum design rates.

An equation by Hajek and Ludwig (*Petro/Chem. Eng.* pp. C31–38, June 1960; pp. C44–51, July 1960) may be used to determine the distance required between a flare flame and a point of exposure where thermal radiation must be limited. [See API RP 521 (2007).]

A flame under the influence of wind will tilt in the direction that the wind is blowing and will change the location of the flame center. API RP 521 (2007) presents two methods for estimating the required flare height when the flame is tilted.

Other methods for estimating the required flare height are presented in *Guidelines for Pressure Relief and Effluent Handling Systems* (AIChE-CCPS, 1998).

Purge gas requirements Purging is used to prevent the formation of explosive mixtures in flare systems by preventing the admission of air into the system through leaks, backflow of air at the flare tip at very low flows, and back diffusion of air into the flare tip. Combustible gases such as methane or natural gas, or inert gases such as nitrogen or carbon dioxide, are frequently used for purging flare systems. API RP 521 (section 7.3.3.3) presents equations for calculating the purge rate as a function of the stack diameter, the desired oxygen level, and the type of purge gas.

Air (purge reduction) seals Air seals (also called purge reduction or gas seals) are often installed near the top of flare stacks. Others are located near the base of the flare. They are used to prevent air from entering the stack and are often recommended to prevent flashbacks and explosions. They also greatly reduce the amount of purge gas required. There are two basic types of air seals primarily used: diffusion (sometimes called molecular or labyrinth seals) and velocity types.

The diffusion seal consists of a baffled concentric cylinder arrangement, which uses the difference in molecular weight of the flare gas and ambient air to prevent air from entering the stack. Diffusion seals, combined with the purge gas flow rate, are intended to form a secondary line of defense against the entry of air into the flare stack. In a vertical flare stack, gravity exerts a driving force to increase the diffusion of air into the stack. Diffusion seals create an inverted flow field to reverse the gravitational effect and reduce air entry. Use of lighter-than-air purge gas (e.g., methane or natural gas) creates a pocket of light gas at the top of the air entry path. Use of diffusion seal and lighter-than-air purge gas allows a much lower purge gas rate. Installation of a diffusion seal at the top of the flare, and immediately below the burner tip, reduces the purge gas required to 1/50 of the volume required if the diffusion seal is not used (Niemeyer and Livingston, *Chem. Eng. Prog.*, pp. 39–44, December 1993).

Velocity seals are more recent developments in air seal design. They use conical baffles to redirect and focus the purge gas flow field just below the flare tip to sweep air from the flare stack. Some velocity seal designs can reduce the purge gas flow rate requirement to about 1/10 of the rate needed without the seal. Also, some velocity seal designs reportedly require only about 25 to 33 percent of the purge gas used in diffusion seals (AIChE-CCPS, 1998). More details about air (purge reduction) seals may be found in API RP 521 (2007).

Liquid seal drums Emergency vent streams are usually passed through a liquid seal, commonly water, before going to the flare stack. The liquid seal drum is usually located downstream of the knockout drum, and some vendors' designs include them in the base of the flare stack. A liquid seal drum is used to maintain a positive pressure in the vent header system and upstream system. It also reduces the possibility of flame flashbacks, caused when air is inadvertently introduced into the flare system and the flame front pulls down into the stack; it also acts as a mechanical damper on any explosive shock wave in the flare stack. Figure 23-58 is a schematic of a typical flare stack liquid seal drum, designed per API RP 521 criteria.

A properly designed and operated liquid seal drum should allow gas to pass through the seal with minimum surging in gas flow and/or upstream gas pressure. The design of the liquid seal internals and the design of the vessel can have a significant impact on the ability of the seal to meet the performance objective. For example, a common design for the end of a dipleg pipe uses V notches cut into the end of the pipe wall. This design is less effective than the proprietary designs developed by vendors of liquid seal drums. These proprietary designs, which use alternative design guidelines, may offer

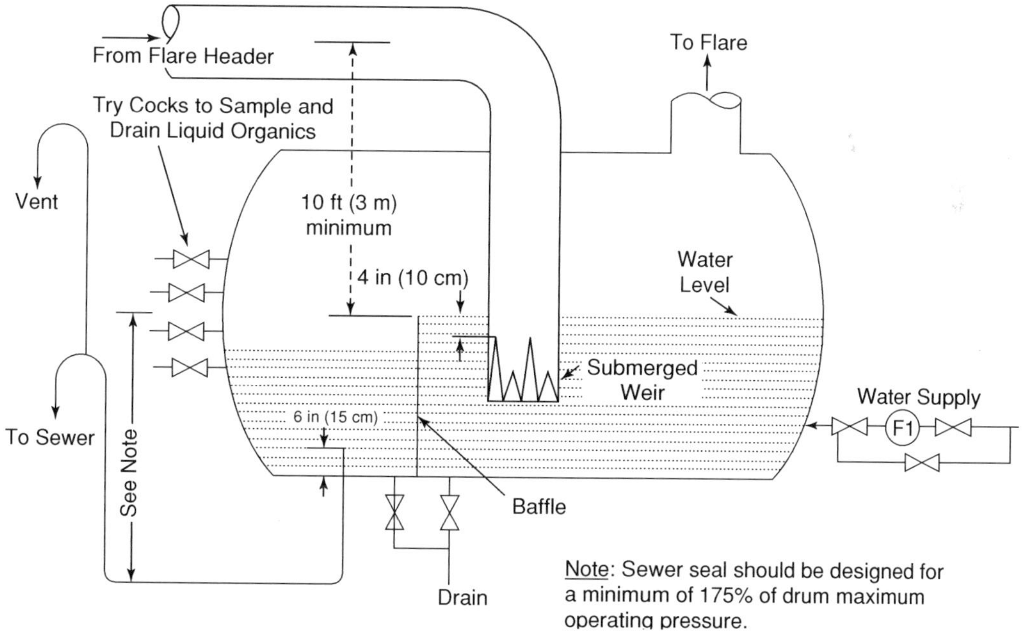

FIG. 23-58 Schematic for typical flare stack seal drum. (*Adapted from API RP 521.*) (Guidelines for Pressure Relief and Effluent Handling Systems, *Center for Chemical Process Safely (CCPS) of the American Institute of Chemical Engineers (AIChE); copyright 1988 AIChE and reproduced with permission.*)

economic or operational advantages. Details about several types of proprietary designs for liquid seal drums are presented by Grossel (*Deflagration and Detonation Flame Arresters*, AIChE-CCPS, New York, 2002).

More detailed discussions of flare systems and their components are presented in API RP 521 (2007), API STD 537 (2003), and AIChE-CCPS (1998).

Emergency Scrubbers (Absorbers) Emergency relief discharges are often passed through scrubbers (also called absorbers, or absorption columns or towers) for removal of flammable, corrosive, or toxic chemicals. The removal mechanism in some scrubbers involves physical absorption in a solvent, whereas in others chemical absorption (reactive scrubbing) is required.

Unlike quench pools for treatment of emergency relief effluent, scrubbers are not passive. For reliable performance, all critical components, including instruments, must be working correctly. There must be a reliable and sufficient supply of solvent and utilities (cooling water, electricity, etc.), with all process variables maintained under design conditions at all times. This includes proper liquid levels, liquid concentrations, flow rates, and reliable operation of ancillary equipment (e.g., pumps, blowers, etc.). General practice is to provide redundancy in pumps and blowers and utilities to ensure reliable scrubber operation (see Fig. 23-57). In the preferred mode of operation for hazardous materials, especially toxic gases, the scrubber is in a standby mode, with solvent flow maintained at all times. An alternate mode is to have the scrubber shut down (solvent not circulating) and then started up after detecting a release from the pressure relieving system. The consequences of the delay in starting the scrubber up after it has been shut down must be considered before accepting this mode of operation.

A discussion of emergency scrubber design and operating criteria and practices is given in *Guidelines for Pressure Relief and Effluent Handling Systems* (AIChE-CCPS, 1998).

FLAME ARRESTERS

GENERAL REFERENCES: [1] Howard, W. B., Precautions in Selection, Installation and Use of Flame Arresters, *Chem. Eng. Prog.*, April 1992. [2] Piotrowski, T., Specification of Flame Arresting Devices for Manifolded Low Pressure Storage Tanks, *Plant/Operations Progress*, vol. 10, no. 2, April 1991. [3] Roussakis and Lapp, A Comprehensive Test Method for In-Line Flame Arresters, *Plant/Operations Progress*, vol. 10, no. 2, April 1991. [4] Deflagration and Detonation Flame Arresters, Chapter 13, *Guidelines for Engineering Design for Process Safety*, Center for Chemical Process Safety, American Institute of Chemical Engineers, 1993. [5] Ibid. Chapter 15, Effluent Disposal Systems. [6] Howard, W. B., Flame Arresters and Flashback Preventers, *Plant/Operations Progress*, vol. 1, no. 4, 1982. [7] Thibault, P., Britton, L., and Zhang, F., Deflagration and Detonation of Ethylene Oxide Vapor in Pipelines, *Process Safety Progress*, vol. 19, no. 3, 2000. [8] Grossel, S. S., *Deflagration and Detonation Flame Arresters*, AIChE-CCPS Concept Book, 2002. [9] Britton, L. G., Using Maximum Experimental Safe Gap to Select Flame Arresters, *Process Safety Progress*, vol. 19, no. 3, 2000. [10] Britton, L. G., Using Heats of Oxidation to Evaluate Flammability Hazards, *Process Safety Progress*, vol. 21, no. 1, 2002. [11] Davies, M., and Heidermann, T., Investigation of Common Application Failures Proven by Life Field Testing of Endurance Burning Tested End-of-Line Flame Arresters, American Institute of Chemical Engineers, 37th Loss Prevention Symposium, 2003. [12] Brandes, E., and Redeker, T., Maximum Experimental Safe Gap of Binary and Ternary Mixtures, Fourth International Symposium on Hazards, Prevention and Mitigation of Industrial Explosions, IV ISHPMIE October 2002, pp. 207–213. [13] Förster, H., and Kersten, C., Investigation of Deflagrations and Detonations in Pipes and Flame Arresters by High Speed Framing, 4th International Symposium on Hazards, Prevention and Mitigation of Industrial Explosions (ISHPMIE), Bourges, France, October 21–25, 2002. [14] Lunn, G. A., The Maximum Experimental Safe Gap: The Effect of Oxygen Enrichment and the Influence of Reaction Kinetics, *Journal of Hazardous Materials*, 8 (1984), pp. 261–270, Elsevier Science Publisher B.V. Amsterdam. [15] European Standard EN 12874, *Flame Arrester Performance Requirements, Test Methods and Limits of Use*, European Committee for Standardization, Brussels, 2001.

General Considerations In 2002, the Center for Chemical Process Safety (CCPS) of the American Institute of Chemical Engineers (AIChE) published a Concept Book on this topic [8]. The book was intended to expand and update the coverage given in [4], providing extensive information on the history, technology, practice, and regulatory aspects of deflagration and detonation flame arresters.

Flame arresters are passive devices designed to prevent propagation of gas flames through pipelines. Typical applications are to prevent flames entering a system from outside (such as via a tank vent) or propagating within a system (such as from one tank to another). Flame arrestment is achieved by a permeable barrier, usually a metallic matrix containing narrow channels, which removes heat and free radicals from the flame fast enough to both quench it within the matrix and prevent reignition of the hot gas on the protected side of the arrester. These metallic matrices are known as *elements*. Some preliminary considerations for arrester selection and placement are as follows:

1. Identify the at-risk equipment and the potential ignition sources in the piping system to determine where arresters should be placed and what general type (deflagration or detonation, unidirectional or bidirectional) is needed.
2. Determine the worst-case gas mixture combustion characteristics, system pressure, and permissible pressure drop across the arrester, to help select the most appropriate element design. Not only does element design impact pressure drop, but also the rate of blockage due to particle impact, liquid condensation, and chemical reaction (such as monomer polymerization) can make some designs impractical even if in-service and out-of-service arresters are provided in parallel.
3. The possibility of a stationary flame residing on the arrester element surface should be evaluated, and the need for additional safeguards, should such an event occur, should be evaluated (see the subsection Endurance Burn)
4. Consider any material of construction limitations due to reactive or corrosive stream components, plus the impact of elevated temperature and pressure during flame events.
5. Consider upset conditions that could exceed the test conditions at which the arrester was certified. These include the gas composition with regard to concentration of sensitive constituents such as ethylene or hydrogen, maximum system pressure during an emergency shutdown, plus the maximum temperature and oxygen concentration. Under certain upset conditions such as a high-pressure excursion and/or elevated oxygen concentration [14], there may be no flame arrester available for the task.
6. Consider the type and location of the arrester with respect to ease of maintenance, particularly for large in-line arresters.

These questions address the type of arrester needed, the appropriate location, and the best design with respect to flow resistance, maintainability, and cost. It should be recognized that while flame arrester effectiveness is high, it is not 100 percent. To maximize effectiveness, attention should be given to proper selection, application, and maintenance of the device. Since arresters may fail on demand, it is good engineering practice to conduct layers of protection analysis to determine what additional mitigation may be required (Förster, H., Flame Arresters—The New Standard and Its Consequences, *Proceedings of the International ESMG Symposium, Part 2: Industrial Explosion Protection,* Nürnberg, Germany, March 27–29, 2001). For marine vapor control systems in the United States, flame arrester applications are regulated by the U.S. Coast Guard. For other applications, alternative test protocols and procedures have been developed by different agencies. Some arresters, such as hydraulic arresters and in-line types used to stop decomposition flames, have specialized applications for which general design and testing information is scarce. Where flame arresters are impractical, alternative strategies such as fast-acting valves, vapor suppression, and flammable mixture control should be considered. Dynamic systems such as fast-acting valves and suppression should be tested and certified as suitable for the required task.

Combustion: Deflagrations and Detonations A deflagration is a combustion wave propagating at less than the speed of sound as measured in the unburned gas immediately ahead of the flame front. Flame speed relative to the unburned gas is typically 10 to 100 m/s although owing to expansion of hot gas behind the flame, several hundred meters per second may be achieved relative to the pipe wall. The combustion wave propagates via a process of heat transfer and species diffusion across the flame front, and there is no coupling in time or space with the weak shock front generated ahead of it. Deflagrations typically generate maximum pressures in the range of 8 to 12 times the initial pressure. The pressure peak coincides with the flame front although a marked pressure rise precedes it; thus the unburned gas is compressed as the deflagration proceeds, depending on the flame speed and vent paths available. The precompression of gas ahead of the flame front (also known as *cascading* or *pressure piling*) establishes the gas conditions in the arrester when the flame enters it, and hence affects both the arrestment process and the maximum pressure generated in the arrester body. A severe deflagration arrestment test involves placing a restricting orifice behind the arrester, which increases the degree of precompression. This is known as *restricted-end* deflagration testing. As discussed later, the *maximum experimental safe gap* (MESG) of the gas mixture may underestimate its tendency to defeat the arrester under such conditions. This is so because the current MESG test method minimizes precompression effects and hence the likelihood of autoignition on the "protected side" of the arrester. It is most important to consider the influences of pressure, temperature, and oxygen concentration on the safe gap. In addition the safe gap of mixtures is influenced by the chemical interaction of the different gases [12, 14].

As the deflagration flame travels through piping, its speed increases due to flow-induced turbulence and compressive heating of the unburned gas ahead of the flame front. Turbulence is especially enhanced by flow obstructions such as valves, elbows, and tees. Once the flame speed has attained the order of 100 m/s, a *deflagration-to-detonation transition* (DDT) can occur, provided that the gas composition is within the detonable limits, which lie inside the flammable limits. The travel distance for this to occur is referred to as the *run-up* distance for detonation. This distance varies with the gas mixture sensitivity and increases with pipe diameter. It is often difficult to estimate the run-up distance at which a DDT may occur in a piping system, and when to specify either a deflagration or a detonation flame arrester. However, some indication of this can be obtained from [13]. If the actual plant piping installation is different from that of the experimental configurations, then it is recommended that tests be conducted simulating the actual plant piping configurations to establish the DDT run-up length. The European Standard EN12 874 [15] limits the use of in-line deflagration arresters to $L/D = 50$ for group D and C applications and $L/D = 30$ for group B. Tabulated run-up distances are generally for straight pipe runs, and DDT can occur for much smaller distances in pipe systems containing flow obstructions. At the instant of transition, a transient state of *overdriven detonation* is achieved and persists for a distance of a few pipe diameters. Overdriven detonations propagate at speeds greater than the speed of sound (as measured in the burned gas immediately behind the flame front), and side-on pressure ratios (at the pipe wall) in the range of 50 to 100 have been measured. The peak pressure is variable depending on the amount of precompression during deflagration. A severe test for detonation-type flame arresters is to arrange for the arrester to encounter a series of overdriven detonations.

After the abnormally high velocities and pressures associated with DDT have decayed, a state of stable detonation is attained. A detonation is a combustion-driven shock wave propagating at the speed of sound as measured in the burned gas immediately behind the flame front. Since the speed of sound in this hot gas is much larger than that in the unburned gas or the ambient air, and the flame front speed is augmented by the burned gas velocity, stable detonations propagate at supersonic velocities relative to an external fixed point. A typical velocity for a stoichiometric gas mixture is 1800 m/s. The wave is sustained by chemical energy released by shock compression and ignition of the unreacted gas. The flame front is coupled in space and time with the shock front with no significant pressure rise ahead of the shock front. The high velocities and pressures associated with detonations require special element design to quench the high-velocity flames plus superior arrester construction to withstand the associated impulse loading. Since this entails narrower and/or longer element channels plus bracing of the element facing, both inherent pressure drop and the possibility of fouling of detonation arresters should be considered.

The problem of flame arrestment, of either deflagrations or detonations, depends on the properties of the gas mixture involved plus the initial temperature, pressure, and oxygen concentration. Gas mixture combustion properties cannot be quantified for direct use in flame arrester selection, and only general characteristics can be assigned. For this reason flame arrester performance must be demonstrated by realistic testing. Such testing has demonstrated that arresters capable of stopping even overdriven detonations may fail under restricted end deflagration test conditions. It is important to understand the significance of the test conditions addressed and their possible limitations. It is recommended that users request a detailed test report from the flame arrester manufacturer and only an approved arrester tested to a recognized test standard be installed.

Combustion: Gas Characteristics and Sensitivity Combustion thermodynamic calculations allow determination of peak deflagration and detonation pressures, plus stable detonation velocity. The peak pressure calculation may be used to determine combustion product venting requirements although a conservative volume increase of 9:1 may be used for essentially closed systems. Other relevant gas characteristics are entirely experimental. The sensitivity to detonation depends on the detonable range and fundamental burning velocity, although no specific correlations or measures of sensitivity exist based on fundamental properties. National Electrical Code (NEC) groups are commonly used to rank gases for the purposes of flame arrester selection. By this method, group A gases (acetylene) are considered most sensitive to detonation (and most difficult to arrest) while group D gases (such as saturated hydrocarbons) are considered least sensitive and easiest to arrest. As formerly applied, successful testing of an arrester using one gas in a NEC Electrical Group was assumed to mean that the arrester would be suitable for all other gases in that group. As presently applied in various codes, a representative test gas from each group (such as hydrogen in group B, ethylene in group C, or propane in group D) is typically used for arrester certification, and its ability to arrest a different gas or gas mixture is deduced by comparing the respective maximum experimental safe gaps. The arrester is assumed to be suitable for any single gas or mixture having an MESG greater than or equal to that of the representative test gas. Although the arrester may be directly tested using the gas or gas mixture of interest, this is rarely carried out owing to cost. The MESG comparison procedure is currently applied to both deflagration and detonation arresters. Specific testing protocols are specified by the codes used. Since MESGs provide an independent ranking for arrester selection, reference to NEC groups is not essential. Nevertheless, arresters continue to be described in terms of the NEC group of the representative test gas used for certification. For example, an arrester might be described as a "group C (ethylene)" type. A detailed discussion of the use of MESGs is given in [9].

It is cautioned that there have been no systematic studies proving that arrester performance can be directly correlated with MESG, especially if the MESG for a mixture is estimated by using the Le Chatelier rule. An alternative way for calculating the MESG and safe gap of mixtures is given in [12]. Furthermore, major revisions to the MESG test method have caused many "historic" MESG values to increase significantly; the current test method minimizes compression of the unburned gas mixture and may therefore underestimate the likelihood of arrester failure via autoignition.

Corrosion Consideration should be given to possible corrosion of both the element material and the arrester housing, since corrosion may weaken the structure, increase the pressure drop, and decrease the effectiveness of the element. While the housing might be designed to have a corrosion allowance, corrosion of the element must be avoided by proper material specification. Common materials of construction include aluminum, carbon steel, ductile iron and 316 stainless steel housings, and aluminum or 316 stainless steel elements. While special materials such as Hastelloy might be used for situations such as high HCl concentrations, it may be more cost-effective to use a hydraulic arrester made of carbon steel and lined with a suitable polymeric lining in such applications. Also, at least one flame arrester manufacturer can provide flame arresters with PTFE (Teflon) arresting elements and lined housings up to a nominal size of 4 in for very corrosive service [8].

Directionality To select an arrester for any service, the potential sources of ignition must be established in relation to the pipe system and the equipment to be protected. The pipe connecting an arrester with an identified ignition source is the *unprotected side* of the arrester.

The pipe connecting the arrester with at-risk equipment is the *protected side*. If the arrester will encounter a flame arriving only from one direction, a *unidirectional* arrester can be used. If a flame may arrive from either direction a *bidirectional* arrester is needed. The latter either are symmetrically constructed or are certified by testing. Back-to-back use of unidirectional arresters will not usually be cost-effective unless testing reveals a specific advantage such as increased allowable operating pressure during restricted end deflagration testing.

Endurance Burn Under certain conditions a successfully arrested flame may stabilize on the unprotected side of an arrester element. Should this condition not be corrected, the flame will eventually penetrate the arrester as the channels become hot. An endurance burn time can be determined by testing, which specifies that the arrester has withstood a stabilized flame without penetration for a given period. The test should address either the actual or the worst-case geometry since heat transfer to the element will depend on whether the flame stabilizes on the top, bottom, or horizontal face. In general the endurance burn time identified by test should not be regarded as an accurate measure of the time available to take remedial action, since test conditions will not necessarily approximate the worst possible practical case. Temperature sensors may be incorporated at the arrester to indicate a stabilized flame condition and either alarm or initiate appropriate action, such as valve closure. It is very important to install an endurance burning flame arrester in the same way as tested to avoid malfunction [11].

Installation End-of-line arresters should be protected by using appropriate weather hoods or cowls. In-line arresters (notably detonation arresters and in-line deflagration arresters) must be designed to withstand the highest line pressure that might be seen, including upset conditions. The design should be verified by hydrostatic and pneumatic pressure tests. The piping system should be designed with adequate supports and should allow routine access to the arrester for inspection and maintenance.

Maintenance It is important to provide for arrester maintenance by both selection of the most suitable arrester type and judicious location. Inspection and maintenance should be performed on a regular basis depending on experience with the particular arrester in the service involved. It should also be carried out after successful function of the arrester. Some in-line designs allow removal, inspection, and cleaning of the element without the need to expand the line. Unit designs featuring multiple elements in parallel can reduce downtime by extending the period between cleaning. For systems that cannot be shut down during maintenance, parallel arresters incorporating a three-way valve may be used. Detonation arrester elements are especially prone to damage during dismantling, cleaning, and reassembly. Maintenance must be carefully done, avoiding sharp objects that could disable the delicate channels in the element. Spare elements should be available to reduce downtime and provisions made for storing, transporting, and cleaning the elements without damage.

Monitoring The differential pressure across the arrester element can be monitored to determine the possible need for cleaning. The pressure taps must not create a flame path around the arrester. It can be important to provide temperature sensors such as thermocouples at the arrester to detect flame arrival and stabilization. Since arrester function may involve damage to the arrester, the event of successful function (flame arrival) may be used to initiate inspection of the element for damage. If the piping is such that flame stabilization on the element is a realistic concern, action must be taken immediately upon indication of such stabilization (see also Endurance Burn). Such action may involve valve closure to shut off gas flow or inert gas purging.

Operating Temperature and Pressure Arresters are certified subject to maximum operating temperatures and absolute pressures normally seen at the arrester location. Arrester placement in relation to heat sources such as incinerators must be selected so that the allowable temperature is not exceeded, with due consideration for the detonation potential as run-up distance is increased. Flame arrester manufacturers can provide recommended distances from heat sources, such as open flames, to avoid thermal damage to a flame arrester element. If heat tracing is used to prevent condensation of liquids, the same temperature constraint applies. In the case of in-line arresters, there may be certain upset conditions that produce unusually large system pressures outside the stipulated operating range of the arrester. Since the maximum operating pressure for a detonation arrester may be in the range of 16 to 26 psia, depending on the gas sensitivity and arrester design, it may be impossible to find a suitable arrester to operate during such an upset. The situation may be exacerbated by pressure drop across the device caused by high flow and/or fouling.

Pressure Drop Flow resistance depends on flame arrester channel arrangement and on a time-dependent fouling factor due to corrosion, or accumulation of liquids, particles or polymers, depending on the system involved. Monomer condensation is a difficult problem since inhibitors will usually be removed during monomer evaporation and catalysis might occur over particulates trapped in the element. Pressure drop can be a critical factor in operability, and cleaning may represent a large hidden cost. Sizing for pressure drop must be based on worst-credible-case operating conditions rather than normal operating conditions.

Fouling may be mitigated in a number of ways. First, the least sensitive element design can be selected, and in the case of end-of-line arresters, weather hoods or cowls can be used to protect against water or ice accumulation. Second, a fouling factor (20 percent or greater) might be estimated and an element with a greater tested flow capacity selected to reduce the pressure drop. This should be further increased if liquid condensation might occur. It is important that certified flow curves for the arrester be used rather than calculated curves since the latter can be highly optimistic. Condensation and polymerization may be mitigated by geometry (minimizing liquid accumulation in contact with the element) and provision for drainage. Alternatively, the arrester may be insulated and possibly heat-traced. Drains should not provide flame paths around the arrester or leak in either direction when closed. If heat tracing is used, the temperature must be limited to the certified operating range of the arrester.

In addition to using an arrester element with greater flow capacity, it is common to use two arresters in parallel where frequent cleaning is required, with one arrester in standby. A three-way valve can be used to allow uninterrupted operation during changeover. Where elements have an intrinsically high pressure drop, such as sintered metal elements used in acetylene service, multiple parallel elements can be used.

Deflagration Arresters The two types of deflagration arrester normally encountered are the *end-of-line arrester* (Figs. 23-59 and 23-60) and the *tank vent deflagration arrester,* which is mounted close to the end of the line going direct to the atmosphere. Neither type of arrester is designed to stop in-line deflagrations or detonations. If mounted sufficiently far from the atmospheric outlet of a piping system, which constitutes the unprotected side of the arrester, the flame can accelerate sufficiently to cause these arresters to fail. Failure can occur at high flame speeds even without a run-up to detonation.

If atmospheric tanks are equipped with flame arresters on the vents, fouling or blockage by extraneous material can inhibit gas flow to the degree that the tank can be damaged by underpressure. API standards allow the use of pressure vacuum (P/V) valves without flame arresters for free-venting tanks on the basis that the high vapor velocity in the narrow gap between pressure pallet (platter) and valve body will prevent flashback. However, it is important to ensure the pallet is not missing or stuck open since this will remove the protection. Absence of the pallet was a listed factor in the 1991 Coode Island fire (State Coroner Victoria, Case No. 2755/91, Inquest into Fire at Coode Island on August 21 and 22, 1991, Finding). Whether or not flame arresters are used, proper inspection and maintenance of these vent systems is required.

End-of-line arresters are mounted at the outlet of a pipe system and go directly to the atmosphere, so there is no potential for significant flame acceleration in the pipe. **Tank vent deflagration arresters** are strictly limited by the approval agency, but for group D gases they are typically mounted no more than 20 ft from the end of a straight pipe that vents directly to the atmosphere. The allowed distance must be established by proper testing with the appropriate gas mixture and the pipe diameter involved. Turbulence-promoting irregularities in the flow (bends, tees, elbows, valves, etc.) cannot be used unless testing has addressed the exact geometry. It is essential that run-up to detonation not occur in the

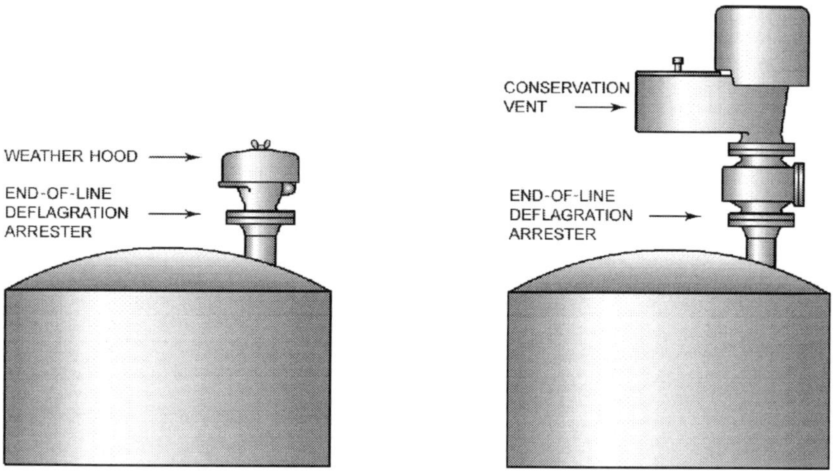

FIG. 23-59 Typical end-of-line deflagration arrester installations. (*Courtesy of PROTEGO®.*)

available piping system, and run-up distance can be less than 2 ft for some fast-burning gases such as hydrogen in air (group B). Thus the NEC grouping of the gas mixture must be considered. More importantly, it must be emphasized that even if run-up to detonation does not occur, a deflagration arrester can fail if the flame speed is great enough. Thus the run-up distance is not an adequate criterion for acceptable location, and this limitation can be determined only by realistic testing. A number of explosions have occurred due to misapplication of end-of-line or tank vent deflagration arresters where detonation arresters or in-line deflagration arresters should have been used. The latter are described in the next subsection.

Detonation and Other In-Line Arresters If the point of ignition is remote from the arrester location, the arrester is an in-line type such as might be situated in a vapor collection system connecting several tanks. Due to the possibility of DDT, most in-line arresters are designed to stop both deflagrations and detonations (including overdriven detonations) of the specified gas mixture. These are known as **detonation arresters**. Figure 23-61 shows a typical design. Detonation arresters may be further delineated into those types that will arrest only stable detonations and those that will arrest both stable and unstable detonations (i.e., overdriven detonations or DDT events).

In some cases, in-line arresters need to stop deflagrations only. However, in such cases it must be demonstrated that detonations cannot occur in the actual pipework system; unless the gas mixture is intrinsically not capable of detonation, this requires full-scale testing using the exact pipe geometry to be used in practice, which must not be changed after installation. In certain exceptional cases, an in-line deflagration arrester may be mounted without regard to run-up distance. This can be done only where the system is known to be incapable of detonation. Examples are the decomposition flames of ethylene and ethylene oxide, which are briefly discussed under Special Arrester Types and Alternatives.

Detonation arresters are typically used in conjunction with other measures to decrease the risk of flame propagation. For example, in vapor control systems the vapor is often enriched, diluted, or inerted, with appropriate instrumentation and control [5]. In cases where ignition sources are present or predictable (such as most vapor destruct systems), the detonation arrester is used as a last-resort method anticipating possible failure of vapor composition control. Where vent collection systems have several vapor/oxidant sources, stream compositions can be highly variable and this can be additionally complicated when upset conditions are considered. It is often cost-effective to perform hazard analyses such as HAZOP or fault

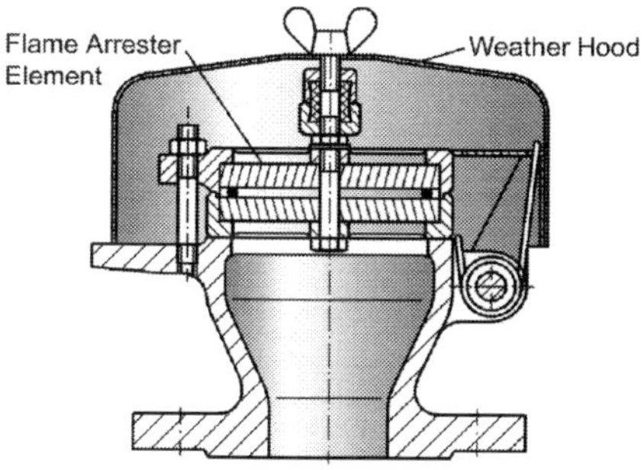

FIG. 23-60 Typical end-of-line deflagration arrester design. (*Courtesy of PROTEGO®.*)

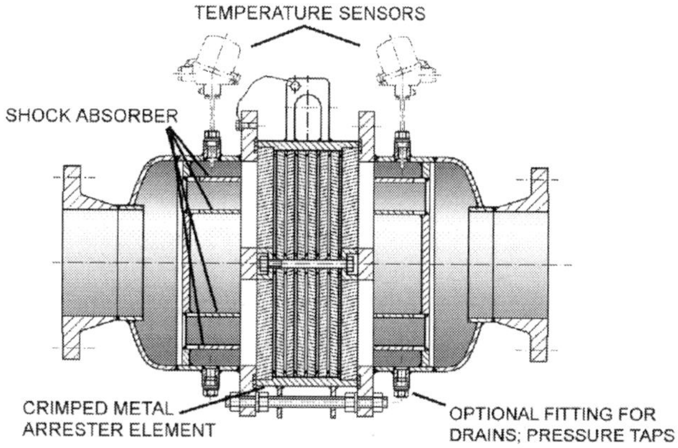

FIG. 23-61 Typical detonation arrester design (crimped ribbon type). (*Courtesy of PROTEGO®.*)

tree analysis to determine whether such vent streams can enter the flammable region and, if so, what composition corresponds to the worst credible case. Such an analysis is also suitable to assess alternatives to arresters.

Effect of Pipe Diameter Changes Arrester performance can be impaired by local changes in pipe diameter. It was shown that a minimum distance of 120 pipe diameters should be allowed between the arrester and any increase in pipe diameter; otherwise, a marked reduction in maximum allowable operating pressure would occur. This impairment was observed during detonation testing but was most pronounced during restricted-end deflagration testing (Lapp and Vickers, *Int. Data Exchange Symp. on Flame Arresters and Arrestment Technology,* Banff, Alberta, October 1992). As a rule, arresters should be mounted in piping either equal to or smaller than the nominal size of the arrester.

Venting of Combustion Products As gas deflagrates or detonates in the piping system, there is a volume expansion of the products and an associated pressure increase. In some instances where the pipe system volume involved is relatively large, a significant overpressure might be developed in the vapor spaces of connected tanks, especially when vapor space is minimal due to high liquid level. It can be assumed that all the gas on the unprotected side of the arrester is converted to equilibrium products; the pressure is relieved via gas expansion into the entire system volume and to the atmosphere via any vent paths present. If heat losses are neglected by the assumption of high flame speeds or detonation, and atmospheric venting paths are neglected, a conservative approach is that storage vessels be designed with a capacity to handle 9 times the pipe volume on the unprotected side of the arrester. With regard to the high pressures associated with detonations, it has been shown (Lapp, Independent Liquid Terminal Association Conference, Houston, June 23, 1992) that detonation arresters attenuate the peak detonation pressure by up to 96 percent depending on the arrester design, and therefore protect from much of the pressure pulse. To further reduce the pressure pulse, relief devices may be provided at the arrester.

Arrester Testing and Standards Regulatory and approval agencies and insurers impose *acceptance testing* requirements, sometimes as part of certification standards. The user may also request testing to demonstrate specific performance needs just as the manufacturer can help develop standards. These interrelationships have resulted in several new and updated performance test procedures. *Listing* of an arrester by a testing laboratory refers only to performance under a defined set of test conditions. The flame arrester user should develop specific application requirements based on the service involved and the safety and risk criteria adopted.

As discussed in [8], a variety of test procedures and use guidelines have been developed. General considerations are given in Chapter 9.3 of National Fire Protection Association Standard 69. The Federal Register, 33 CFR, Part 154, contains the USCG requirements for detonation arresters in marine vapor control systems. Other U.S. procedures are given in ASTM F 1273-91, UL 525, FM Procedure Classes 6061 and 7371, plus API Publications 2028 and 2210. For U.S. mining applications, the Mine Safety and Health Administration (MSHA) provides regulation and guidance; e.g., in CFR Title 30, Part 36. The International Maritime Organization (IMO) Standard MSC/Circ. 677 (1994) provides testing procedures for end-of-line deflagration and in-line detonation arresters for use on tanker ships. In Canada, CSA-Z343 is followed while in Europe the CEN Standard EN 12874 [15] has replaced previous European National Standards such as BS 7244.

Deflagration Arrester Testing For end-of-line, tank vent, and in-line deflagration flame arresters, approval agencies may require manufacturers to provide users with data for flow capacity at operating pressures, proof of success during an endurance burn or continuous flame test, evidence of flashback test results (for end-of-line arresters) or explosion test results (for in-line or tank vent arrester applications), hydraulic pressure test results, and results of a corrosion test.

Endurance burn testing generally implies that the ignited gas mixture and flow rate are adjusted to give the worst-case heating (based on temperature observations on the protected side of the element surface), that the burn continues for a specified duration, and flame penetration does not occur. Continuous flame testing implies a gas mixture and flow rate are established at specified conditions and burn on the flame arrester for a specified duration. The endurance burn test is usually a more severe test than the continuous burn. In both cases the flame arrester attachment configuration and any connecting piping or valves should be installed in the same configuration used for testing. General reference [11] gives additional information.

Flashback tests incorporate a flame arrester on top of a tank with a large plastic bag surrounding the flame arrester. A specific gas mixture (e.g., propane, ethylene, or hydrogen at the most sensitive composition in air) flows through and fills the tank and the bag. Deflagration flames initiated in the bag (three at different bag locations) must not pass through the flame arrester into the protected vessel. On the unprotected side, piping and attachments such as valves are included as intended for installation; a series of tests, perhaps 10, is conducted.

Whatever the application, a user should be aware that not all test procedures are the same, or of the same severity, or use the same rating designations. Therefore, it is important to review the test procedure and determine whether the procedure used is applicable to the intended installation and potential hazard the flame arrester is meant to prevent.

Detonation Arrester Testing Requirements are described by various agencies in the documents listed above (UL 525, etc). For installations governed by the USCG in Appendix A of 33 CFR, Part 154 (Marine Vapor Control Systems), the USCG test procedures must be followed. These are similar but not identical to those of other agencies listed. The European Union mandates arrester testing by an approved testing laboratory according to the EN 12874 Standard. Reference [8] discusses differences between the requirements of disparate agencies.

Detonation arresters are extensively tested for proof of performance against deflagrations, detonations, and endurance burns. In the United States, arrester manufacturers frequently test detonation arresters according to the USCG protocol; other test standards might alternatively or additionally be met. Under this protocol, the test gas must be selected to have either the same or a lower MESG than the gas in question. Typical MESG benchmark gases are stoichiometric mixtures of propane, hexane, or gasoline in air to represent group D gases having an MESG equal to or greater than 0.9 mm, and ethylene in air to represent group C gases with an MESG no less than 0.65 mm. Commercially available arresters are typically certified for use with one or another of these benchmark gas types. An ethylene-type arrester is selected, should the gas in question have an MESG of less than 0.9 mm but not less than 0.65 mm. Five low and five high overpressure deflagration tests are required with and without a flow restriction on the protected side. Of these 20 tests, the restricted-end condition is usually the most severe and often limits the maximum initial pressure at which the arrester will be suitable. Five detonation tests and five overdriven detonation tests are also required, which may involve additional run-up piping and turbulence promoters to achieve DDT at the arrester. If these tests are successful, an endurance burn test is required. This test does not use propane for group D gases but hexane or gasoline, owing to their lower autoignition temperatures. For group C tests, ethylene can be used for all test stages.

Care must be taken when applying the MESG method [4, 9, 12]. The user has the option to request additional tests to address such concerns and may wish to test actual stream compositions rather than simulate them on the basis of MESG values.

Special Arrester Types and Alternatives

Hydraulic (Liquid Seal) Flame Arresters Hydraulic (liquid seal) flame arresters are most commonly used in large-pipe-diameter systems where fixed-element flame arresters are either cost-prohibitive or otherwise impractical (e.g., very corrosive gas or where the gas contains solid particles that would quickly plug a conventional arrester element). These arresters contain a liquid, usually water-based, to provide a flame barrier. Figure 23-62 shows one design. Realistic tests are needed to ensure performance, as described in EN 12874 [15]. Note that hydraulic flame arresters may fail at high flow rates, producing a sufficiently high concentration of gas bubbles to allow transmission of flame. This is distinct from the more obvious failure mode caused by failure to maintain adequate liquid level.

Alternatives to Arresters Alternatives to the use of flame arresters include fast-acting isolation valves, vapor suppression systems,

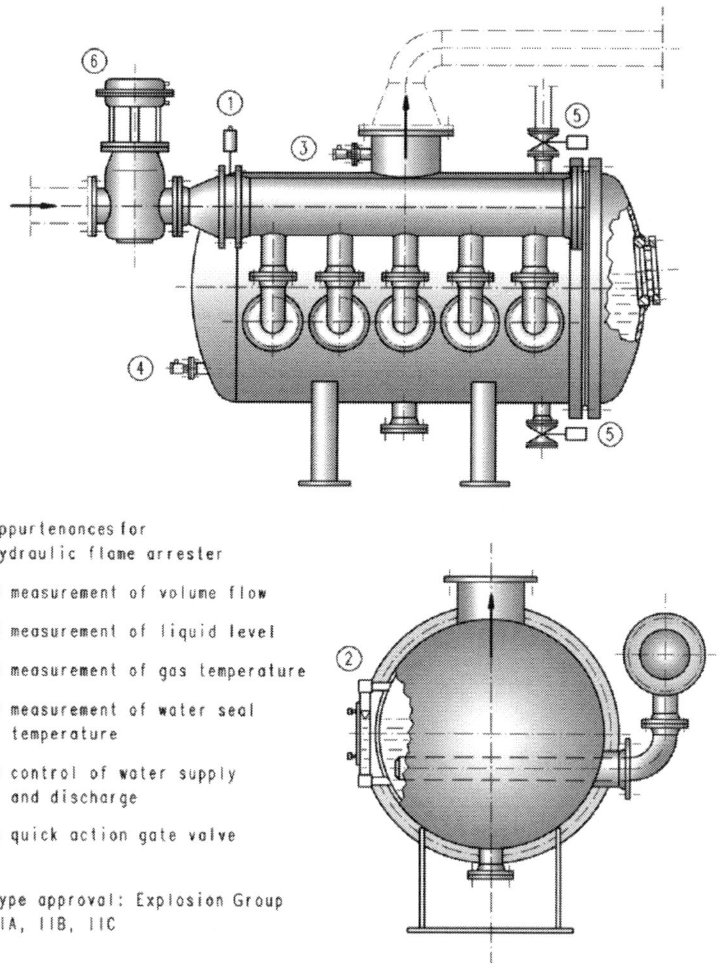

FIG. 23-62 Tested and approved hydraulic (liquid seal) flame arrester. (*Courtesy of PROTEGO®.*)

velocity-type devices in which gas velocity is designed to exceed flashback velocity, and control of the flammable mixture (NFPA 69 Standard, "Explosion Prevention Systems"). The latter alternative frequently involves reduction of oxygen concentration to less than the limiting oxygen concentration (LOC) of the gas stream.

STORAGE AND HANDLING OF HAZARDOUS MATERIALS

General References: API-620, *Design and Construction of Large, Welded, Low-Pressure Storage Tanks,* American Petroleum Institute, Washington. AP-40, *Air Pollution Engineering Manual,* 2d ed., U.S. Environmental Protection Agency, Office of Air Quality Planning and Standards, 1973. AP-42, *Compilation of Emission Factors for Stationary Sources,* 5th ed., U.S. Environmental Protection Agency, Office of Transportation and Air Quality, 1995. *API Standards,* American Petroleum Institute, Washington ASME, *Process Piping: The Complete Guide to ASME B31.3,* 2d ed., American Society of Mechanical Engineers, New York, 2004. ASME, *ASME Boiler and Pressure Vessel Code; ASME Code for Pressure Piping; ASME General and Safety Standards; ASME Performance Test Codes,* American Society of Mechanical Engineers, New York. *Chemical Exposure Index,* 2d ed., AIChE, New York, 1994. *Code of Federal Regulations,* Protection of Environment, Title 40, Parts 53 to 80, Office of the Federal Register, Washington. CGA, *Handbook of Compressed Gases,* 4th ed., Compressed Gas Association, Chantilly, Va., 1999. CCPS, *Guidelines for Chemical Process Quantitative Risk Analysis,* 2d ed., CCPS-AIChE, New York, 2000. CCPS, *Guidelines for Engineering Design for Process Safety.* CCPS, *Guidelines for Facility Siting and Layout,* CCPS-AIChE, New York, 2003. CCPS, *Guidelines for Process Safety in Batch Reaction Systems,* CCPS-AIChE, New York, 1999. CCPS, *Guidelines for Safe Storage and Handling of High Toxic Hazard Materials,* CCPS-AIChE, New York, 1988. CCPS, *Guidelines for Safe Storage and Handling of Reactive Materials,* CCPS-AIChE, New York, 1995. CCPS, *Guidelines for Mechanical Integrity Systems,* Wiley, New York, 2006. Englund, "Opportunities in the Design of Inherently Safer Chemical Plants," in J. Wei et al., eds., *Advances in Chemical Engineering,* vol. 15, Academic Press, 1990. Englund, "Design and Operate Plants for Inherent Safety," *Chem. Eng. Prog.,* pts. 1 and 2, March and May 1991. Englund, Mallory, and Grinwis, "Preventing Backflow," *Chem. Eng. Prog.,* February 1992. Englund and Grinwis, "Redundancy in Control Systems," *Chem. Eng. Prog.,* October 1992. Fisher et al., "Emergency Relief System Design Using DIERS Technology: The Design Institute for Emergency Relief Systems (DIERS) Project Manual," AIChE, New York, 1992. Grossel and Crowl, *Handbook of Highly Toxic Materials Handling and Management,* Marcel Dekker, New York, 1995. Hendershot, "Alternatives for Reducing the Risks of Hazardous Material Storage Facilities," *Environ. Prog.,* 7, August 1988, pp. 180ff. Kletz, *An Engineer's View of Human Error,* Institution of Chemical Engineers, VCH Publishers, New York, 1991. Kletz, "Friendly Plants," *Chem. Eng. Prog.,* July 1989, pp. 18-26. Kletz, *Plant Design for Safety: A User Friendly Approach,* Hemisphere Publishing, London, 1991. Kohan, *Pressure Vessel Systems: A User's Guide to Safe Operations and Maintenance,* McGraw-Hill, New York, 1987. Mannan, *Lees' Loss Prevention in the Process Industries,* 3d ed., Elsevier, Amsterdam, 2005. Prokop, "The Ashland Tank Collapse," *Hydrocarbon Processing,* May 1988. Russell and Hart, "Underground Storage Tanks, Potential for Economic Disaster," *Chemical Engineering,* March 16, 1987, pp. 61-69. *Ventsorb for Industrial Air Purification,*

Bulletin 23-56c, Calgon Carbon Corporation, Pittsburgh, Pa., 1986. White and Barkley, "The Design of Pressure Swing Adsorption Systems," *Chem. Eng. Prog.*, January 1989.

Introduction The inherent nature of most chemicals handled in the chemical process industries is that they each have physical, chemical, and toxicological hazards to a greater or lesser degree. This requires that these hazards be contained and controlled throughout the entire life cycle of the facility, to avoid loss, injury, and environmental damage. The provisions that will be necessary to contain and control the hazards will vary significantly depending on the chemicals and process conditions required.

Established Practices Codes, standards, regulatory requirements, industry guidelines, recommended practices, and supplier specifications have all developed over the years to embody the collective experience of industry and its stakeholders in the safe handling of specific materials. These should be the engineer's first resource in seeking to design a new facility.

The ASME Boiler and Pressure Vessel Code, Section VIII, is the standard resource for the design, fabrication, installation, and testing of storage tanks and process vessels rated as pressure vessels (i.e., above 15-psig design). ASME B31.3 is a basic resource for process piping systems.

Examples of established practices and other resources—some of which pertain to the safe storage and handling of specific hazardous chemicals, classes of chemicals, or facilities—include those listed in Table 23-29 from the publications of two U.S. organizations, the NFPA and the Compressed Gas Association (CGA). Other organizations that may have pertinent standards include the International Standards Organization (ISO), the American National Standards Institute (ANSI), ASTM International (Conshohocken, Pa; *www.astm.org*), and other well-established national standards such as British Standards and Deutsches Institut für Normung e.V. (DIN) standards. Local codes and regulations should be checked for applicability, and the latest version should always be used when employing established practices.

Basic Design Strategies The storage and handling of hazardous materials involve risks that can be reduced to very low levels by good planning, design, and management practices. Facilities that handle hazardous materials typically represent a variety of risks, ranging from small leaks, which require prompt attention, to large releases, which are extremely rare in well-managed facilities but which have the potential for widespread impact (CCPS, 1988). It is essential that good techniques be developed for identifying significant hazards and mitigating them where necessary. Hazards can be identified and evaluated by using approaches discussed in the section on hazard and risk analysis.

Loss of containment due to mechanical failure or improper operation is a major cause of chemical process incidents. The design of storage and piping systems should be based on minimizing the likelihood of loss of containment, with the accompanying release of hazardous materials, and on limiting the amount of the release. An effective emergency response program that can reduce the impacts of a release should be available.

Thus, the basic design strategy for storing and handling hazardous materials can be summarized as follows, with reference to other parts of this section in parentheses:

1. Understand the hazardous properties of the materials to be stored and handled (Flammability, Reactivity, Toxicity, Other Hazards), as well as the physical hazards associated with the expected process design.
2. Reduce or eliminate the underlying hazards as much as is feasible (Inherently Safer and More User-Friendly Design).
3. Evaluate the potential consequences associated with major and minor loss-of-containment events and other possible emergency situations involving the hazardous materials and energies; and take this information into account in the process of site selection and facility layout and the evaluation of the adequacy of personnel, public, and environmental protection (Source Models, Atmospheric Dispersion, Estimation of Damage Effects).
4. Design and build a robust and well-protected primary containment system following codes, standards, regulations, and other established practices (Security).

TABLE 23-29 Examples of Established Practices Related to Storage and Handling of Hazardous Materials

Designation	Title
National Fire Protection Association (Quincy, Mass.; *www.nfpa.org*)	
NFPA 30	Flammable and Combustible Liquids Code
NFPA 30B	Code for the Manufacture and Storage of Aerosol Products
NFPA 36	Standard for Solvent Extraction Plants
NFPA 45	Standard on Fire Protection for Laboratories Using Chemicals
NFPA 53	Recommended Practice on Materials, Equipment and Systems Used in Oxygen-Enriched Atmospheres
NFPA 55	Standard for the Storage, Use, and Handling of Compressed Gases and Cryogenic Fluids in Portable and Stationary Containers, Cylinders, and Tanks
NFPA 58	Liquefied Petroleum Gas Code
NFPA 59A	Standard for the Production, Storage, and Handling of Liquefied Natural Gas (LNG)
NFPA 68	Guide for Venting of Deflagrations
NFPA 69	Standard on Explosion Prevention System
NFPA 318	Standard for the Protection of Semiconductor Fabrication Facilities
NFPA 326	Standard for the Safeguarding of Tanks and Containers for Entry, Cleaning, or Repair
NFPA 329	Recommended Practice for Handling Releases of Flammable and Combustible Liquids and Gases
NFPA 400	Hazardous Chemical Code
NFPA 430	Code for the Storage of Liquid and Solid Oxidizers
NFPA 432	Code for the Storage of Organic Peroxide Formulations
NFPA 434	Code for the Storage of Pesticides
NFPA 484	Standard for Combustible Metals, Metal Powders, and Metal Dusts
NFPA 490	Code for the Storage of Ammonium Nitrate
NFPA 495	Explosive Materials Code
NFPA 497	Recommended Practice for the Classification of Flammable Liquids, Gases, or Vapors and of Hazardous (Classified) Locations for Electrical Installations in Chemical Process Areas
NFPA 499	Recommended Practice for the Classification of Combustible Dusts and of Hazardous (Classified) Locations for Electrical Installations in Chemical Process Areas
NFPA 654	Standard for the Prevention of Fire and Dust Explosions from the Manufacturing, Processing, and Handling of Combustible Particulate Solids
NFPA 655	Standard for Prevention of Sulfur Fires and Explosions
NFPA 704	Standard System for the Identification of the Hazards of Materials for Emergency Response
Compressed Gas Association (Chantilly, Va.; *www.cganet.com*)	
CGA G-1	Acetylene
CGA G-2	Anhydrous Ammonia
CGA G-3	Sulfur Dioxide
CGA G-4	Oxygen
CGA G-5	Hydrogen
CGA G-6	Carbon Dioxide
CGA G-8.1	Standard for Nitrous Oxide Systems at Consumer Sites
CGA G-12	Hydrogen Sulfide
CGA G-14	Code of Practice for Nitrogen Trifluoride (EIGA Doc. 92/03)
CGA P-1	Safe Handling of Compressed Gases in Containers
CGA P-8	Safe Practices Guide for Cryogenic Air Separation Plants
CGA P-9	The Inert Gases: Argon, Nitrogen, and Helium
CGA P-12	Safe Handling of Cryogenic Liquids
CGA P-16	Recommended Procedures for Nitrogen Purging of Tank Cars
CGA P-32	Safe Storage and Handling of Silane and Silane Mixtures
CGA P-34	Safe Handling of Ozone-Containing Mixtures Including the Installation and Operation of Ozone-Generating Equipment
CGA S-1.1	Pressure Relief Device Standards—Part 1—Cylinders for Compressed Gases
CGA S-1.2	Pressure Relief Device Standards—Part 2—Cargo and Portable Tanks for Compressed Gases
CGA S-1.3	Pressure Relief Device Standards—Part 3—Stationary Storage Containers for Compressed Gases

NOTE: Always check the latest edition when using established practices.

5. Design and implement a reliable and fault-tolerant basic process control system to ensure the design limitations of the primary containment system are not exceeded.

6. Include provisions for detecting abnormal process conditions and bringing the process to a safe state before an emergency situation occurs (Safety Instrumented Systems).

7. Design, install, and maintain reliable and effective emergency relief systems, as well as mitigation systems such as secondary containment, deluge, and suppression systems, to reduce the severity of consequences in the event an emergency situation does occur (Pressure Relief Systems; Emergency Relief Device Effluent Collection and Handling).

8. Evaluate the risks associated with the process and its safety systems taken as a whole, including consideration of people, property, business, and the environment, that could be affected by loss events; and determine whether the risks have been adequately reduced (Hazard Analysis, Risk Analysis, Source Models, Atmospheric Dispersion, Estimation of Damage Effects).

9. Take human factors into account in the design and implementation of the control system and the facility procedures (Human Error, Key Procedures).

10. Ensure staffing, training, inspections, tests, maintenance, and management of change are all adequate to maintain the integrity of the system throughout the facility lifetime (Key Procedures, Audit Process).

Designers and operating companies will address these items in different ways, according to their established procedures. The steps that are addressed elsewhere in this section are not repeated here.

Site Selection, Layout, and Spacing Facility siting decisions that will have critical, far-reaching implications are made very early in a new facility's life cycle, or in the early planning stages of a site expansion project. The degree of public and regulatory involvement in this decision-making process, as well as the extent of prescriptive requirements and established practices in this area, varies considerably among countries, regions, and companies. Insurance carriers are also generally involved in the process, particularly with regard to fire protection considerations.

From the perspective of process safety, key considerations with respect to site selection, layout, and spacing can be summarized as

- Where on-site personnel (including contractors and visitors), critical equipment, the surrounding public, and sensitive environmental receptors are located with respect to hazardous materials and processes
- Whether the design and construction of control rooms and other occupied structures, as well as detection, warning, and emergency response provisions, will provide adequate protection in the event of a major fire, explosion, or toxic release event

Recommended distances for spacing of buildings and equipment for fire protection were issued as IRI IM.2.5.2, *Plant Layout and Spacing for Oil and Chemical Plants* (Industrial Risk Insurers, Hartford, Conn). These are referenced in "Typical Spacing Tables" included as Appendix A of the CCPS *Guidelines for Facility Siting and Layout* (2003). Other resources pertaining to siting and layout include

- *Dow's Fire & Explosion Index Hazard Classification Guide,* 7th ed. (AIChE, New York, 1994), which gives an empirical radius of exposure and damage factor based on the quantity and characteristics of the material being stored and handled
- API RP 752, "Management of Hazards Associated With Location of Process Plant Buildings," 2d ed. (American Petroleum Institute, Washington, 2003), which gives a risk-based approach to evaluating protection afforded by occupied structures

Storage

Storage Facilities Dating back to at least 1974, when a vapor cloud explosion in Flixborough, U.K., claimed 28 lives and destroyed an entire chemical plant (Mannan, 2005), a major emphasis in the safe storage and handling of hazardous materials has been to reduce hazardous material inventories. Inventory reduction can be accomplished not only by using fewer and smaller storage tanks and vessels but also by eliminating any nonessential intermediate storage vessels and batch process weigh tanks and generating hazardous materials on demand when feasible. Note, however, that reduction of inventory may require more frequent and smaller shipments and improved management.

There may be more chances for errors in connecting and reconnecting with small shipments. Quantitative risk analysis of storage facilities has revealed solutions that may run counter to intuition. [Schaller, *Plant/Oper. Prog.* 9(1), 1990]. For example, reducing inventories in tanks of hazardous materials does little to reduce risk in situations where most of the exposure arises from the number and extent of valves, nozzles, and lines connecting the tank. Removing tanks from service altogether, on the other hand, generally helps. A large pressure vessel may offer greater safety than several small pressure vessels of the same aggregate capacity because there are fewer associated nozzles and lines. Also, a large pressure vessel is inherently more robust, or it can economically be made more robust by deliberate overdesign than can a number of small vessels of the same design pressure. On the other hand, if the larger vessel has larger connecting lines, the relative risk may be greater if release rates through the larger lines increase the risk more than the inherently greater strength of the vessel reduces it. In transporting hazardous materials, maintaining tank car integrity in a derailment is often the most important line of defense in transportation of hazardous materials.

Safer Storage Conditions The hazards associated with storage facilities can often be reduced significantly by changing storage conditions. The primary objective is to reduce the driving force available to transport the hazardous material into the atmosphere in case of a leak (Hendershot, 1988). Some methods to accomplish this follow.

Dilution Dilution of a low-boiling hazardous material reduces the hazard in two ways:

1. The vapor pressure is reduced. This has a significant effect on the rate of release of material boiling at less than ambient temperature. It may be possible to store an aqueous solution at atmospheric pressure, such as aqueous ammonium hydroxide instead of anhydrous ammonia.

2. In the event of a spill, the atmospheric concentration of the hazardous material will be reduced, resulting in a smaller hazard downwind of the spill.

Refrigeration Loss of containment of a liquefied gas under pressure and at atmospheric temperature causes immediate flashing of a large proportion of the gas. This is followed by slower evaporation of the residue. The hazard from a gas under pressure is normally much less in terms of the amount of material stored, but the physical energy released if a confined explosion occurs at high pressure is large.

Refrigerated storage of hazardous materials that are stored at or below their atmospheric boiling points mitigates the consequences of containment loss in three ways:

1. The rate of release, in the event of loss of containment, will be reduced because of the lower vapor pressure in the event of a leak.

2. Material stored at a reduced temperature has little or no superheat, and there will be little flash in case of a leak. Vaporization will be mainly determined by liquid evaporation from the surface of the spilled liquid, which depends on weather conditions.

3. The amount of material released to the atmosphere will be further reduced because liquid entrainment from the two-phase flashing jet resulting from a leak will be reduced or eliminated.

Refrigerated storage is most effective in mitigating storage facility risk if the material is refrigerated when received.

The economics of storage of liquefied gases are such that it is usually attractive to use pressure storage for small quantities, pressure or semirefrigerated storage for medium to large quantities, and fully refrigerated storage for very large quantities. Quantitative guidelines can be found in Mannan (2005).

It is generally considered that there is a greater hazard in storing large quantities of liquefied gas under pressure than at low temperatures and low pressures. The trend is toward replacing pressure storage by refrigerated low-pressure storage for large inventories. However, it is necessary to consider the risk of the entire system, including the refrigeration system, and not just the storage vessel. The consequences of failure of the refrigeration system must be considered. Each case should be carefully evaluated on its own merits.

Preventing Leaks and Spills from Accumulating under Tanks or Equipment Around storage and process equipment, it is a good idea to design dikes that will not allow toxic and flammable materials to accumulate around the bottom of tanks or equipment in case of a spill. If liquid is spilled and ignites inside a dike where there are storage tanks or process equipment, the fire may be continuously supplied with fuel and the consequences can be severe. It is usually much better to direct possible spills and leaks to an area away from the tank or equipment and provide a firewall to shield the equipment from most of the flames if a fire occurs. Figure 23-63 shows a diking design for directing leaks and spills to an area away from tanks and equipment.

The surface area of a spill should be minimized for hazardous materials that have a significant vapor pressure at ambient conditions, such as acrylonitrile or chlorine. This will make it easier and more practical to collect vapor from a spill or to suppress vapor release with foam or by other means. This may require a deeper nondrained dike area than normal or some other design that will minimize surface area, in order to contain the required volume. It is usually not desirable to cover a diked area to restrict loss of vapor if the spill consists of a flammable or combustible material.

Minimal Use of Underground Tanks The U.S. Environmental Protection Agency's (EPA) Office of Underground Storage Tanks defines underground tanks as those with 10 percent or more of their volume, including piping, underground. An aboveground tank that does not have more than 10 percent of its volume (including piping) underground is excluded from the underground tank regulations. Note, however, that a 5000-gal tank sitting wholly atop the ground but having 1400 ft of 3-in buried pipe or 350 ft of 6-in buried pipe is considered an underground storage tank.

At one time, burying tanks was recommended because it minimized the need for a fire protection system, dikes, and distance separation. At many companies, this is no longer considered good practice. Mounding, or burying tanks above grade, has most of the same problems as burying tanks below ground and is usually not recommended.

Problems with buried tanks include

- Difficulty in monitoring interior and exterior corrosion (shell thickness)
- Difficulty in detecting leaks
- Difficulty of repairing a tank if the surrounding earth is saturated with chemicals from a leak
- Potential contamination of groundwater due to leakage

Government regulations concerning buried tanks have become stricter. This is so because of the large number of leaking tanks that have been identified as causing adverse environmental and human health problems.

Design of Tanks, Piping, and Pumps Six basic tank designs are used for the storage of organic liquids: (1) fixed-roof, (2) external floating-roof, (3) internal floating-roof, (4) variable vapor space, (5) low-pressure, and (6) high-pressure tanks. The first four tank designs listed are not generally considered suitable for highly toxic hazardous materials.

Low-Pressure Tanks (up to 15 psig) Low-pressure storage tanks for highly hazardous toxic materials should meet, as a minimum, the American Petroleum Institute (API) 620 Standard, "Recommended Rules for the Design and Construction of Large Welded, Low-Pressure Storage Tanks" (*API Standards*). This standard covers aboveground tanks designed for all pressures less than or equal to 15 psig and metal temperatures less than or equal to 250°F (121°C). There are no specific requirements in API 620 for highly hazardous toxic materials.

API 650, "Welded Steel Tanks for Oil Storage" (*API Standards*), has limited applicability to storage of highly hazardous toxic materials because it prohibits refrigerated service and limits pressures to 2.5 psig and only if designed for certain conditions. Most API 650 tanks have a working pressure approaching atmospheric pressure, and hence their pressure-relieving devices must generally vent directly to the atmosphere. Its safety factors and welding controls are less stringent than required by API 620.

Horizontal and vertical cylindrical tanks are used to store highly toxic liquids and other hazardous materials at atmospheric pressure. Horizontal, vertical, and spherical tanks are used for refrigerated liquefied gases that are stored at atmospheric pressure. The design pressure of tanks for atmospheric pressure and low-pressure storage at ambient temperature should not be less than 100 percent of the vapor pressure of the material at the maximum design temperature. The maximum design metal temperature to be used takes into consideration the maximum temperature of material entering the tank; the maximum ambient temperature, including solar radiation effects; and the maximum temperature attainable by expected or reasonably foreseeable abnormal chemical reactions.

Since discharges of vapors from highly hazardous materials cannot simply be released to the atmosphere, the use of a weak seam roof is not normally acceptable. It is best that tanks in low-pressure hazardous service be designed and stamped for 15 psig to provide maximum safety, and pressure relief systems must be provided to vent relieved overpressure to equipment that can collect, contain, and treat the effluent.

The minimum design temperature should be the lowest temperature to which the tank will be subjected, taking into consideration the minimum temperature of material entering the tank, the minimum temperature to which the material may be autorefrigerated by rapid evaporation of low-boiling liquids or mechanically refrigerated, the minimum ambient temperature of the area where the tank is located, and any possible temperature reduction by endothermic physical processes or chemical reactions involving the stored material. API 620 provides for installations in areas where the lowest recorded 1-day mean temperature is 50°F (10°C).

While either rupture disks or relief valves are allowed on storage tanks by Code, rupture disks by themselves should not be used on tanks for the storage of toxic or other highly hazardous materials since they do not close after opening and may lead to continuing release of hazardous material to the atmosphere.

The API 620 Code requires a combined pneumatic hydrotest at 125 percent of design tank loading. In tanks designed for low-density liquid, the upper portion is not fully tested. For highly hazardous materials, consideration should be given for hydrotesting at the maximum specified design liquid level. It may be required that the lower shell thickness be increased to withstand a full head of water and that the foundation be designed such that it can support a tank full of water or the density of the liquid, if it is greater than water. Testing in this manner not only tests the containment capability of the tank, but also provides an overload test for the tank and the foundation similar to the overload test for pressure vessels. API 620 also requires radiography.

Proper preparation of the subgrade and grade is extremely important for tanks that are to rest directly on grade. Precautions should be taken to prevent ground freezing under refrigerated tanks, as this can

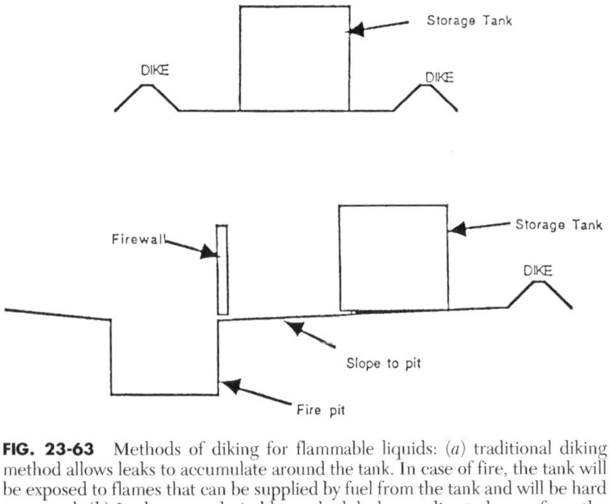

FIG. 23-63 Methods of diking for flammable liquids: (*a*) traditional diking method allows leaks to accumulate around the tank. In case of fire, the tank will be exposed to flames that can be supplied by fuel from the tank and will be hard to control. (*b*) In the more desirable method, leaks are directed away from the tank. In case of fire, the tank will be shielded from most flames and fire will be easier to fight. (*From England, in* Advances in Chemical Engineering, *vol. 15, Academic Press, San Diego, 1990, pp. 73–135, by permission.*)

cause the ground to heave and damage the foundation or the tank. Designing for free air circulation under the tank is a method for passive protection from ground freezing.

Steels lose their ductility at low temperatures and can become subject to brittle failure. There are specific requirements for metals to be used for refrigerated storage tanks in API 620, Appendices Q and R.

Corrosive chemicals and external exposure can cause tank failure. Materials of construction should be chosen so that they are compatible with the chemicals and exposure involved. Welding reduces the corrosion resistance of many alloys, leading to localized attack at the heat-affected zones. This may be prevented by the use of the proper alloys and weld materials, in some cases combined with annealing heat treatment.

External corrosion can occur under insulation, especially if the weather barrier is not maintained or if the tank is operating at conditions at which condensation is likely. This form of attack is hidden and may be unnoticed for a long time. Inspection holes and plugs should be installed in the insulation to monitor possible corrosion under the insulation.

Pressure Vessels (above 15 psig) The design of vessels above 15 psig falls within the scope of the American Society of Mechanical Engineers (ASME) Boiler and Pressure Vessel Code, Section VIII, "Pressure Vessels, Division I," and should be designated as lethal service if required. *Lethal service* means containing substances that are "poisonous gases or liquids of such a nature that a very small amount of the gas or vapor of the liquid mixed or unmixed with air is dangerous to life when inhaled. This class includes substances which are stored under pressure or may generate a pressure if stored in a closed vessel." This is similar to, but not exactly the same as, the same definition as that for "Category M" fluid service of the ASME Pressure Piping Code (see below). Pressure vessels for the storage of highly hazardous materials should be designed in accordance with requirements of the ASME code even if the vessels could be exempted because of high pressure or size. The code requires that the corrosion allowance be adequate to compensate for the more or less uniform corrosion expected to take place during the life of the vessel and not weaken the vessel below design strength.

Venting and Drainage Low-pressure storage tanks are particularly susceptible to damage if good venting practices are not followed. A vent that does not function properly at all times may cause damage to the tank from pressure that is too high or too low. Vapors should go to a collection system, if necessary, to contain toxic and hazardous vents.

Piping Piping systems for toxic fluids fall within Chapter VIII of the ASME Pressure Piping Code, "Piping for Category M Fluid Service." Category M fluid service is defined as "fluid service in which the potential for personnel exposure is judged to be significant and in which a single exposure to a small quantity of a toxic fluid, caused by leakage, can produce serious irreversible harm to persons on breathing or bodily contact, even when prompt restorative measures are taken."

Piping systems should meet the requirements for both Category M fluid service and for "severe cyclic conditions." Piping systems should be subjected to a flexibility analysis, and if they are found to be too rigid, flexibility should be added. Severe vibration pulsations should be eliminated. Expansion bellows, flexible connections, and glass equipment should be avoided. Pipelines should be designed with the minimum number of joints, fittings, and valves. Joints should be flanged or butt-welded. Threaded joints should not be used.

Instrumentation (CCPS, 1986.) Instrument systems are an essential part of the safe design and operation of systems for storing and handling hazardous materials. They are key elements of systems to eliminate the threat of conditions that could result in loss of containment. They are also used for early detection of releases so that mitigating action can be taken before these releases result in serious effects on people in the plant or in the public sector, or on the environment.

Pumps and Gaskets The most common maintenance problem with centrifugal pumps is with the seals. Mechanical seal problems account for most of the pump repairs in a chemical plant, with bearing failures a distant second. The absence of an external motor (on canned pumps) and a seal is appealing to those experienced with mechanical seal pumps.

Sealless pumps are very popular and are widely used in the chemical industry. Sealless pumps are manufactured in two basic types: canned-motor and magnetic-drive. Magnetic-drive pumps have thicker "cans," which hold in the process fluid, and the clearances between the internal rotor and can are greater compared to canned-motor pumps. This permits greater bearing wear before the rotor starts wearing through the can. Many magnetic-drive pump designs now have incorporated a safety clearance, which uses a rub ring or a wear ring to support the rotating member in the event of excessive bearing wear or failure. This design feature prevents the rotating member (outer magnet holder or internal rotating shaft assembly) from accidentally rupturing the can, as well as providing a temporary bearing surface until the problem bearings can be replaced. Because most magnetic-drive pumps use permanent magnets for both the internal and external rotors, there is less heat to the pumped fluid than with canned-motor pumps. Some canned-motor pumps have fully pressure-rated outer shells, which enclose the canned motor; others do not. With magnetic-drive pumps, containment of leakage through the can to the outer shell can be a problem. Even though the shell may be thick and capable of holding high pressures, there is often an elastomeric lip seal on the outer magnetic rotor shaft with little pressure containment capability.

Canned-motor pumps typically have a clearance between the rotor and the containment shell or can, which separates the fluid from the stator, of only 0.008 to 0.010 in (0.20 to 0.25 mm). The can has to be thin to allow magnetic flux to flow to the rotor. It is typically 0.010 to 0.015 in (0.25 to 0.38 mm) thick and made of Hastelloy. The rotor can wear through the can very rapidly if the rotor bearing wears enough to cause the rotor to move slightly and begin to rub against the can. The can may rupture, causing uncontrollable loss of the fluid being pumped.

It should not be assumed that just because there is no seal, sealless pumps are always safer than pumps with seals, even with the advanced technology now available in sealless pumps. Use sealless pumps with considerable caution when handling hazardous or flammable liquids.

Sealless pumps rely on the process fluid to lubricate the bearings. If the wear rate of the bearings in the fluid being handled is not known, the bearings can wear unexpectedly, causing rupture of the can.

Running a sealless pump dry can cause complete failure. If there is cavitation in the pump, hydraulic balancing in the pump no longer functions and excessive wear can occur, leading to failure of the can. *The most common problem with sealless pumps is bearing failure, which occurs either by flashing the fluid in the magnet area because of a drop in flow below minimum flow or by flashing in the impeller eye as it leaves the magnet area.* It is estimated that 9 out of 10 conventional canned-motor pump failures are the result of dry running. Canned pumps are available which, their manufacturer claims, can be operated dry for as long as 48 h.

It is especially important to avoid deadheading a sealless pump. Deadheaded sealless pumps can cause overheating. The bearings may be damaged, and the pump may be overpressurized. The pump and piping systems should be designed to avoid dead spots when pumping monomers. Monomers in dead spots may polymerize and plug the pump. There are minimum flow requirements for sealless pumps. It is recommended that a recirculation system be used to provide internal pump flow whenever the pump operates. Inlet line filters are recommended, but care must be taken not to cause excessive pressure drop on the suction side. Typical inlet filters use sieve openings of 0.0059 in (0.149 mm).

A mistreated sealless pump can rupture with potentially serious results. The can can fail if valves on both sides of the pump are closed and the fluid in the pump expands, either due to heating up from a cold condition or if the pump is started up. If the pump is run dry, the bearings can be ruined. The pump can heat up and be damaged if there is insufficient flow to take away heat from the windings. Sealless pumps, especially canned-motor pumps, produce a significant amount of heat, since nearly all the electric energy lost in the system is absorbed by the fluid being pumped. *If this heat cannot be properly dissipated, the fluid will heat up with possibly severe consequences.* Considerable care must be used when installing a sealless pump to be sure that improper operations cannot occur.

The instrumentation recommended for sealless pumps may seem somewhat excessive. However, sealless pumps are expensive, and they can be made to last for a long time, compared to conventional centrifugal pumps where seals may need to be changed frequently. Most failures of sealless pumps are caused by running them dry and damaging the bearings. Close monitoring of temperature is necessary in sealless pumps. Three temperature sensors (resistance temperature devices, or RTDs) are recommended: (1) in the internal fluid circulation loop, (2) in the magnet, or shroud, area, and (3) in the pump case area.

It is very important that sealless pumps be flooded with liquid before starting, to avoid damage to bearings from imbalance or overheating. Entrained gases in the suction can cause immediate imbalance problems and lead to internal bearing damage. Some type of liquid sensor is recommended. Sealless pumps must not be operated deadheaded (pump liquid full with inlet and/or outlet valves closed). Properly installed and maintained, sealless pumps, both canned-motor and magnetic-drive, offer an economical and safe way to minimize hazards and leaks of hazardous liquids.

Loss-of-Containment Causes The list in Table 23-30 indicates four basic ways in which containment can be lost. These cause categories can be used both as a checklist of considerations during the design process and as a starting point for evaluating the adequacy of safeguards as part of a process hazard and risk analysis.

Maintaining the Mechanical Integrity of the Primary Containment System The second main category in the above list pertains to containment failure under design operating conditions due to imperfections in the equipment. This group of causes is the main focus of a facility's mechanical integrity (MI) program. The MI program should also detect other imperfections such as previous periods of operating outside design limits, or improper process materials or impurities that cause accelerated corrosion, chemical attack of seals or gaskets, stress corrosion cracking, embrittlement, etc. MI programs include quality assurance of the initial construction and of maintenance materials; routine preventive maintenance activities; regular inspections and nondestructive testing (NDT) of vessels, tanks, and piping to detect corrosion, pitting, erosion, cracking, creep, etc.; functional testing of standby equipment including alarms, safety instrumented systems, and emergency relief systems; and correcting problems that are identified while inspecting, testing, or maintaining the equipment and instrumentation. CCPS (2006) provides guidance on developing and implementing a mechanical integrity program.

TABLE 23-30 Summary of Loss-of-Containment Causes in the Chemical Industry

I. Containment lost via an "open-end" route to atmosphere
- A. Due to genuine process relief or dumping requirements
- B. Due to maloperation of equipment in service; e.g., spurious relief valve operation
- C. Due to operator error; e.g., drain or vent valve left open, misrouting of materials, tank overfilled, unit opened up under pressure

II. Containment failure under design operating conditions due to imperfections in the equipment
- A. Imperfections arising prior to commissioning and not detected before start-up
- B. Imperfections due to equipment deterioration in service and not detected before the effect becomes significant
- C. Imperfections arising from routine maintenance or minor modifications not carried out correctly, e.g., poor workmanship, wrong materials

III. Containment failure under design operating conditions due to external causes
- A. Impact damage, such as by cranes, road vehicles, excavators, machinery associated with the process
- B. Damage by confined explosions due to accumulation and ignition of flammable mixtures arising from small process leaks, e.g., flammable gas buildup in analyzer houses, in enclosed drains, around submerged tanks
- C. Settlement of structural supports due to geologic or climatic factors or failure of structural supports due to corrosion, etc.
- D. Damage to tank trucks, railcars, containers, etc., during transport of materials on- or off-site
- E. Fire exposure
- F. Blast effects from a nearby explosion (unconfined vapor cloud explosion, bursting vessel, etc.), such as blast overpressure, projectiles, structural damage
- G. Natural events (acts of God) such as windstorms, earthquakes, floods, lightning

IV. Containment failure due to deviations in plant conditions beyond design limits
- A. Overpressurizing of equipment
- B. Underpressurizing of non-vacuum-rated equipment
- C. High metal temperature (causing loss of strength)
- D. Low metal temperature (causing cold embrittlement and overstressing)
- E. Wrong process materials or abnormal impurities (causing accelerated corrosion, chemical attack of seals or gaskets, stress corrosion cracking, embrittlement, etc.)

SOURCE: Summarized from Appendix A of Prugh and Johnson, *Guidelines for Vapor Release Mitigation*, CCPS-AIChE, New York, 1988.

Release Detection and Mitigation **Mitigation** means reducing the severity of consequences of an emergency situation such as a major release, fire, and/or explosion. The choice of mitigation strategies will depend on the nature of the hazardous materials and energies that can be released and the degree to which risk reduction is needed to ensure people, property, and the environment are adequately protected. The latter will be affected by the proximity of populations and sensitive environments surrounding the facility. An unstaffed remote natural gas facility will obviously not warrant the same mitigation measures as a facility using large quantities of high-toxic-hazard materials with other industry or residences nearby.

To be effective, a mitigation strategy will need to be capable of
- Detecting either an incipient or an actual emergency situation
- Deciding on and initiating the proper course of action to mitigate the situation
- Reducing the severity of consequences at the source, in transit, and/or at the receptor locations
- Preventing domino effects that could have even more severe consequences

Each of these steps might be performed either by direct action of operations or emergency response personnel or by automatic systems. An example of the latter might be an array of toxic or flammable gas detectors that might trip an emergency shutdown system that closes remotely actuated block valves and vents off the process pressure to a flare if two adjacent sensors read above a predetermined vapor concentration.

Mitigation measures can also be *passive* safeguards, meaning that they require no human intervention and no engineered sensing and actuation system to work. Examples of passive mitigation measures are secondary containment systems, blast-resistant and fire-resistant structures, insulated or low-heat-capacity spill surfaces to reduce the rate of evaporation, and an increased distance between the hazardous materials and energies and the sensitive receptors.

SAFETY INSTRUMENTED SYSTEMS

REFERENCES: *Guidelines for Safe and Reliable Instrumented Protective Systems*, American Institute of Chemical Engineers, New York, 2007; ISA TR84.00.04, *Guidelines for the Implementation of ANSI/ISA 84.00.01-2004 (IEC 61511)*, Instrumentation, Systems, and Automation Society, N.C., 2005; ANSI/ISA 84.00.01-2004, *Functional Safety: Safety Instrumented Systems for the Process Industry Sector*, Instrumentation, Systems, and Automation Society, N.C., 2004; IEC 61511, *Functional Safety: Safety Instrumented Systems for the Process Industry Sector*, International Electrotechnical Commission, Geneva, Switzerland, 2003.

GENERAL REFERENCES: *Guidelines for Hazard Evaluation Procedures, Second Edition with Worked Examples*, American Institute of Chemical Engineers, New York, 1992; *Layer of Protection Analysis: A Simplified Risk Assessment Approach*, American Institute of Chemical Engineers, New York, 2001; ISA TR84.00.02, *Safety Instrumented Functions (SIF)—Safety Integrity Level (SIL) Evaluation Techniques*, Instrumentation, Systems, and Automation Society, N.C., 2002.

Glossary

Basic process control system (BPCS) System that responds to input signals from the process, its associated equipment, other programmable systems, and/or an operator and generates output signals, causing the process and its associated equipment to operate in the desired manner. The BPCS is commonly referred to as the *control system.*

Compensating measures Planned means for managing process risk during periods of process operation with known faults or problems that increase risk.

Control layer Protection layer that is used to maintain the process within the normal operating limits, such as standard operating procedures, basic process control system, and process alarms.

Core attribute Fundamental underlying property of a protection layer. The core attributes are independence, functionality, integrity, reliability, auditability, management of change, and access security.

Independent protection layer An IPL is a device, system, or action that is capable of preventing a hazard scenario from proceeding to the undesired consequence regardless of the initiating cause occurrence (or its consequences) or the failure of any other protection layer.

Safety instrumented function (SIF) A safety function allocated to the safety instrumented system with a safety integrity level necessary to achieve the desired risk reduction for an identified process hazard.

Safety instrumented system (SIS) Any combination of separate and independent devices (sensors, logic solvers, final elements, and support systems) designed and managed to achieve a specified safety integrity level. An SIS may implement one or more safety instrumented functions.

Safety integrity level (SIL) Discrete level (one out of a possible four SIL categories) used to specify the probability that a safety instrumented function will perform its required function under all operational states within a specified time.

Introduction The chemical processing industry relies on many types of instrumented systems, e.g., the basic process control systems (BPCSs) and safety instrumented system (SIS). The BPCS controls the process on a continuous basis to maintain it within prescribed control limits. Operators supervise the process and, when necessary, take action on the process through the BPCS or other independent operator interface. The SIS detects the existence of unacceptable process conditions and takes action on the process to bring it to a safe state. In the past, these systems have also been called emergency shutdown systems, safety interlock systems, and safety critical systems.

In 1993, the Center for Chemical Process Safety (CCPS) published *Guidelines for Safe Automation of Chemical Processes* (referred to henceforth as *Safe Automation*). *Safe Automation* provides guidelines for the application of automation systems used to control and shut down chemical and petrochemical processes. The popularity of one of the hazard and risk analysis methods presented in *Safe Automation* led to the publication of the 2001 Concept Series book from CCPS, *Layer of Protection Analysis: A Simplified Risk Assessment Approach*. This method builds upon traditional process hazards analysis techniques. It uses a semiquantitative approach to define the required performance for each identified protective system.

The Instrumentation, Systems, and Automation Society (ISA) published the Standard ANSI/ISA 84.01-1996, documenting the good engineering practice for the design, operation, maintenance, and testing of SIS. The standard established a numerical benchmark for the SIS performance known as the safety integrity level (SIL) and provided requirements on how to design and manage the SIS to achieve the target SIL.

Safe Automation and ANSI/ISA 84.01-1996 served as significant technical references for the first international standard, IEC 61511, issued by the International Electrotechnical Commission (IEC). In the United States, IEC 61511 was accepted by ISA as ISA 84.00.01-2004, replacing the 1996 standard. In 2004, the European Committee for Electrotechnical Standardization (CENELEC) and the American National Standards Institute (ANSI) recognized IEC 61511 as a consensus standard for the process industry. IEC 61511 covers the complete process safety management life cycle. With its adoption, this standard serves as the primary driving force behind the work processes followed to achieve and maintain safe operation using safety instrumented systems.

It is important that personnel understand how to achieve safe operation, but not at the exclusion of other important considerations, such as reliability, operability, and maintainability. The chemical industry has also found significant benefit to plant productivity and operability when SIS work processes are used to design and manage other instrumented protective systems (IPS), such as those mitigating potential economic and business losses. The CCPS book (2007) *Guidelines for Safe and Reliable Instrumented Protective Systems* discusses the activities and quality control measures necessary to achieve safe and reliable operation throughout the IPS lifecycle.

Hazard and Risk Analysis Consideration should be given to identifying process hazards as early as possible in the process equipment design, so that measures can be taken to reduce or eliminate the hazards. Inherently safer design strategies, such as minimize, substitute, moderate, and simplify, should be implemented.

When it is no longer practical to reduce the risk further by process design modification, protection layers are used to mitigate the remaining process risk. IPLs must meet the necessary rigor associated with seven core attributes: independence, functionality, integrity, reliability, auditability, access security, and management of change. There are two critical activities to be completed during the risk assessment phase. First, the safety functions (i.e., those functions that detect and respond to process hazards) are identified by using an accepted hazard and risk analysis (H&RA) methodology. Second, each safety function is allocated to a protection layer that is designed and managed to achieve the required risk reduction.

An H&RA involves a review of the process design and its control, operation, and maintenance practices. The review is conducted by a multidisciplinary team with expertise in the design and operation of the process unit. The team uses a systematic screening process to determine how deviations from normal operation lead to process hazards. The H&RA identifies areas where the process risk is too high, requiring the implementation of safety functions. The team's objective is to reduce the risk to below the owner/operator's risk criteria.

Process risk is defined by the frequency of the occurrence and the potential consequence severity of the process hazard. To define the frequency, the initiating causes (e.g., single causes or multiple causes and conditions) are identified for each process hazard, and their frequency of occurrence is estimated. The consequence severity is the logical conclusion to the propagation of the process hazard if no protection layers are implemented as barriers to the event.

The gap between the process risk and the owner/operator's risk criteria establishes the requirements for risk reduction. The risk gap can be managed by a single safety function or by multiple functions allocated to protection layers. The team defines the risk reduction that must be provided by each safety function and allocates the safety function to a protection layer that is designed and managed to achieve the allocated risk reduction.

When the safety function is allocated to the SIS, it is a safety instrumented function (SIF). The risk reduction allocated to the SIF defines its target safety integrity level (SIL). This target is related to the SIF probability of failure on demand (PFD), e.g., SIL 1 (PFD range: 0.01 to 0.1), SIL 2 (PFD range: 0.001 to 0.01), SIL 3 (PFD range: 0.0001 to 0.001), and SIL 4 (PFD range: 0.00001 to 0.0001).

The identification of safety functions continues until the process risk associated with the hazard is reduced to meet the risk criteria. When there is insufficient risk reduction provided by the current or planned design, the team makes recommendations for process design changes (e.g., inherently safer design), improvement to existing functions, or the design and implementation of new functions. These recommendations are generally prioritized based on the magnitude of the gap between the mitigated process risk (i.e., risk considering the presence of existing functions) and the risk criteria.

Design Basis In the design phase, the project team works together to create an SIS design basis that achieves the risk reduction strategy established in the risk reduction phase. This strategy relies, in part, on the implementation of SIFs to address identified process risk. The SIF uses dedicated devices, including process sensors that detect the process hazard, a logic solver that decides what to do, and final elements that take action on the process. Often, a single logic solver implements multiple SIFs, so the potential for common-cause failures between SIFs should be considered during design.

The SIS is normally designed to fail-safe on loss of power and takes action only when the process demands that it do so. These demands often occur when safe operating limits are exceeded due to BPCS failures. Therefore, the SIS is designed and managed to be independent of the BPCS in terms of its hardware and software and its user interfaces, such as operator, maintenance, and engineering interfaces.

Systematic errors can occur anywhere in the design and implementation process or during the operational life of an SIS device. These errors put the SIS on the path to failure in spite of the design elements incorporated to achieve robust hardware and software systems. Systematic errors are minimized using work processes that address potential human errors in the SIS design and management (e.g., programming errors or hardware specification errors).

Random hardware failure can occur throughout the device life as components age in the environmental conditions of the process unit. These failures can cause a device to fail dangerously; i.e., it cannot perform as required. These failures are estimated by examining the dangerous failure modes of each device and their frequency of occurrence. The resulting failure rate is used to estimate the PFD of the SIS considering its specific devices, redundancy, diagnostics, common-cause failure potential, and proof test interval. The PFD is then compared to the target SIL assigned during the risk assessment phase to determine whether the design is adequate.

The design basis includes the process requirements specification and the safety requirements specification. The process requirements specification is typically developed by process engineering, with input from operations personnel. The process requirements are provided to the instrumentation, electrical, or controls systems personnel to develop the safety requirements specification with input from operations and maintenance personnel.

Process Requirements Specification Process engineering uses the H&RA findings, process design information, and operations input to

- Define safe state, including safety and nonsafety actions.
- Define reliability requirements necessary to achieve desired process unit uptime performance.
- Define operability requirements for modes of operation, such as start-up, reduced rates, maintenance modes, and shutdown.
- Identify windows of opportunity for SIS testing.
- Define process-related parameters.
- Define human-related parameters.

Guidance can be found in the CCPS book (2007) *Guidelines for Safe and Reliable Instrumented Protective Systems* related to the development of the process requirements specification.

Safety Requirements Specification The instrumentation and electrical (I&E) requirements are developed to meet the intent of any H&RA findings and the process requirements. The design documentation should establish a clear connection between each process hazard and the design of its SIFs. I&E personnel should meet with the process engineering representative responsible for the process requirements to ensure that the intent is understood.

I&E design focuses on achieving the target SIL through careful selection of the devices (e.g., user approved for safety), use of redundancy, on-line diagnostics, and frequent proof testing. The ISA technical report TR84.00.04 gives extensive guidance on design requirements for the hardware and software systems used to implement the SIS. Application-specific standards by organizations such as American Society of Mechanical Engineers (ASME), American Petroleum Institute (API), and the National Fire Protection Association (NFPA) may provide additional requirements and guidance.

There is often quite a bit of give and take between the process requirements and I&E requirements in the early stages of the project. For example, the ideal process measurement may not be practical in the existing installation. At all times, it should be recognized that the goal of the design is to prevent the process hazard from propagating to an incident.

Engineering, Installation, Commissioning, and Validation (EICV) This phase involves the physical realization of the design basis, which is developed in response to process risk identified in an H&RA study. The bulk of the work in this phase is not a process engineering effort. Detailed engineering, installation, and commissioning is generally an I&E function. However, this is where the assumptions and requirements developed by the process engineer are put into practice and validated.

Validation of the SIS functionality is performed as part of a site acceptance test (SAT). Validation involves a full functional test that demonstrates the SIS actually works in the real-world installation. It proves the SIS devices execute the logic according to the specification and ensures that the SIS and its devices interact as intended with other systems, such as the BPCS and operator interface. From a systematic error standpoint, the SAT also provides an opportunity for a first-pass validation of the procedures developed for the operating basis (see next subsection).

Pre-start-up Safety Review (PSSR) approval of the SIS establishes the point where the SIS design and construction is considered complete. All documentation should be formally updated to as-built status, incorporating any modifications made since the last formal drawing or document revision. Once the PSSR has approved the SIS for process unit start-up, formal management of change procedures should be followed to address proposed modification to the SIS or its associated documentation. Any deviation from the approved design basis should be reviewed and approved by appropriate parties prior to change implementation.

Operating Basis As the SIS engineering design nears completion, the resources and skills of plant operations should be considered. At some point, the SIS is turned over to operations and maintenance personnel, who must be trained on the new SIS and on their responsibilities. Consequently, thought should be given to the content and depth of the information that must be communicated to various personnel. This is especially important as the responsibility for the SIS transitions from the project team to operations and maintenance control.

The process engineer is responsible for defining the content of SIS operating procedures, which should cover SIS specific information (e.g., set points, SIS actions, and the hazard that is being prevented with SIS), the correct use of bypasses and resets, the operator response to SIS alarms and trips, when to execute a manual shutdown, and provisions for operation with detected faults (e.g., compensating measures). These procedures, along with analogous ones developed by maintenance and reliability engineering for maintenance activities, make up the backbone of the operating basis.

Since a device can fail at any time during its life, periodic proof tests are performed to demonstrate the functionality of the SIS. Proof tests are covered by operation and maintenance procedures that ensure that the test is done correctly, consistently, and safely and that the device is returned to a fully operational state after test. Each test serves as an opportunity for personnel to see the SIS in action and to validate the procedures associated with its operation.

Proof testing is required for all SISs. It is used to demonstrate that the devices are operating as specified and are maintained in "as good as new" condition. Failures found during testing indicate gaps in the mechanical integrity program, necessitating root-cause investigation and corrective action.

SECURITY

Definition of Terms (American Petroleum Institute/National Petroleum Refiner's Association, *Security Vulnerability Assessment Methodology for the Petroleum Industry,* 2004.)

Adversary Any individual, group, organization, or government that conducts activities, or has the intention and capability to conduct activities detrimental to critical assets. An adversary could include intelligence services of host nations, or third-party nations, political and terrorist groups, criminals, rogue employees, and private interests. Adversaries can include site insiders, site outsiders, or the two acting in collusion.

Alert levels A progressive, qualitative measure of the likelihood of terrorist actions, from negligible to imminent, based on government or company intelligence information. Different security measures may be implemented at each alert level based on the level of threat to the facility.

Asset Any person, environment, facility, material, information, business reputation, or activity that has a positive value to an owner. The asset may have value to an adversary, as well as an owner, although the nature and magnitude of those values may differ. Assets in the SVA include the community and the environment surrounding the site.

Asset category Assets may be categorized in many ways. Among these are by people, hazardous materials (used or produced), information, environment, equipment, facilities, activities and operations, and company reputation.

Countermeasures An action taken or a physical capability provided whose principal purpose is to reduce or eliminate one or more vulnerabilities. The countermeasure may also affect the threat(s) (intent and/or capability) as well as the asset's value. The cost of a countermeasure may be monetary, but may also include nonmonetary costs such as reduced operational effectiveness, adverse publicity, unfavorable working conditions, and political consequences.

Cyber security Protection of critical information systems including hardware, software, infrastructure, and data from loss, corruption, theft, or damage.

Delay A countermeasures strategy that is intended to provide various barriers to slow the progress of an adversary in penetrating a site, to prevent an attack or theft, or in leaving a restricted area to assist in apprehension and prevention of theft.

Detection A countermeasures strategy that is intended to identify an adversary attempting to commit a security event or other criminal activity in order to provide real-time observation as well as postincident analysis of the activities and identity of the adversary.

Deterrence A countermeasures strategy that is intended to prevent or discourage the occurrence of a breach of security by means of fear or doubt. Physical security systems such as warning signs, lights, uniformed guards, cameras, and bars are examples of countermeasures that provide deterrence.

Hazard A situation with the potential for harm.

Intelligence Information to characterize specific or general threats including the motivation, capabilities, and activities of adversaries.

Intent A course of action that an adversary intends to follow.

Likelihood of adversary success (LAS) The potential for causing a catastrophic event by defeating the countermeasures. LAS is an estimate that the security countermeasures will thwart or withstand the attempted attack, or if the attack will circumvent or exceed the existing security measures. This measure represents a surrogate for the conditional probability of success of the event.

Physical security Security systems and architectural features that are intended to improve protection. Examples include fencing, doors, gates, walls, turnstiles, locks, motion detectors, vehicle barriers, and hardened glass.

Response The act of reacting to detected or actual criminal activity either immediately following detection or after the incident.

Risk The potential for damage to or loss of an asset. Risk, in the context of process security, is the potential for a catastrophic outcome to be realized. Examples of the catastrophic outcomes that are typically of interest include an intentional release of hazardous materials to the atmosphere, or the theft of hazardous materials that could later be used as weapons, or the contamination of hazardous materials that may later harm the public, or the economic costs of the damage or disruption of a process.

Risk assessment Risk (R) assessment is the process of determining the likelihood of an adversary (T) successfully exploiting vulnerability (V) and the resulting degree of consequences (C) on an asset. A risk assessment provides the basis for rank-ordering risks and thus establishing priorities for the application of countermeasures.

Security layers of protection Also known as concentric "rings of protection," a concept of providing multiple independent and overlapping layers of protection in depth. For security purposes, this may include various layers of protection such as countersurveillance, counterintelligence, physical security, and cyber security.

Security vulnerability assessment (SVA) The process of determining the likelihood of an adversary successfully exploiting vulnerability, and the resulting degree of damage or impact. SVAs are not quantitative risk analyses, but are performed qualitatively using the best judgment of security and safety professionals. The determination of risk (qualitatively) is the desired outcome of the SVA, so that it provides the basis for rank ordering of the security-related risks and thus establishing priorities for the application of countermeasures.

Target attractiveness An estimate of the value of a target to an adversary based on the factors shown below. Experience has shown that, particularly for terrorist attacks, certain targets better accomplish the objectives of the adversaries than do others. Since the SVA is a risk-based analytical approach, consideration must be given to these factors in defining the threat and in determining the need for any enhanced countermeasures.

- Potential for mass casualties and fatalities
- Extensive property damage
- Proximity to national assets or landmarks
- Possible disruption or damage to critical infrastructure
- Disruption of the national, regional, or local economy
- Ease of access to target
- Media attention or possible interest of the media
- Company reputation and brand exposure

Threat Any indication, circumstance, or event with the potential to cause the loss of, or damage to, an asset. Threat can also be defined as the intention and capability of an adversary to undertake actions that would be detrimental to critical assets.

Threat categories Adversaries may be categorized as occurring from three general areas:

- Insiders
- Outsiders
- Insiders working in collusion with outsiders

Vulnerability Any weakness that can be exploited by an adversary to gain access to an asset. Vulnerabilities can include, but are not limited to, building characteristics; equipment properties; personnel behavior; locations of people, equipment, and buildings; or operational and personnel practices.

Introduction Prior to September 11, 2001, known as 9/11, chemical process safety activities primarily focused on accidental release risks and excluded most considerations of intentional releases. Security was provided mostly for lesser threats than such extreme acts of violence, and terrorism was generally not provided for except in high-security areas of the world. Exceptions to this included general concerns for sabotage. This was due to a perception that these risks were managed adequately, and that the threat of a terrorist attack, particularly on U.S. chemical manufacturing facilities or transportation system, was remote.

Following 9/11 it became apparent that the threat of intentional harm to infrastructure, especially where hazardous materials were manufactured, stored, processed, or transported, had to be considered a credible concern. Security for the chemical industry took on increased emphasis as a result, and such organizations as the American Institute of Chemical Engineers recognized the paradigm shift and published guidelines on analyzing these threats.* The concerns of international terrorism have spread to many countries around the world, and addressing this concern is now a permanent part of the requirements of the chemical engineering profession. Chemical engineers now must include *chemical process security* as a critical element of the management of a process facility.

Chemical process security management has as its objectives

1. To minimize the risk of harm to the public or employees from intentional acts against a process facility
2. To protect the assets (including employees) of the process facility to maintain the ongoing integrity of the operation and to preserve the value of the investment

Process security and process safety have many parallels and use many common programs and systems for achieving their ends. Process security management requires a systems approach to develop a comprehensive security program, which shares many common elements with process safety management.

Chemical process security includes, but goes beyond, traditional *physical security.* Physical security includes such considerations as guards, barriers, surveillance equipment, and other physical system considerations. Physical security is an element of chemical process security, but physical security alone is not always adequate to address the new challenges of security against extreme acts of violence, such as terrorism.

**Guidelines for Managing and Analyzing the Security Vulnerabilities of Fixed Chemical Sites*, American Institute of Chemical Engineers, August 2002.

Effective chemical process security must also consider integration of broader process elements including technology, chemical usage and quantities, procedures, administrative controls, training, and cyber interface with those traditional physical security elements.

The chemical engineer has an opportunity to influence these considerations in all stages of a process life cycle, including concept, engineering, construction, commissioning, operations, modification, and decommissioning. Security issues that are recognized in the concept and design phases of a project may allow for cost-effective considerations that can eliminate or greatly minimize security risks. For example, if a buffer zone can be provided between the public areas and a plant fence, and then again between a plant fence and critical process equipment, those two zones can effectively provide such benefits as

- *Detection zone(s)*, given they are free of obstacles and have sufficient depth to allow for adversaries to be detected while attempting unlawful entry
- *Standoff zone(s)*, given they have sufficient depth to keep adversaries from using explosives or standoff weapons effectively from the perimeter
- *Delay zone(s)* allowing intervening force the time to respond or time for operators to take evasive action before an adversary reaches a target following detection

A chemical engineer may have a choice of inherent safety variables, such as quantity stored or process temperatures and pressures, or process safety measures such as emergency isolation valves or containment systems, all of which may greatly reduce the vulnerabilities or the consequences of intentional loss. These are in addition to traditional security measures, which may include physical security, background checks, administrative controls, access controls, or other protective measures. For a more complete discussion of the options, refer to the AIChE Center for Chemical Process Safety *Guidelines for Analyzing and Managing the Security Vulnerabilities of Fixed Chemical Sites** and other references.†

Threats of Concern Terrorist acts can be the most problematic to defend against since they may be more extreme or malevolent than other crimes focused on monetary gains or outcomes with less malicious intent. Plus terrorists may use military tactics not often provided for in base chemical facility design. Chemical facility security must be considered in context with local and national homeland security and law enforcement activities, as well as with emergency response capabilities. There is a practical limit to the ability of a chemical site to prevent or mitigate a terrorist act. Above a certain level of threat, the facility needs to rely on law enforcement and military services to provide physical security against extreme acts of intentional harm. The security posture must be risk-based, and so extremely robust security measures are not always applicable or necessary.

The acts of concern for terrorism can be generally defined as involving the four motives shown in Table 23-31.

Other adversaries that must be considered as applicable include those capable and interested in perpetrating a full spectrum of security acts. These may include outside parties or insiders or a combination of the two working in collusion.

The threats that are applicable and the adversaries that may be culpable are characterized to understand their capabilities, intent, and therefore potential targets and tactics. The targets and acts of interest to various adversaries will vary with the group. For example, a terrorist may be interested in destroying a process through violent means, such as by the use of an explosive device. An activist may be interested only in a nonviolent protest or in causing some limited physical damage, but not in harming the environment or the public in the process. The various adversaries and strategies of interest form the basis of the vulnerability assessment, which is the foundation of a chemical process security management system specific to address the anticipated threats.

**Guidelines for Managing and Analyzing the Security Vulnerabilities of Fixed Chemical Sites,* American Institute of Chemical Engineers, August 2002.

†*Counterterrorism and Contingency Planning Guide,* special publication from Security Management Magazine and American Society for Industrial Security, 2001; Dalton, D., *Security Management: Business Strategies for Success,* Butterworth-Heinemann Publishing, Newton, Mass., 1995; Walsh, Timothy J., and Richard J. Healy, eds., *Protection of Assets Manual,* Merritt Co., Santa Monica, Calif. (four-volume loose-leaf reference manual, updated monthly).

TABLE 23-31 Security Issues of Concern with Example Applications to Terrorism

Security motives of concern*	Example terrorist means and objectives
Intentional loss of containment	By causing a release of chemicals to the atmosphere and potential toxic release, fire, or explosion to harm the public, workers, or the environment, or to destroy the facility
Theft of chemicals	For their eventual reuse as primary or secondary improvised weapons against a third party
Contamination or spoilage of a process	To cause immediate or delayed harm to people or the environment, or to cause severe economic injury
Degradation of the asset	By causing mechanical damage or physical or cyber disruption, for purposes of causing severe direct or indirect economic damages

*Adapted from *Guidelines for Managing and Analyzing the Security Vulnerabilities of Fixed Chemical Sites,* American Institute of Chemical Engineers, August 2002.

Overall Objectives of Terrorism Terrorists attempt to cause change to accomplish their goals by creating fear and uncertainty in the population they are targeting through the use of violent acts. The underlying goals include fundamentalist objectives, such as purity of religion or idealistic goals, but they may include power struggles, such as trying to overthrow a government, or reparations, such as revenge for past actions. The reason for a chemical plant being targeted may be that it serves an adversary of the terrorist (economic or military significance) or that it can be weaponized to cause third-party harm (health and safety consequences from intentional release of hazardous materials).

Security Vulnerability Assessment A security vulnerability assessment is intended to identify security vulnerabilities from a wide range of threats ranging from vandalism to terrorism. With the recognition of threats, consequences, and vulnerabilities, the risk of security events can be evaluated, and a security management system can be organized that will effectively mitigate those risks.

SVA Methodologies There are several SVA techniques and methods available to the industry, all of which share common elements. The following is a list of some available SVA methodologies published by various governments, private, and trade and professional organizations. Some are merely chapters or sections of documents that address security or risk assessment/risk management in broader terms. Some are SVA or VA publications by themselves. Some of these "methods" are complete, systematic analytical techniques, and others are mere checklists.

- American Institute of Chemical Engineers Center for Chemical Process Safety: *Guidelines for Analyzing and Managing the Security Vulnerabilities of Fixed Chemical Sites,* 2002.
- American Petroleum Institute/National Petroleum Refiner's Association, *Security Vulnerability Assessment Methodology for the Petroleum Industry,* 2003.
- National Institute of Justice, *Chemical Facility Vulnerability Assessment Methodology,* July 2002 (Sandia VAM).
- Synthetic Organic Chemical Manufacturers Association, Inc. (SOCMA), *Manual on Chemical Site Security Vulnerability Analysis Methodology and Model,* 2002.

One approach to conducting an SVA is shown in Fig. 23-64. This methodology was published by the American Institute of Chemical Engineers, Center for Chemical Process Safety, in 2002. The CCPS SVA is founded on a risk-based approach to managing chemical facility security. To begin the process, companies may perform an enterprise-level screening methodology to sort out significant risks among multiple sites and to determine priorities for analysis and implementation of any recommended changes. The screening, if performed, would result in a prioritized list of sites and forms the foundation of the choice of specific SVAs required. The book covers how to integrate chemical security

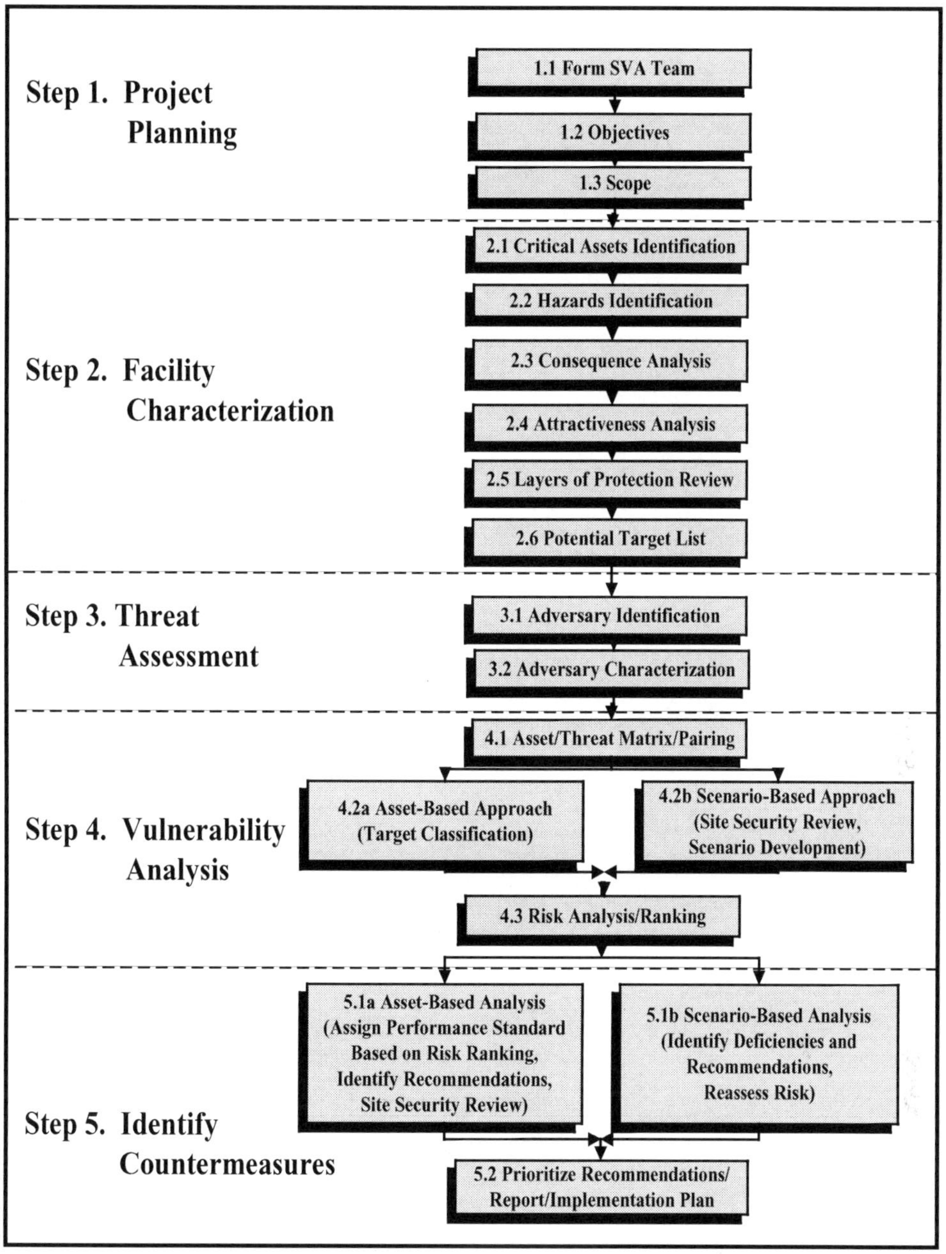

FIG. 23-64 CCPS SVA process.

management and process safety management strategies into a comprehensive process safety and security strategy. Security risk reduction opportunities during the process life cycle are explained, as well as various process risk management strategies (including inherent safety) that are applicable.

In the appendices, the book contains a set of tools including an enterprise-level screening tool, reference information available to conduct the CCPS SVA, and a workbook with worksheets for conducting the CCPS SVA, examples of enhanced security measures, and checklists for assessing security measures at a site.

Defining the Risk to Be Managed For the purposes of an SVA, the definition of risk is shown in Fig. 23-65. The risk that is being analyzed for the SVA is defined as an expression of the likelihood that a defined threat will target and successfully attack a specific security vulnerability of a particular target or combination of targets to cause a given set of consequences. This is contrasted with the usual accidental risk definitions. The risk variables are defined as shown in Table 23-32.

A challenge for security vulnerability analysis is that the accurate prediction of the frequency and location of terrorist acts is not considered credible. As such, the analyst has a choice of assuming a frequency of a certain attack or assuming the attack frequency is 1, thereby focusing solely on the conditional likelihood of success of the adversary who attempts an attack. While the latter approach provides a baseline for making decisions about vulnerability, it does not fully

Intentional release risk is a function of	Accidental release risk is a function of
• Consequences of a successful attack against an asset • Likelihood of a successful attack against an asset	• Consequences of an accidental event • Likelihood of the occurrence of the event
Likelihood is a function of • The attractiveness to the adversary of the asset • The degree of threat posed by the adversary • The degree of vulnerability of the asset	Likelihood is a function of • The probability of an event cascading from initiating event to the consequences of interest and the frequency of the events over a given period

FIG. 23-65 Intentional release vs. accidental release.

answer the question of cost/benefit of any countermeasures. Certain crimes other than terrorism may be more predictable or frequent, allowing for statistical analysis to help frame the risks and justify countermeasure expenditures. Due to this limitation, the factor of attractiveness is considered along with consequences, threat, and vulnerability, to determine the priorities for and design of security measures for the industry.

Security Strategies A basic premise is that not all security risks can be completely prevented. Appropriate strategies for managing security can vary widely depending on the circumstances including the type of facility and the threats facing the facility. As a result, it is difficult to prescribe security measures that apply to all facilities in all industries. Instead, it is suggested to use the SVA as a means of identifying, analyzing, and reducing vulnerabilities. The specific situations must be evaluated individually by local management using best judgment of applicable practices. Appropriate security risk management decisions must be made commensurate with the risks. This flexible approach recognizes that there isn't a uniform approach to security in the chemical process industry, and that resources are best applied to mitigate high-risk situations primarily.

Security strategies for the process industries are generally based on the application of four key concepts against each threat: deterrence, detection, delay, and response.[*]

[*]*Managing and Analyzing the Security Vulnerabilities of Fixed Chemical Sites*, AIChE, August 2002.

A complete security design includes these four concepts in layers of protection or a defense in depth arrangement. The most critical assets should be placed in the center of conceptual concentric levels of increasingly more stringent security measures. In the concept of rings of protection, the spatial relationship between the location of the target asset and the location of the physical countermeasures is important.

In the case of malicious acts, the layers or rings of protection must be particularly robust because the adversaries are intentionally attempting to breach the protective features and can be counted on to use whatever means are available to be successful. This could include explosions or other initiating events that result in widespread common-cause failures. Some particularly motivated adversaries might commit suicide while attempting to breach the security layers of protection.

Countermeasures and Security Risk Management Concepts Countermeasures are actions taken to reduce or eliminate one or more vulnerabilities. Countermeasures include hardware, technical systems, software, interdictive response, procedures, and administrative controls.

Security risk reduction at a site can include the following strategies:

- Physical security
- Cyber security
- Crisis management and emergency response plans
- Policies and procedures
- Information security
- Intelligence
- Inherent safety

TABLE 23-32 SVA Risk Variables

Concequences	Potential Impact of the Event
Likelihood	Likelihood is a function of the chance of being targeted for attack, and the conditional chance of mounting a successful attack (both planning and executing), given the threat and existing security measures. This is a function of three variables below.
Threat	Threat, is a function of the adversary's existence, intent, motivation, capabilities, and known patterns of potential adversaries. Different adversaries may pose different threats to various assets within a given facility.
Vulnerability	This is weakness that can be exploited by an adversary to gain access and damage or steal an asset or disrupt a critical function. This is a variable that indicates the likelihood of a successful attack, given the intent to attack an asset.
Target attractiveness	Target attractiveness is a surrogate measure for likelihood of attack. This factor is a composite estimate of the perceived value of a target to the adversary and the adversary's degree of interest in attcking the target.

SOURCE: *Managing and Analyzing the Security Vulnerabilities of Fixed Chemical Sites*, AIChE, August 2002.

TABLE 23-33 American Chemistry Council's Responsible Care® Security Code Process Security Management System*

1.	Leadership commitment
2.	Analysis of threats, vulnerabilities, and consequences
3.	Implementation of security measures
4.	Information and cyber security
5.	Documentation
6.	Training, drills, and guidance
7.	Communications, dialogue, and information exchange
8.	Response to security threats
9.	Response to security incidents
10.	Audits
11.	Third-party verification
12.	Management of change
13.	Continuous improvement

**Site Security Guidelines for the U.S. Chemical Industry*, American Chemistry Council, October 2001.

Security Management System A comprehensive process security management system must include management program elements that integrate and work in concert with other management systems to control security risks. The 13 management practices shown in Table 23-33 are an example of a management system developed by the American Chemistry Council.

The purpose of a security management system is to ensure the ongoing, integrated, and systematic application of security principles and programs to protect personnel and assets in a dynamic security environment to ensure the continuity of the operation and supporting or dependent infrasturcture. Traditional industrial facility security management tended to focus on protection of persons and property from crime (e.g., theft of property, workplace violence) and crime prevention, response, and investigation. While that is still an element of facility security, a management system allows incorporation of broader security concerns relating to intentional attack on fixed assets, such as by terrorists. To develop and implement a *security* management system not only provides a more thorough, dynamic, risk-based, and proactive approach, but also allows security management to be integrated into a facility's overall EH&S management systems.

The American Chemistry Council's Responsible Care® Security Code is designed to encourage continuous improvement in security performance by using a risk-based approach to identify, assess, and address vulnerabilities; prevent or mitigate incidents; enhance training and response capabilities; and maintain and improve relationships with key stakeholders. As a condition of membership in the council, each member company must implement the Security Code for facilities, transportation and value chain, and cyber security.

KEY PROCEDURES

Safety by design should always be our aim but is often impossible or too expensive, and then we have to rely on procedures. Key features of all procedures are as follows:

- They should be as simple as possible and described in simple language, so as to help the reader rather than protect the writer.
- They should be explained to and discussed with those who will have to carry them out, not just sent to them through the post.
- Regular checks and audits should be made to confirm that they are being carried out correctly.

Many accidents have occurred because the three procedures discussed below were unsatisfactory or were not followed.

Preparation of Equipment for Maintenance The essential feature of this procedure is a permit-to-work system: The operating team members prepare the equipment and write down on the permit the work to be done, the preparation carried out, the remaining hazards, and the precautions necessary. The permit is then accepted by the person or group who will carry out the work and is returned when the work is complete. The permit system will not make maintenance 100 percent safe, but it reduces the chance that hazards will be overlooked, lists ways of controlling them, and informs those doing the job what precautions they should take. The system should cover such matters as who is authorized to issue and accept permits-to-work, the training they should receive (not forgetting deputies), and the period of time for which permits are valid. It should also cover the following:

Isolation of the Equipment under Maintenance Poor or missing isolation has been the cause of many serious accidents. Do not rely on valves except for quick jobs; use blinds or disconnection and blanking unless the job is so quick that blinding (or disconnection) would take as long and be as hazardous as the main job. Valves used for isolation (including isolation while fitting blinds or disconnecting) should be locked shut (e.g., by a padlock and chain). Blinds should be made to the same standard (pressure rating and material of construction) as the plant. Plants should be designed so that blinds can be inserted without difficulty; i.e., there should be sufficient flexibility in the pipework, or a slip-ring or figure 8 plate should be used. Electricity should be isolated by locking off or removal of fuses. Do not leave the fuses lying around for anyone to replace. Always try out electrical equipment after defusing to check that the correct fuses have been withdrawn.

Identification of the Equipment Many accidents have occurred because maintenance workers opened up the wrong equipment. Equipment that is under repair should be numbered or labeled unambiguously. Temporary labels should be used if there are no permanent ones. Pointing out the correct equipment is not sufficient. "The pump you repaired last week is leaking again" is a recipe for an accident.

Freeing from Hazardous Materials Equipment that is to be repaired should be freed as far as possible from hazardous materials. Gases can be removed by sweeping out with nitrogen (if the gases are flammable) or air; water-soluble liquids, by washing with water; and oils, by steaming. Some materials such as heavy oils and materials that polymerize are very difficult or impossible to remove completely. Tests should be carried out to make sure that the concentration of any hazardous material remaining is below an agreed level. Machinery should be in the lowest energy state. Thus the forks of fork lift trucks should be lowered, and springs should not be compressed or extended. For some machinery the lowest energy state is less obvious. Do not work under heavy suspended loads.

Special Jobs Certain jobs, such as entry to vessels and other confined spaces, hot work, and responsibilities of contractors, raise special problems.

Handover Permits should be handed over (and returned when the job is complete) person to person. They should not be left on the table for people to sign when they come in.

Change of Intent If there is a change in the work to be done, the permit should be returned and a new one issued (Crowl and Grossel, eds., *Handbook of Toxic Materials Handling and Management*, Marcel Dekker, 1994, Chap. 12).

Control of Modifications to Plants, Processes, and Organization Many accidents have occurred when such modifications had unforeseen and unsafe side effects (Sanders, *Chemical Process Safety Learning from Case Histories*, 3d ed., Gulf Professional, 2005). No such modifications should therefore be made until they have been authorized by a professionally qualified person who has made a systematic attempt to identify and assess the consequences of the proposal, by hazard and operability study or a similar technique. When the modification is complete, the person who authorized it should inspect it to make sure that the design intention has been followed and that it "looks right." What does not look right is usually wrong and should at least be checked.

Unauthorized modifications are particularly liable to occur

- During start-ups, as changes may be necessary to get the plant on-line.
- During maintenance, as the maintenance workers may be tempted to improve the plant as well as repair it. They may suggest modifications, but should put the plant back as it was unless a change has been authorized.
- When the modification is cheap and no financial authorization is necessary. Many seemingly trivial modifications have had tragic results.
- When the modification is temporary. Twenty-eight people were killed by the temporary modification at Flixborough, one of the

most famous of all time (Mannan, *Lees' Loss Prevention in the Process Industries*, 3d ed., Elsevier, Amsterdam, 2005, Appendix A1; Kletz, *Learning from Accidents*, 3d ed., Gulf Professional, Boston, 2001, Chap.8).

- When one modification leads to another, and then another (Kletz, *Plant/Operations Progress*, vol. 5, 1986, p. 136).
- When organizations are changed often, especially when staffing is reduced. Such changes should be studied as thoroughly as changes to equipment or processes.

Inspection and Testing of Protective Equipment All protective equipment should be scheduled for regular inspection and for testing if failure is latent (hidden); e.g., we do not know if an interlock, trip, alarm, or relief valve is in working order unless we test it. The frequency of testing or inspection depends on the failure rate and the length of time considered tolerable if it fails. Relief valves fail about once per 100 years on average, and testing every 1 or 2 years is usually adequate. Protective systems based on instruments, such as trips and alarms, fail more often, about once every couple of years on average; so more frequent testing is necessary, about once per month. Pressure systems (vessels and pipework) on noncorrosive duties can go for many years between inspections, but on some duties they may have to be inspected annually or even more often.

All protective equipment should be designed so that it can be tested or inspected, and access should be provided. Audits should include a check that the tests are carried out and the results acted on. The supervisor, manager, or engineer responsible should be reminded when a test or inspection is due, and senior managers should be informed if it has not been carried out by the due date. Test and inspection schedules should include guidance on the methods to be used and the features that should be inspected. For example, if the time of response is critical, it should be checked.

Test results should be displayed for all to see, e.g., on a board in the control room.

Tests should be like real life. For example, a high-temperature trip failed to work despite regular testing. It was removed from its case before testing so the test did not disclose that the pointer rubbed against the case. This prevented it from indicating a high temperature.

Operators sometimes regard tests and inspections as a nuisance, interfering with the smooth operation of the plant. Operator training should emphasize that protective equipment is there for their protection and they should "own" it.

Key Performance Indicators Preparation for maintenance, the control of modifications, and the testing of protective equipment are examples of key performance indicators; i.e., taken together, they indicate the quality of the plant's and company's process safety. If they are below standard, the plant is at risk. The usual measure of safety, the lost-time accident (LTA) rate, does not measure process safety. Many companies that had a low LTA rate and assumed that their process safety was therefore under control have experienced serious fires and explosions.